Statistics in Plain English

P9-DYE-113

Statistics in Plain English

with Computer Applications

Robert S. Schulman

Department of Statistics
Virginia Polytechnic Institute and State University

CHAPMAN & HALL
London · Weinheim · New York · Tokyo · Melbourne · Madras

Published by Chapman & Hall, 2—6 Boundary Row, London SE1 8HN

Chapman & Hall, 2—6 Boundary Row, London SE1 8HN, UK

Chapman & Hall GmbH, Pappelallee 3, 69469 Weinheim, Germany

Chapman & Hall USA, 115 Fifth Avenue, New York, NY 10003, USA

Chapman & Hall Japan, ITP-Japan, Kyowa Building, 3F, 2-2-1 Hirakawacho, Chiyoda-ku, Tokyo 102, Japan

Chapman & Hall Australia, 102 Dodds Street, South Melbourne, Victoria 3205, Australia

Chapman & Hall India, R. Seshadri, 32 Second Main Road, CIT East, Madras 600 035, India

---

First edition 1992
Reprinted 1996

Printed in Great Britain by Galliard, Great Yarmouth

ISBN 0 412 06731 5

A catalogue record for this book is available from the British Library

Library of Congress Cataloging-in-Publication Data available

∞ Printed on permanent acid-free text paper, manufactured in accordance with ANSI/NISO Z39.48-1992 and ANSI/NISO Z39.48-1984 (Permanence of Paper).

To my parents, Cy and Maggie Schulman, who can show this book in response when asked what their statistician-son does for a living.

Contents

Preface

"The prescribed calm was eventually found for him in Statistics, and it drove him nearly mad."

From *A Perfect Spy*, p. 367, by John LeCarre.

Plain English. Honest. It's easy to understand statistics, because it's really just common sense. The main purpose of statistics is to help us focus our common sense to form more precise and accurate conclusions. While there is lots of mathematics that proves this common sense, the concepts themselves are perfectly reasonable *without the math. Statistics in Plain English* presents these concepts from a logical point of view, with a minimum of mathematics or formulas. It uses a large number of examples, graphs, figures, and other common sense methods to explain and demonstrate the logical concepts used in statistics. It does not hesitate to use the jargon of the field, but that jargon is explained in plain English. A unique feature of this book is the inclusion of instructions for computer applications of statistics. Thus, along with explaining the concepts, we also demonstrate how to perform statistics in the practical world.

This book is primarily intended for practicing professionals whose formal education did not include adequate preparation in statistics, but who find they must use statistics in their work. Most people suffered through a statistics course in college, not fully comprehending the techniques and probably not ready to appreciate the practical utility of statistical analysis. Others have learned a little about statistics on their own, perhaps by running computer programs whose purpose and results seemed somewhat mysterious. *Statistics in Plain English* will give both groups an appreciation of statistical logic, while providing guidance on practical applications using computer programs. Discussion begins with the most fundamental concepts and progresses in a logical fashion through basic statistics to more advanced topics, such as factorial designs and multiple regression. As a result of this breadth of coverage, however, the true novice may find the pace of this book too demanding. Those who have never been exposed to statistics, either through a formal course or by experimenting on their own, might appreciate the material presented here more thoroughly after first reading a book devoted entirely to introductory concepts.

Many people who have been exposed to statistics are not confident that they understand the subject, yet are called upon to perform or interpret statistical analyses in their work. This uncomfortable position is common and somewhat natural, since the advantages of statistical logic have only recently been recognized in many fields. In part, this is due to the computer revolution. We are living in the information age. Recent advances in computer technology have enabled us to collect huge masses of data. Now we need a way to simplify and analyze all this information so that we can understand what it is telling us. This is the role of statistics. Thus, along with the rapid growth of computers comes an increasing need for knowledge of statistics. Many professionals trained only a few years ago now find the need to educate or reeducate themselves in statistics.

Some important features of this book include:

- Explanations of *statistical reasoning* in plain English;
- *Practical examples* fully explained from raw data to final results;
- Complete annotated *computer inputs and outputs*; and
- A comprehensive *glossary* and list of *symbols* at the rear of the book.

Statistical reasoning is the application of common sense to questions involving interpretation of data. The ability to think logically is the only talent needed to understand statistics. Consequently, statistics can be explained in plain English. We do not claim that *Statistics in Plain English* is totally devoid of mathematics. Some mathematics is employed to describe the finer aspects of this logic, and some formulas are presented to explain the calculations performed by computers. But the amount and level of mathematics employed in this book are kept to a minimum. Mathematics is not used to make points that can be discussed in plain English, and no mathematics beyond high school algebra is used at all. We caution, however, that minimizing mathematics does not eliminate the burden of thinking. Understanding statistics requires logical reasoning and concentration.

Practical examples, chosen from a wide variety of fields, are employed in each chapter to motivate the statistical techniques presented. Each example begins with a research question and pertinent data. Appropriate statistical methods are demonstrated for each example, linking the data, analysis, and results to the question. Where alternative techniques exist for a single goal, they are demonstrated using the same example to provide contrast and comparison of the various strategies.

In the modern world, virtually no statistical computations are performed by hand. In practice, statistical analysis is conducted using computer programs written expressly for this purpose. Unfortunately, manuals for most of these programs assume that the reader already understands statistics. Conversely, many statistics books assume that the reader already knows how to use the computer programs. Rarely are the two discussed together. While some statistics books present snippets of computer printouts, none provides detailed instructions on how to input data and request the desired analysis. *Statistics in Plain English* bridges this gap.

Three of the most commonly used programs for statistical analysis are SAS®, SPSS®, and MINITAB®. Every example presented in this book is demonstrated using each of these programs, with *complete annotated inputs and outputs*. Instructions for using each program tell how the data must be entered and which commands are necessary to produce the desired analyses. Outputs are examined in detail so that users can find and interpret the relevant results. While we still recommend that computer users become familiar with official program manuals, the templates provided here relieve most of the burden of learning the "language" of each program. If you have a computer program other than one of these three, you can use the examples to help you learn your own program by running the example data on your program and matching its results with the results discussed in this book. This should allow you to feel confident that you have input the data correctly and understand the output of your own program.

For general reference, a *glossary* and list of *symbols* appear at the end of the book. The glossary provides a brief description of each technical word or phrase, and serves as a dictionary of statistical terminology. Similarly, the list of symbols explains each element of statistical notation used throughout the book.

A couple of things you will *not* find in this book are computational formulas and exercises. Computational formulas are designed to simplify calculations and minimize round-off errors. But they are rarely logical and often lead to confusion about reasonable concepts. Exercises provide practice with computational formulas. Since the emphasis of this book is on the *practical* use of statistics, which requires computer analysis, neither computational formulas nor exercises are appropriate. Those wishing to become proficient in hand calculations will find an abundance of computational formulas and practice exercises in conventional textbooks. Professionals generally have neither the time nor need for extensive drill. *Statistics in Plain English* provides them with the explanation and computer instructions for performing practical statistics in the real world.

SAS is a registered trademark of SAS Institute Inc., Cary, North Carolina. Inputs and outputs from SAS procedures are printed with permission of SAS Institute Inc.

SPSS is a registered trademark of SPSS Inc., Chicago, Illinois. Inputs and outputs from SPSS procedures are printed with permission of SPSS Inc.

MINITAB is a registered trademark of MINITAB Inc., State College, Pennsylvania. Inputs and outputs from MINITAB procedures are printed with permission of MINITAB Inc.

Statistics in Plain English

1

Basic Concepts

1.1 POPULATION AND SAMPLE

Many of the most important questions in life cannot be answered with absolute certainty. If we are interested in the average reading level for all Americans, we recognize that it is a practical impossibility to test each person individually. If we want to know which drug is better in treating a certain illness, we can compare the drugs only on actual patients, not on all people who might potentially suffer from the illness. In general, most of our questions deal with populations that are infinite or nearly so, while most of our information pertains to samples from these populations. The information contained in a finite sample provides some help in answering our questions, but can never completely remove the uncertainty in our answer. One of the functions of statistics is to limit the degree of uncertainty, while providing objective bounds on this uncertainty. Statistics enables us to find answers that are *probably* correct and to specify precisely how much faith to put in these answers.

In order to use the information in a sample effectively to draw conclusions about a population, it is necessary that the sample be representative of the entire population. In assessing the national average reading level, we do not want to look only at depressed areas of the country, but at all areas. We might stratify the country in terms of general economic level, or amount of money spent per capita on education, or both. Then we could draw our sample carefully so that all such strata are represented proportionally. But we might not have included as stratifying variables all the factors that influence reading level. The simplest way to ensure that our sample represents the whole population is to draw the sample at random from the population. If the sample is drawn completely at random, it is likely that all relevant strata are at least somewhat proportionally represented, whether we thought about these strata in advance or not. Most statistical procedures assume that the data represent a random sample from the population of interest. Specialized procedures, not covered in this book, are available for stratified random samples.

A distinction is often made between descriptive and inferential uses of statistics. **Descriptive statistics** refers to describing the data at hand, typically a random sample from some population. For example, the sample average is a descriptive statistic. In most applications, however, we are only interested in the sample to the extent that it gives us information about the population. **Inferential statistics** pertains to using sample results to infer what we think is true in the population. Clearly, we must first describe what is found in the sample before we can infer as to the whole population. Inferential uses of statistics are therefore based on descriptive statistics, and our discussions will involve both simultaneously.

The term **parameter** refers to a quantity that characterizes a population. For instance, the true national average reading level is a parameter. Usually, a parameter is represented by a Greek letter. The letter μ (mu) is used to represent the average in a population. Typically, parameters are not only unknown, but *unknowable*, since we cannot test the entire population. The term **statistic** refers to a quantity that characterizes a sample, such as the sample average. Any quantity that is based on the sample is a statistic. English letters are used to represent statistics.

1.2 VARIABLES

An **experiment** is any process of measurement or observation. This definition is deliberately ambiguous enough to encompass anything that produces data for analysis. The experiment may be as simple as asking whether a person is married or not, or it may be so complex that it takes months of careful laboratory work to provide the basic data for study. Some experiments are carefully planned and must be conducted by experts in the field. Others are as simple as drawing data from an existing file. All experiments, however, are undertaken with a purpose in mind. We gather data to help answer questions about the world around us. Statistics enables us to use these data to provide reasonable answers to our questions.

Generally, we do not analyze the experiment itself, but rather the data produced by the experiment. We refer to such data as random variables. A **random variable** is a variable that takes on values according to the outcome of some experiment. For example, the experiment may be a study of a chemical reaction, and the random variable might be the quantity of a particular compound produced. Or the experiment might be asking people how they feel about a new policy, and the random variable might be the number of people who agree with this policy.

There are many classification schemes for random variables. We will discuss two classification schemes—one that is useful from a conceptual point of view, and another that is more pragmatic for actual data analyses. From a conceptual viewpoint, we classify random variables as either discrete or continuous. A **discrete random variable** is limited in the values it can assume, typically to a finite number of possible values, but always to a countable number of values. If the experiment is rolling a die, for example, and the random variable of interest is the number of spots on the top face, only the values 1, 2, 3, 4, 5, and 6 can be observed. If the experiment involves asking people for their religious affiliation, the random variable might take on values according to whether the response is Baptist, Methodist, Catholic, Jewish, or Muslim. In both these examples there is a finite number of possible values for the random variable. However, for some discrete random variables there is an infinite number of possible values. For instance, if the experiment consists of inspecting a sheet of plexiglass for imperfections, and the random variable is the number of imperfections seen, we could observe values 0, 1, 2, 3, 4, and so on. There is no upper limit to the number of imperfections possible, but they are still countable.

A **continuous random variable**, on the other hand, takes values along a continuum. A continuous variable is said to be dense, like the points on a line. Between any two given points, it is always possible to insert another point. Following this logic repeatedly, we can say that between any two points there is an infinite number of other points. Examples of continuous variables are length, weight, density, specific gravity, and time. If we are measuring weights, and two observed values are 64.7 pounds and 64.8 pounds, it is still possible to find another value between these two, such as 64.73 pounds. Note that this is not always possible with a discrete variable; we cannot observe 3.5 imperfections in a sheet of plexiglass.

Most of the measurements we deal with in life represent continuous variables. The distinction between discrete and continuous variables, however, is not always obvious. Rounding often makes a continuous variable look discrete. For example, when asked their height, a person may say 5 feet and 11 inches, or 5 feet and 2 inches, but would be unlikely to respond that they are 5 feet and 7.045 inches. From the observed data, it may appear that people only come in heights evenly divisible by 1 inch, but this is merely a result of rounding height to the nearest whole inch. In fact, all measurements are rounded, since it is impossible to measure anything infinitely accurately. As a result, continuous variables frequently appear discrete. It is generally not necessary to resolve the distinction between discrete and continuous variables, since a different classification scheme is advantageous for applications of statistics.

For data analysis purposes, it is pragmatic to sort variables into three types: numeric, categorical, and ordered categories. **Numeric** variables, as the name implies, refer to all measurements that naturally result in numeric quantities. All continuous random variables are numeric; some discrete random variables are numeric. The number of imperfections in a sheet of plexiglass, while discrete, is clearly numeric. The questions we typically ask for numeric variables are the same whether the variables are continuous or discrete. We might be interested in the average value of a numeric variable or in how spread out the values are. Since the questions we ask are the same, the same statistical procedures are used for all numeric variables, whether they are discrete or continuous.

Discrete variables that are not numeric are said to be **categorical**. Religious affiliation, for example, is categorical. The questions we ask about categorical variables are frequently different from those we ask about numeric variables, so different statistical procedures are needed for categorical variables. For some categorical variables, the categories are ordered. If we measure level of education on a crude scale, each person might be placed in a category, such as high school education, some college but no degree, college degree, master's degree, or doctoral degree. In this case, the categories are clearly ordered and thus contain some additional information. Such variables are said to be on a scale of **ordered categories**. We frequently use different statistical methods when the categories are ordered, to take advantage of the additional information inherent in knowing the order of the categories.

Finally, it should be noted that not all variables are random. Random variables correspond to the *outcome* of some experiment. But other variables are frequently involved in the experiment. Suppose, for example, that we wish to compare a new drug with an old drug. Patients afflicted with the illness would be given either the new drug or the old drug, and we would observe the degree of relief from symptoms of the illness for each patient. The measure of relief, since it depends on the experiment, is a random variable. But each observation on this variable also comes from one of two conditions, namely the new drug or the old drug. "Drug" (either old or new) is a useful categorical variable, but it is *not* random. Rather, it is part of the design of the experiment to have two drugs. Variables built into the design of the experiment are called **design variables**. We might be interested in determining if there is any relationship between the two variables—relief from symptoms and drug. That is, does one drug tend to produce higher relief from symptoms of the illness than the other? Degree of relief is a random variable, while drug is a design variable. Statistical procedures for looking at the relationship between two variables can be classified in terms of whether the variables are both numeric, both categorical, both ordered categories, or various combinations. Usually, the choice of statistical procedure does not depend on whether the variables are random variables or design variables.

The terms **independent variable** and **dependent variable** are often used instead of design variable and random variable, respectively. A design variable, such as which drug each patient receives, may be determined by the researcher *independently* for each patient. The random variable, relief from symptoms, will generally *depend*, at least partly, on the independent variable—which drug the patient receives.

1.3 PROBABILITY

Probability is fundamental to statistics, since we are interested in reaching decisions that are *probably* correct. We use the concepts of probability to determine how likely it is that our decision is correct or how sure we are that an estimate is within a certain distance of the true parameter. While we will not discuss the theory of probability, it is still necessary to have some understanding of the concepts involved.

In common usage, probability is often expressed as a percentage, such as a 20 percent chance of rain tomorrow. Since the word "percent" means "hundredth," 20 percent is the same as 20 hundredths or 0.20. In statistics, we generally speak of probabilities on a 0 to 1 scale, which is equivalent to using percentages on a 0 to 100 percent scale. Obviously, probabilities cannot be less than 0, nor greater than 1. No matter what any coach may have said to you, it is impossible to give a 150 percent effort!

An experiment is conducted to learn something about a population, and random variables are used to describe the outcome of the experiment. For example, if we are interested in reading levels, the experiment might be to select a person at random from the population and administer a reading test. An associated random variable would be the score on the test. If we repeat this experiment by selecting another person at random, the value of the random variable (test score) will probably be different. If we perform the experiment repeatedly, we will find that some values of the random variable occur more frequently than others. Every random variable has a **probability distribution**, that is, a list of possible values and how likely each value is. The probability distribution paints a picture of the population, at least with respect to this random variable. Generally, the probability distribution is unknown, and its estimation is one of the purposes of the experiment.

We can represent a probability distribution graphically, but different forms of graphs are needed for discrete and continuous random variables. To illustrate a probability distribution for a discrete random variable, consider Example 1.1, whose probability distribution is displayed in Figure 1.1. As shown in this figure, we represent the probability distribution for a discrete random variable by a line graph, where the height of each line is the probability of the corresponding value. In this example, the six values of the random variable X are equally likely and, therefore, the lines are equal in height. If we were looking at a distribution of religious affiliations, however, the labels on the horizontal axis would be Baptist, Methodist, Catholic, and so forth, and the lines would not all be the same height. For any discrete random variable, the probabilities must all be positive (no lines extending below the horizontal axis) and the sum of the probabilities (the sum of the heights of the lines) must be 1.0.

For a continuous random variable, however, there are an infinite number of values, and a line graph is not possible. Instead, we draw a curve over the range of

Example 1.1

We perform the simple experiment of rolling a die and considering the random variable of the number of spots on the top face. Let us use the English letter X to represent this random variable. That is, X = number of spots on the top face. For this experiment, the probability distribution is known. The only possible values of X are 1, 2, 3, 4, 5, and 6, and the probability is 1/6 for each of these values.

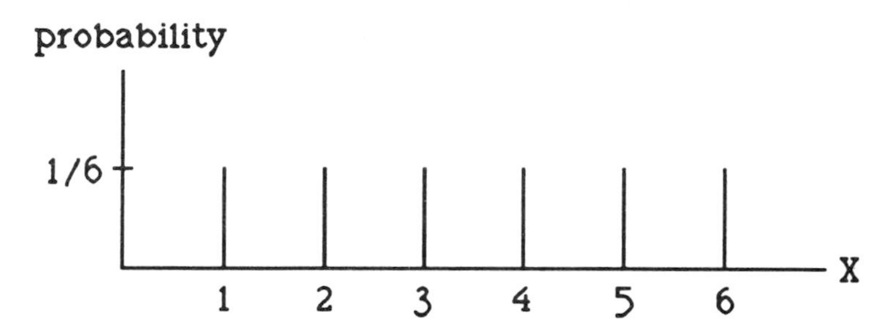

FIGURE 1.1. Probability distribution for number of spots on top of a die.

---

possible values for the random variable and represent probability as area under the curve. (The word "curve" is loosely interpreted to include figures bounded by straight lines as well as those bounded by rounded lines.) An experiment yielding a continuous random variable is presented in Example 1.2, and the corresponding probability distribution is displayed in Figure 1.2.

In the continuous case, probability is represented by the area under the curve, not by height. The fact that probabilities can never be negative means that the curve does not extend below the horizontal axis. Since the total probability is equal to 1, the area under the curve is 1. The area of the triangle in Figure 1.2 is (½)(base)(height) = (½)(1)(2) = 1.0. The probability that the radius of the iron disk lies between two values, say a and b, is given by the area under the curve between these values (see Figure 1.3).

There is one awkward aspect of representing probability by area. The probability of any single value is the area under the curve at that value, namely the area of a line segment. Since a line segment has no width, its area is 0. As a result, the probability of observing any particular value for the radius is 0. For instance, the probability of obtaining a disk with a radius *exactly* ½ inch is the area of a line segment drawn at R = ½, which is 0. At first glance, this seems problematic. But it makes sense when we realize that there are an infinite number of possible radii between 0 and 1 inch, so each single value has probability that is essentially equal to 1 divided by infinity (i.e., 0). For

Example 1.2

Part of a monitoring effort in a mining operation involves melting down nuggets of ore of uniform weight. The components of each nugget are separated, and molten iron is siphoned into a 1-inch radius mold. Since the nuggets are not of uniform purity, the resulting iron disks are not always 1 inch in radius. In practice, the probabilities of various sized disks are not known exactly. For demonstration purposes, however, let us suppose that the probability distribution for the radius of the iron disk, R, is known to have the shape shown in Figure 1.2. Proabability corresponds to the area under the curve. Note that most of the area is close to 1 inch, so large disks are more common than small ones.

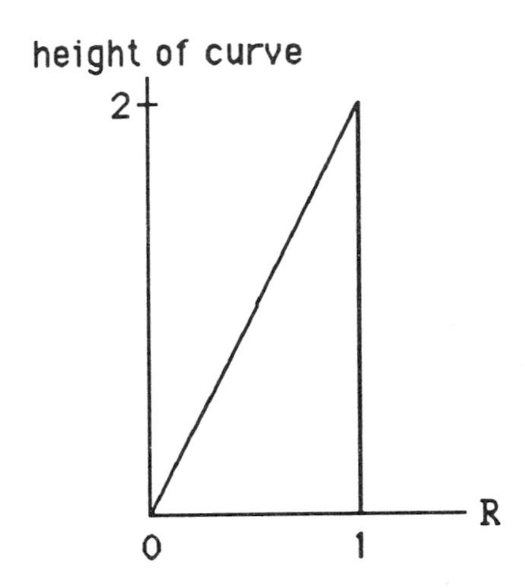

FIGURE 1.2. Probability distribution for radius of iron disks.

---

a continuous distribution, the probability of any particular value for the random variable is 0. The awkward aspect of this is that events with 0 probability can occur. Before the molten iron is siphoned into the mold, each possible value for the radius of the disk has probability 0. Yet, after the drop cools in the mold, some value for the radius will actually occur. To deal with this problem, it is helpful to think of the probability of a single value as *approaching* 0.

1.4 MEAN AND VARIANCE

Consider the experiment of rolling a die, as described in Example 1.1. The population for this experiment consists of all possible rolls of the die. The probability

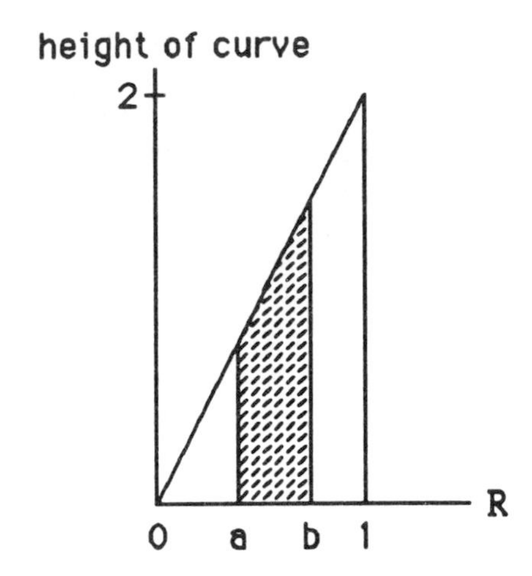

FIGURE 1.3. Probability that radius of iron disk lies between the values a and b.

distribution for X (the number of spots on the top face) is shown in Figure 1.1. This probability distribution describes the population completely, since it lists all possible X values and their probabilities. The probability distribution is a picture of the population. To take a sample from this population, we need only roll the die a few times. Imagine taking a sample and writing down the value of X each time the die is rolled. Suppose we also keep a running average by computing a new average of all the X values after each roll. In the long run, this average would approach 3.5. As shown in Figure 1.4, the number 3.5 is exactly in the center of the probability distribution.

The long-run average describes the center of the probability distribution and, therefore, the average value in the population. This average, since it is a characteristic of the population, is a parameter. As mentioned earlier, we usually use the

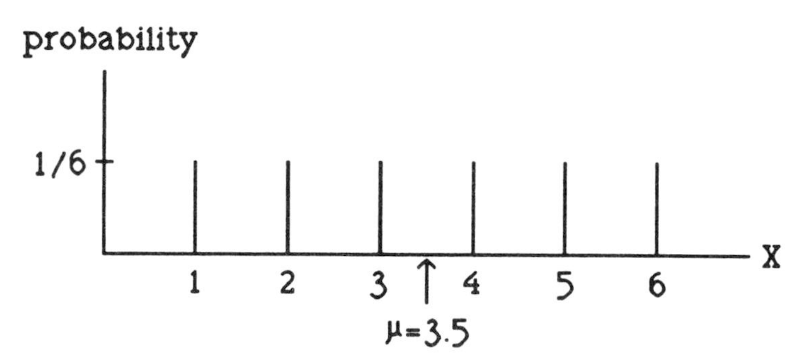

FIGURE 1.4. Mean of probability distribution for number of spots on top of a die.

Greek letter μ to represent the average in the population, which we call the **population mean**. For our example, $\mu = 3.5$. Note that μ does not have to be an observable value. No single roll of the die will yield 3.5 spots on the top face. Nonetheless, μ describes the center of the distribution, in that it represents the long-run average value of the random variable.

The same logic holds for the continuous case. Consider Example 1.2, concerning disks formed from drops of molten iron. The probability distribution for R, the radius of the disk, is shown in Figure 1.2. This curve describes all possible values for the radius and hence represents the population of all possible disks. Suppose we take a sample from this population, by observing the radii of several disks, and compute a running average. If we look at enough disks, the long-run average would approach the population mean, μ. This value can be shown to be ⅔, and again is a measure of the center of the probability distribution (see Figure 1.5). Note that μ is not halfway between 0 and 1, since large disks are more common than small ones. However, it is still a measure of the center of this distribution because it represents the long-run average radius.

The mean describes the *center* of a population. We might also want to describe the amount of *spread* in the population. The quantity used to characterize spread is called the **population variance** and is defined to be the long-run average squared distance from the mean. For each disk, we write down the radius, R, the difference between this value and the mean, $R-\mu$, and the squared distance from the mean, $(R-\mu)^2$. The variance is defined to be the long-run average of $(R-\mu)^2$. If the distribution is narrow, the radii of most disks will be near the mean and $(R-\mu)^2$

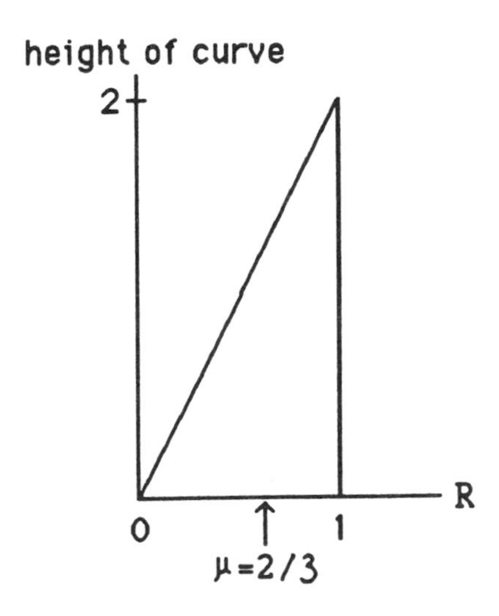

FIGURE 1.5. Mean of probability distribution for radius of iron disks.

will be small for most observations. If the distribution is wide, observed radii will frequently be far from the mean, and $(R-\mu)^2$ will be large for many disks. The variance—the long-run average value of $(R-\mu)^2$—characterizes the amount of spread in the population. If the distribution is very spread out, the variance is large. If the distribution is narrow, the variance is small. The narrowest possible distribution, with all the probability concentrated at a single value, has a variance of 0. The symbol σ^2 (sigma squared) is used to represent the population variance. The long-run average of a square cannot be negative, so σ^2 is positive or 0 for any probability distribution.

A minor inconvenience arises in using σ^2 to characterize spread: it is in the square of the original units. If we measure radii in inches, μ is also in inches. (The long-run average of numbers written in inches is also in inches.) The quantity $(R-\mu)$ is still in inches, since we are subtracting inches from inches. But $(R-\mu)^2$ is in *squared* inches. The variance—the long-run average of $(R-\mu)^2$—measures spread in the square of the original units. To eliminate this problem, we define the **population standard deviation** as the square root of the variance. We use σ to represent the standard deviation, so that $\sigma = \sqrt{\sigma^2}$. By taking the square root, we eliminate the dimensionality problem and σ is back in the original units—in this case, inches. Since σ^2 measures spread of the population, so does σ. Large values of σ represent greater dispersion in the population, while small values represent less dispersion. The value of σ, like σ^2, can never be negative.

The mean and variance of a population help describe the values we are likely to see when we sample from that population. The mean characterizes a typical value, while the variance (or the standard deviation) tells us how far from this value the observations are likely to fall. In order to determine the mean and variance of a population, we must know the probability distribution. However, in practical applications the true probability distribution is unknown. Therefore, the parameters μ and σ^2 *can never be determined exactly*. In practice, all we see is a random sample from the population. One of our objectives, therefore, is to use the information contained in the sample to *estimate* μ and σ^2, without ever knowing the probability distribution. We will return to this estimation problem in the next chapter.

1.5 THE NORMAL DISTRIBUTION

There is an infinite number of possible shapes for probability distributions. In statistics, however, one particular type of curve, called the **normal distribution**, is used extensively. The normal distribution is important in statistics for two main reasons. First, it serves as a good model for data from many real-life experiments. Second, it is magic. The magic of the normal distribution will be explained in the next section. At present, we examine the normal curve and learn how to find probabilities for normal variables.

A normal curve is shown in Figure 1.6. Most of the area (i.e., probability) is in the center of the curve, so the random variable frequently takes a value near the middle. There is less area near the ends of the curve, so the variable is less likely to fall in these regions. This is typical of many real sets of data. Most of the values fall near the center, with occasional values farther away in either direction. The normal distribution is continuous and symmetric about its center, μ. The curve is unimodal (i.e., one lump) and approaches the axis at both ends. The curve never quite hits the axis, although it gets continually closer as we move farther out from the mean. Since this is a probability distribution, the total area under the curve is 1.

Actually, the normal is a whole family of distributions, all sharing the same general properties. Different normal distributions may be centered at different points on the axis (they may have different means), or they may be more or less spread out than the curve shown in Figure 1.6 (they may have different variances). We can identify a particular normal curve by specifying its mean and variance.

Shown in Figure 1.7 are two different normal curves, possibly representing random variables from two different experiments. Let us designate the mean and variance of the first curve as μ_1 and σ_1^2, and the mean and variance of the second curve as μ_2 and σ_2^2. From the figure, we can see that $\mu_1 < \mu_2$ since the first curve is located to the left of the second. We can also tell that $\sigma_1^2 > \sigma_2^2$, since the first curve is more spread out than the second. The second curve, which is less spread out, is taller than the first, since each curve must have area equal to 1.

Tables exist that list areas under the normal curve. The areas correspond to probabilities for variables that have normal distributions. Using these tables, we can find the probability that a normal variable lies in a certain range. This process

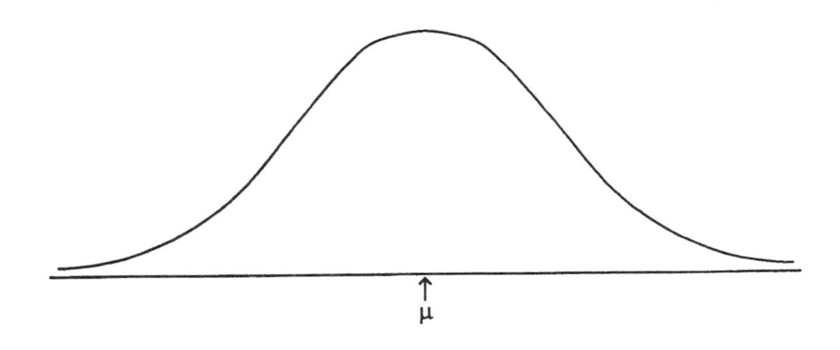

FIGURE 1.6. Normal distribution.

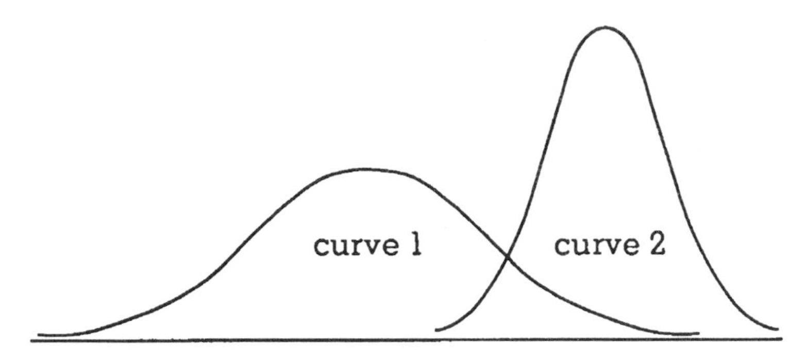

FIGURE 1.7. Comparison of two normal distributions.

is demonstrated in Example 1.3, which asks us to find the probability corresponding to the shaded region in Figure 1.8. At first glance it might appear that we need a table of *this particular normal distribution* to find this area, but this turns out to be unnecessary. Areas under one normal curve can be found from areas under any other normal curve. Because of this relationship, only a single normal distribution need be tabled.

The normal distribution tabled in most books is called the **standard normal**

Example 1.3

Scores on the Stanford-Binet intelligence test are known to be normally distributed with a mean of 100 and a standard deviation of 16 ($\mu = 100, \sigma = 16$). Suppose we want to find the probability of observing a score between 84 and 116 on this test. Using X to represent a score on the Stanford-Binet, the probability that X lies between 84 and 116 is the area shown in Figure 1.8.

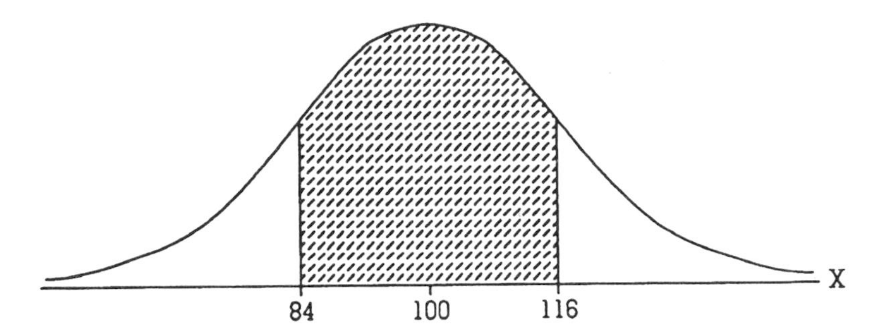

FIGURE 1.8. Probability distribution for scores on the Stanford-Binet.

distribution and has a mean of 0 and a variance of 1. We will first learn how to use the table of the standard normal distribution to find areas under this particular curve. Then we will show how this table can be employed to find areas under any other normal distribution. This will enable us to find the desired probability in Example 1.3.

The letter Z is generally used to represent a variable with the standard normal distribution. On a standard normal curve, the center is at 0, while about 68 percent of the area lies between the Z values of +1 and -1, and about 95 percent of the area lies between the Z values of +2 and -2. Figure 1.9 depicts a standard normal curve.

Table 1 in the Appendix lists upper tail areas for many different Z values. To demonstrate use of this table, suppose we wish to find the area to the right of the Z value of 1.25. In the left margin we locate the row labeled 1.20, and along the top margin we find the column labeled .05, corresponding to the second decimal digit of our Z value. The intersection of this row and column contains the value .1056, which is the area to the right of 1.25 on the Z axis. This area is shown in Figure 1.10.

We stated above that about 68 percent of the area lies between the Z values of +1 and -1, and about 95 percent of the area lies between the Z values of +2 and -2. We can use Table 1 to specify these percentages more precisely. For the Z value of 1.00, the table lists an upper tail area of .1587. By symmetry, the area below Z = -1.00 is also .1587. The total of these two tail areas—above +1 and below -1—is 2(.1587) =.3174. Since the total area under the curve is 1, the area between the Z values of +1 and -1 is 1 - .3174 = .6826, or 68.26 percent. A standard normal variable lies between the values +1 and -1 about 68.26 percent of the time. In similar fashion, we find the area between the Z values of +2 and -2 is .9544, or 95.44 percent.

We will now demonstrate the method for finding areas under other normal

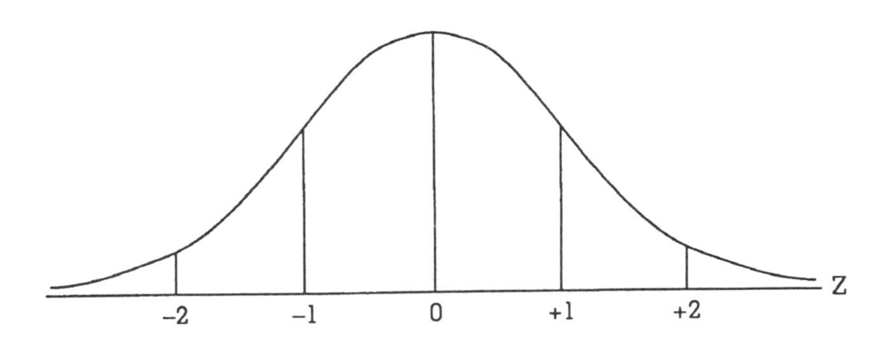

FIGURE 1.9. Standard normal distribution.

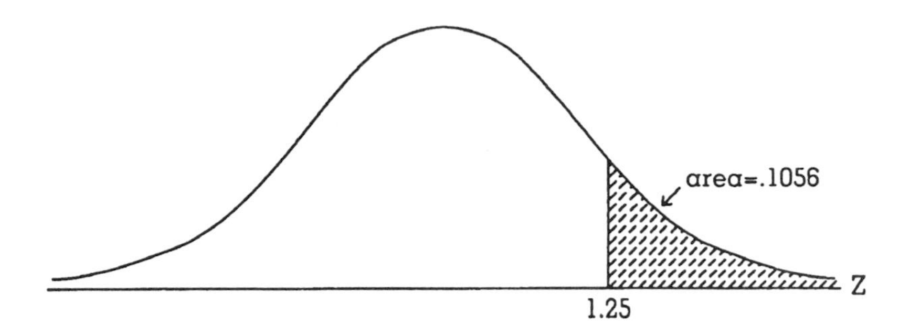

FIGURE 1.10. Upper tail area for Z=1.25.

distributions from the table of standard normal areas. In Example 1.3 we seek the probability that X falls between 84 and 116, where X is normally distributed with mean $\mu = 100$ and standard deviation $\sigma = 16$. To find the desired area, we change the scale of measurement from X to Z. The X values 84 and 116 are changed to the Z scale by subtracting the mean on the X scale and then dividing by the standard deviation on the X scale—that is, each X value is converted to

$$Z = \frac{X-\mu}{\sigma}. \tag{1.5.1}$$

The X value of 84 converts to

$$Z = \frac{X-\mu}{\sigma} = \frac{84 - 100}{16} = -1.$$

Similarly, the X value of 116 converts to

$$Z = \frac{X-\mu}{\sigma} = \frac{116 - 100}{16} = 1.$$

The desired probability can now be found from either the X scale or the Z scale. Figure 1.11 shows the relationship between these two scales. The area on the X scale between 84 and 116 is equivalent to the area on the Z scale between -1 and 1. As we know, this area is 68.26 percent. Thus, the probability of a Stanford-Binet score between 84 and 116 is 0.6826.

The change of scale performed here is similar to processes encountered in everyday life. Fahrenheit and Celsius are two different scales for measuring temperature, and can be thought of as two axes for labeling the amount of heat (see

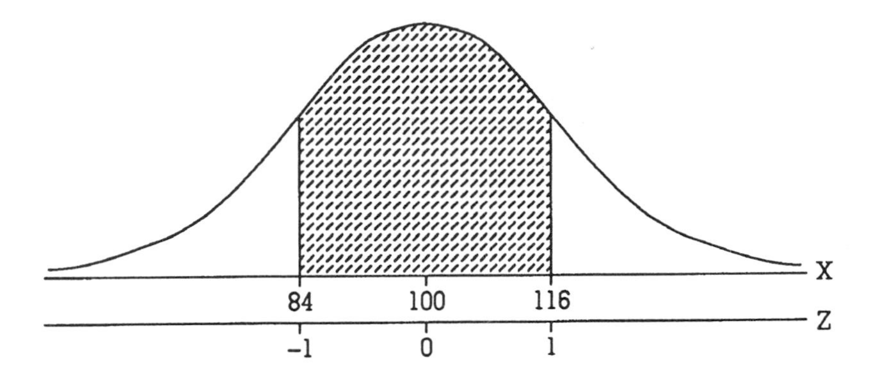

FIGURE 1.11. Original and standardized scales for the Stanford-Binet.

Figure 1.12). The probability that the temperature is between 32 and 212 on the Fahrenheit scale is the same as the probability of it being between 0 and 100 on the Celsius scale. This equivalence is based on the same logic we used in stating that the probability between 84 and 116 on the X scale is identical to the probability between -1 and 1 on the Z scale.

As shown in Figure 1.11, the mean of 100 on the X scale corresponds to the mean on the Z scale, which is 0. When X is equal to μ, then

$$Z = \frac{X-\mu}{\sigma} = \frac{100-100}{16} = 0.$$

A scale with a mean of 0 is called a **centered scale**. If a centered scale has standard deviation equal to 1, as on the Z scale, it is a **standardized scale**.

Table 1 in the Appendix lists areas under the standard normal distribution. For any other normal curve, we can find probabilities by converting X values to Z values, as demonstrated above. With modern computers, however, such calculations are generally unnecessary. Most computer programs calculate and print the desired areas for us. In fact, all of the tables in the Appendix are included only for instruction and reference. They are not typically needed in practical applications.

1.6 SAMPLE MEAN

One of the first things we often do with a sample is to compute the average, also called the **sample mean**. This is probably the single most useful quantity we can obtain from the data. The symbol $\overline{X}$ (read "X-bar") is used to represent the average. The formula for $\overline{X}$ is given by

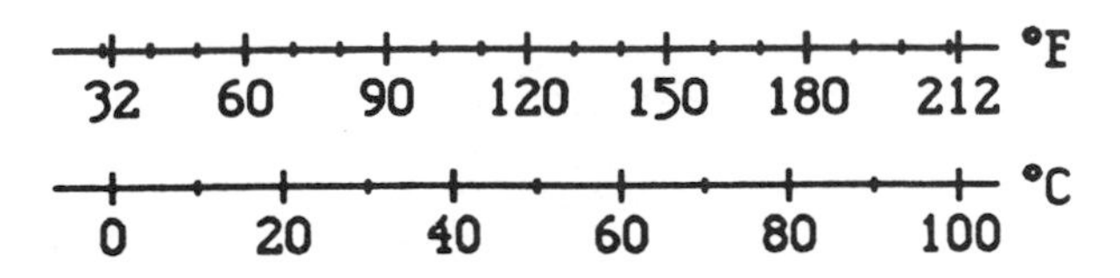

FIGURE 1.12. Fahrenheit and Celsius temperature scales.

$$\overline{X} = \frac{\Sigma X}{n}, \tag{1.6.1}$$

where ΣX stands for the sum of all observations, and n is the number of observations in the sample. The formula for $\overline{X}$ is to add the values of all observations (ΣX) and divide by the number of values (n), yielding the usual average.

The average in our sample serves as an estimate of the population mean. How good is this estimate? Example 1.4 poses a question of this nature, asking for the probability for a range of values for $\overline{X}$. Since this question involves the average, we need to know the probability distribution for the average, $\overline{X}$, rather than the probability distribution for an individual observation. The distribution shown in Figure 1.8 is for an individual observation, since it has the name of the original random variable, X, on the axis. This distribution shows probabilities for values of a single observation. The distribution for $\overline{X}$ has $\overline{X}$ on the axis, and shows probabilities for various values of the sample mean, $\overline{X}$.

To investigate the distribution for $\overline{X}$, consider the experiment of drawing a random sample of size 25 from the population. Each time we perform this experiment, we get a single value for the average, $\overline{X}$. Therefore, $\overline{X}$ is a random variable associated with this experiment. If we perform this experiment repeatedly (i.e., draw lots of random samples), we could plot the obtained $\overline{X}$ values to see which values are more likely and which are less likely. The result would approximate the probability distribution for $\overline{X}$.

Common sense should tell us a great deal about this distribution. First, since the distribution for an individual observation is normal, it is reasonable to expect that

---

Example 1.4

A random sample of 25 persons is tested on the Stanford-Binet, and the average score is computed. What is the probability that the average is within 6.4 points of the population mean?

the shape of the $\overline{X}$ curve is also normal. Second, we would guess that the typical $\overline{X}$ value should be near 100, the mean of the X distribution. Sometimes $\overline{X}$ will be above 100 and sometimes below, but the long-run average value of $\overline{X}$ should be 100. The mean on the $\overline{X}$ scale, denoted by $\mu_{\overline{X}}$, is the same as the mean on the X scale, μ. Finally, we might expect that most of the $\overline{X}$ values will be close to 100. It may be possible to find a *single* Stanford-Binet score far from 100, but it would be most unlikely to find the *average* of 25 scores far from 100. Most of the probability for $\overline{X}$ is concentrated near the mean value of 100. As a result, the variance of the $\overline{X}$ curve, $\sigma^2_{\overline{X}}$, is smaller than the variance of the X curve, σ^2. Figure 1.13 depicts the curve resulting from these logical deductions.

Imagine what this curve would look like if the sample size were 250 instead of 25. If we look at repeated samples of 250 scores and compute $\overline{X}$ for each sample, we still expect the $\overline{X}$ distribution to be normal and centered at 100, but it is even less likely that $\overline{X}$ would be far from 100. The distribution is even narrower. As a result, the variance of $\overline{X}$ becomes smaller as the sample size increases. The precise relationship between variance and sample size is given by

$$\sigma^2_{\overline{X}} = \frac{\sigma^2}{n}. \tag{1.6.2}$$

As sample size increases, the variance of the average, $\sigma^2_{\overline{X}}$ decreases. Taking the square root of both sides in (1.6.2), we write the standard deviation of $\overline{X}$ as

$$\sigma_{\overline{X}} = \frac{\sigma}{\sqrt{n}}. \tag{1.6.3}$$

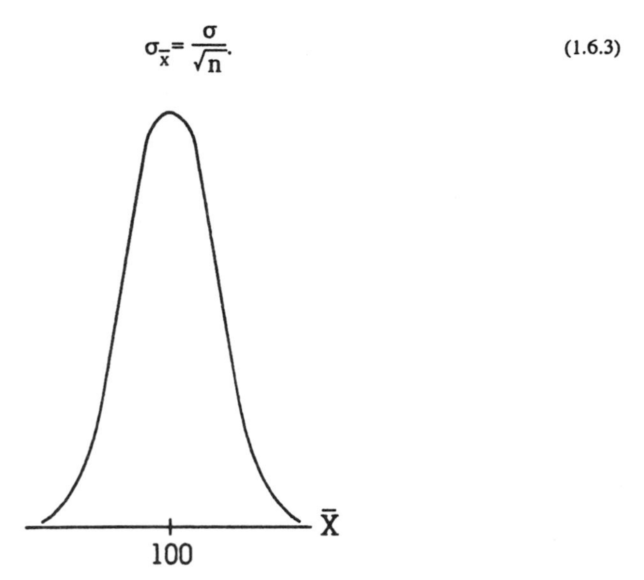

FIGURE 1.13. Probability distribution for the average.

In summary, if we draw a random sample of size n from a normal distribution with mean μ and standard deviation σ, $\overline{X}$ is normally distributed with mean $\mu_{\overline{X}} = \mu$ and standard deviation $\sigma_{\overline{X}} = \sigma/\sqrt{n}$. For a sample of n = 25 scores on the Stanford-Binet, $\mu_{\overline{X}} = \mu = 100$ and

$$\sigma_{\overline{X}} = \frac{\sigma}{\sqrt{n}} = \frac{16}{\sqrt{25}} = 3.2.$$

Now that we know the probability distribution for $\overline{X}$, we can find probabilities for the average of a sample. Example 1.4 asks for the probability that the average of 25 scores is within 6.4 of the mean. Since μ is 100, this probability is the area between 93.6 and 106.4. To find this probability, we must *standardize the* $\overline{X}$ *scale.* In Section 1.5, we learned that a normal variable is converted to the Z scale by subtracting its mean and dividing by its standard deviation. Since $\overline{X}$ is normally distributed, we convert the relevant $\overline{X}$ values to the Z scale by *subtracting the mean on the* $\overline{X}$ *scale* and *dividing by the standard deviation on the* $\overline{X}$ *scale.* An $\overline{X}$ value is converted to a Z value using

$$Z = \frac{\overline{X} - \mu_{\overline{X}}}{\sigma_{\overline{X}}}. \tag{1.6.4}$$

For $\overline{X} = 93.6$ we find

$$Z = \frac{\overline{X} - \mu_{\overline{X}}}{\sigma_{\overline{X}}} = \frac{93.6 - 100}{3.2} = -2,$$

and for $\overline{X} = 106.4$ we find

$$Z = \frac{\overline{X} - \mu_{\overline{X}}}{\sigma_{\overline{X}}} = \frac{106.4 - 100}{3.2} = 2.0.$$

The probability that $\overline{X}$ is between 93.6 and 106.4 is the same as the probability that Z is between -2 and 2. In Section 1.5, this probability was found to be 0.9544. Thus, the probability that the average of 25 Stanford-Binet scores is within 6.4 of the mean is 0.9544.

Any normal variable can be standardized to the Z scale by subtracting its mean and dividing by its standard deviation. In Section 1.5, we standardized the normal variable X. In the example above, we standardized the normal variable $\overline{X}$. Subsequent chapters employ this rule for a variety of other normal variables. In each case, standardizing enables us to find probabilities for normally distributed variables from the table of the standard normal distribution.

1.7 THE MAGIC OF THE NORMAL DISTRIBUTION

When a random variable is *known* to have a normal distribution, the average is also normally distributed and we can find probabilities for $\overline{X}$ by the method above. Since many real-life processes yield data resembling the normal shape, this procedure has fairly wide applicability. Frequently, however, we are interested in a random variable *and have no idea whether its distribution is normal.* Even worse, we may plot our data and find that the plot definitely does *not* resemble the normal curve. Strangely enough, the normal distribution is still appropriate for finding probabilities about the average in such situations. This is the magic of the normal distribution.

In nature, certain quantities seem magical in that they appear in a variety of situations. As an example, consider the number π (pi). For any circle, the ratio of the circumference to the diameter is equal to π. π is also involved in finding the area of a circle, the volume of a sphere, and many other useful quantities. Similar magical constants include e (the base of natural logarithms), Planck's constant (from quantum mechanics), Avogadro's number (chemistry), and the speed of light (physics). In this same sense, the normal distribution is magical in the field of statistics.

The magic of the normal distribution becomes evident anytime we ask questions about the *average* in a sample. It can be shown that the probability distribution of the average approaches the normal curve as the sample size increases. This is true even if the distribution for individual observations is not normal. Were it not for this magical result, most of statistics would be limited to applications where normality of individual observations could be demonstrated.

We can see this magic taking place with a simple example. To illustrate, let us consider the trivial experiment of a coin flip, for which the probability distribution of an individual observation is known exactly. Let the random variable X be assigned the value 0 if the result is heads, and 1 if the result is tails. For a fair coin, X is a discrete random variable with probability distribution:

X	Probability
0	$\frac{1}{2}$
1	$\frac{1}{2}$

Since X is discrete, we represent this probability distribution with a line graph, as shown in Figure 1.14(a). The horizontal axis is labeled $\overline{X}$, although the average of a single observation is simply the observed X value—that is, for a sample of size one, $\overline{X} = X$. Clearly, the distribution for an individual coin flip is not normal. In fact, it is not even continuous.

Now suppose we draw a sample of size n = 2 from this population by flipping the coin twice. Let the result of the first flip be designated X_1 (with value of 0 if heads, 1 if tails), and the result of the second flip be designated X_2 (again with value

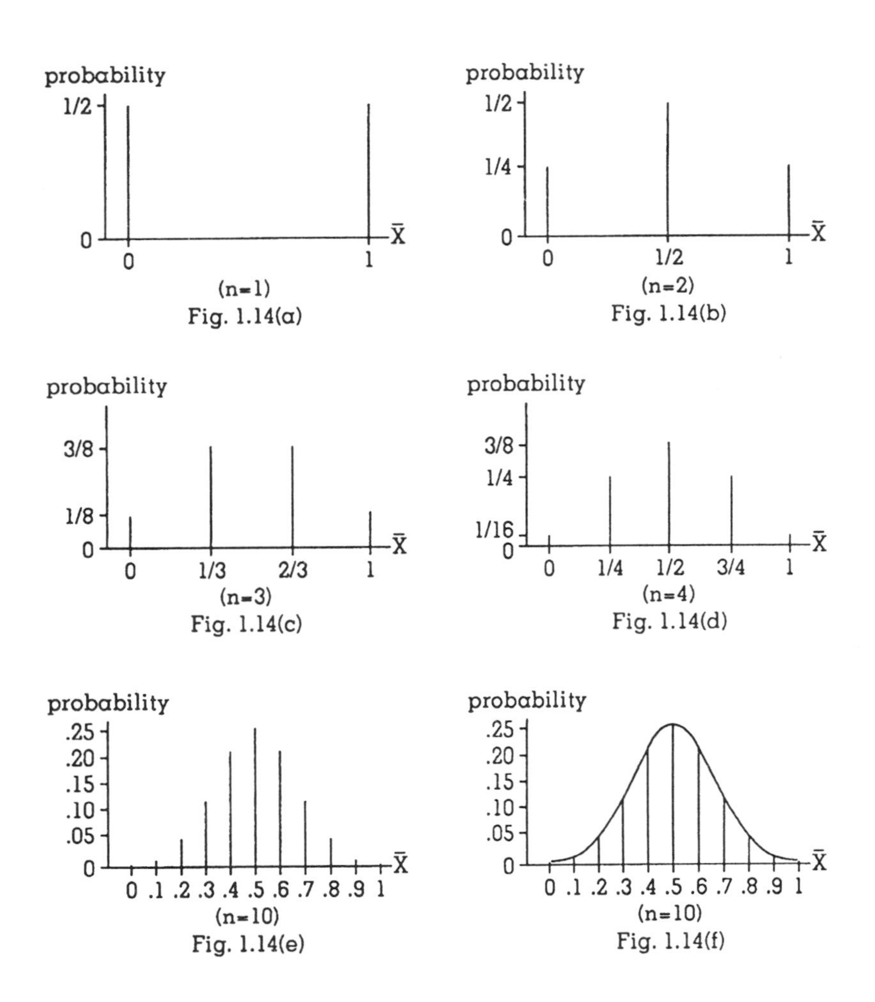

FIGURE 1.14. Probability distributions for the average in coin flipping experiment.

of 0 or 1). The probability of getting heads twice in a row is (½)(½) = ¼. In similar fashion, the probability of getting heads on the first flip and tails on the second is also (½)(½) = ¼. Table 1.1 lists all possible samples of size two (i.e., all possible X_1, X_2 pairs) and their corresponding probabilities. The last column in this display shows the average, $\overline{X}$, for each sample, where

$$\overline{X} = \frac{\Sigma X}{n} = \frac{X_1+X_2}{2}.$$

To obtain a probability distribution for $\overline{X}$, we list each $\overline{X}$ value and its probability. The value $\overline{X} = 0$ occurs with probability ¼, the value $\overline{X} = ½$ occurs with probability ¼ + ¼ = ½ (since $\overline{X} = ½$ with either heads-tails or tails-heads), and the value $\overline{X} = 1$ occurs with probability ¼. The probability distribution for $\overline{X}$ is summarized in Table 1.2. The graph of this distribution (Figure 1.14(b)) has a peak in the middle, but otherwise does not particularly resemble the normal curve.

Continuing in this manner, let us look at all possible samples of size n = 3, denoting the results of the three coin flips by X_1, X_2, and X_3. Table 1.3 displays all possible samples, while Table 1.4 shows the resulting probability distribution for $\overline{X}$. This distribution appears in Figure 1.14(c). Again, we see only a slight similarity to the normal curve.

For a sample of size n = 4, we follow a similar process, obtaining the distribution shown in Figure 1.14(d). By the time we get to a sample of size n = 10, the

TABLE 1.1 Possible Results for Two Coin Flips

X_1X_2	Probability	$\overline{X}$
00	¼	0
01	¼	½
10	¼	½
11	¼	1

TABLE 1.2 Probability Distribution for $\overline{X}$ Based on Two Coin Flips

$\overline{X}$	Probability
0	¼
½	½
1	¼

TABLE 1.3 Possible Results for Three Coin Flips

$X_1X_2X_3$	Probability	$\overline{X}$
000	1/8	0
001	1/8	1/3
010	1/8	1/3
100	1/8	1/3
011	1/8	2/3
101	1/8	2/3
110	1/8	2/3
111	1/8	1

TABLE 1.4 Probability Distribution for $\overline{X}$ Based on Three Coin Flips

$\overline{X}$	Probability
0	1/8
1/3	3/8
2/3	3/8
1	1/8

probability distribution for $\overline{X}$ begins to resemble the normal curve quite closely, as shown in Figures 1.14(e) and 1.14(f).

Since the distribution of $\overline{X}$ approaches a normal curve, we can use the normal distribution to find approximate probabilities for the average. Suppose we want to find the probability that $\overline{X}$ is less than or equal to 0.3 when $n = 10$. This probability is the sum of the heights of the line segments for $\overline{X} = 0.3$, $\overline{X} = 0.2$, $\overline{X} = 0.1$, and $\overline{X} = 0$ in Figure 1.14(e). By enumeration, as above, these heights can be shown to be 0.1172, 0.0439, 0.0098, and 0.0010, respectively. Adding these values yields the exact answer of 0.1719. Since the distribution resembles the normal distribution, however, we could approximate this probability by finding the area of the shaded region in Figure 1.15. We convert $\overline{X}$ to the Z scale and use the table of the standard normal distribution. Omitting the details, we find the approximate probability of 0.1711.

The magic of the normal distribution is that *the distribution of $\overline{X}$ approaches normality as sample size increases.* This is true regardless of the shape of the original distribution for an individual observation. How large a sample size is needed to achieve good accuracy depends on the shape of the original distribution. The closer this distribution is to the normal shape, the faster the approximation

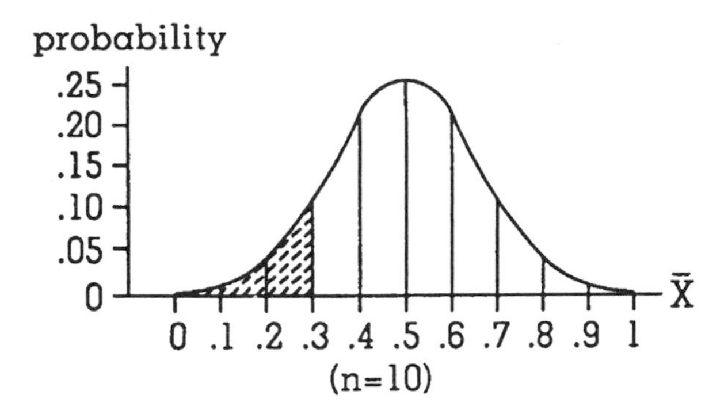

FIGURE 1.15. Normal approximation to the probability distribution of the average in coin flipping experiment.

works. If the original distribution is nearly normal, a small sample size is sufficient. If the distribution of an individual observation is far from normal, a relatively larger sample size is needed to make $\overline{X}$ approximately normal. In general, a sample of size 30 or more is sufficient to make the distribution of $\overline{X}$ extremely close to normal, no matter what the original distribution looks like. However, we do not always need 30 observations. In the coin flipping example, excellent accuracy is achieved for a sample of size 10, despite the fact that the distribution for an individual observation (Figure 1.14(a)) is far from normal.

We will discuss several other sample size issues in later chapters. The problem of determining an appropriate sample size cannot be answered without knowing why it is posed. For the purpose of making the distribution of $\overline{X}$ normal, the answer is that 30 or more observations is always a safe bet.

1.8 SUMMARY

This chapter has introduced the concept of using a random sample to infer conclusions about the population. We obtain data by conducting an experiment, and measure the outcome with a random variable. Every random variable has a probability distribution that reflects the population. The probability distribution for a discrete random variable is drawn as a line graph, in which the height of a line corresponds to probability. For a continuous random variable, the probability distribution is drawn as a curve, and probability is represented by area under the curve. In either case, probability can never be negative and must total to 1.0. Every probability distribution has a mean that describes the center of the population and a variance that describes the spread.

The normal distribution is important in statistics as a model for real data and as a limiting form for the distribution of the sample average. When sampling from a normal population, the sample average has a normal distribution with the same mean, but with variance inversely proportional to sample size. Even if the population is not normal, the probability distribution of the sample average approaches normality as sample size increases. Areas under any normal curve can be found by standardizing the normal variable to the Z scale. Since the sample average is normally distributed, it can be converted to a Z value to find probabilities.

2

Estimation

The population mean, μ, describes the center of the population with respect to a particular random variable and represents a typical value for observations. The population variance, σ^2, describes the spread and indicates how far away from this typical value observations are likely to fall. Knowing these two key parameters would tell us a great deal about what to expect when we observe the random variable. For most experiments, however, the probability distribution of the associated random variable is unknown. Thus, μ and σ^2 cannot be determined exactly. To estimate μ and σ^2, we draw a random sample from the population. Two kinds of estimates are distinguished: point estimates and interval estimates. A **point estimate** is a single value derived from the sample that serves as a "best guess" for the value of a parameter. An **interval estimate**, on the other hand, is a range of possible values for the parameter.

2.1 POINT ESTIMATES OF CENTER

In this section we discuss various point estimates for the center of a population. The population mean, μ, is the long-run average value of the random variable. The most obvious estimate for the long-run average is the sample average. In Example 2.1, the number of observations is $n = 6$, and the sum of these observations is $\Sigma X = 7 + 4 + 10 + 2 + 9 + 4 = 36$. The resulting value of the average, or sample mean, is

$$\overline{X} = \frac{\Sigma X}{n} = \frac{36}{6} = 6.$$

The sample mean is by far the most commonly used estimate for the population mean.

Another frequently used estimate for the center of the population is the **sample median**, or middle value. To find the median, we must first order the sample

Example 2.1

A large company was interested in the productivity of its sales force. Six salespersons were randomly selected, and the number of contracts negotiated in the past month was determined for each. The resulting values were 7, 4, 10, 2, 9, and 4.

---

from smallest to largest. For Example 2.1, the ordered observations are 2, 4, 4, 7, 9, and 10. If the number of observations is odd, there is a single value in the middle, and this middle value is the median. If the number of observations is even, the median is defined to be the average of the two middle values. In our example, n is even and the two middle values are 4 and 7. Thus, the sample median is (4 + 7)/2 = 5.5. One of the drawbacks to the median is that it does not use all the information in the sample. It is based only on the middle value or, if n is even, the two middle values. The rest of the sample is ignored. On the other hand, this can sometimes be advantageous. Suppose, for example, that a recording error was made when the data were collected and the observation with value of 10 was mistakenly listed as 100. The sample mean is radically affected by this error. We now find

$$\overline{X} = \frac{\Sigma X}{n} = \frac{7 + 4 + 100 + 2 + 9 + 4}{6} = \frac{126}{6} = 21.$$

However, the sample median, the average of the two middle observations, is unaffected by altering one of the extremes. The median is still 5.5.

Extreme values, or **outliers**, are not always caused by recording errors. Some populations are highly asymmetric, or skewed, such as the one shown in Figure 2.1. The distribution of personal incomes, for example, has a shape much like the one pictured. Most people have low to moderate incomes, but a few individuals have extremely large incomes. If a random sample is drawn from such a distribution, most of the values are likely to be near the bulk of the distribution, but a few observations may be very large. In this case, the sample average may not adequately represent the income of a typical individual. For describing a typical value, the median is preferable to the sample mean when the population is highly skewed.

Another way of dealing with extreme values is to discard them and average the rest of the sample, yielding the **trimmed mean**. Typically, the same number of observations are deleted at both the low and high ends of the sample. Judging in Olympic events is often based on trimming the lowest and highest scores awarded by the judges, and averaging the rest of the values. With very large samples, 5 percent of the data may be trimmed at each end, resulting in the 5%

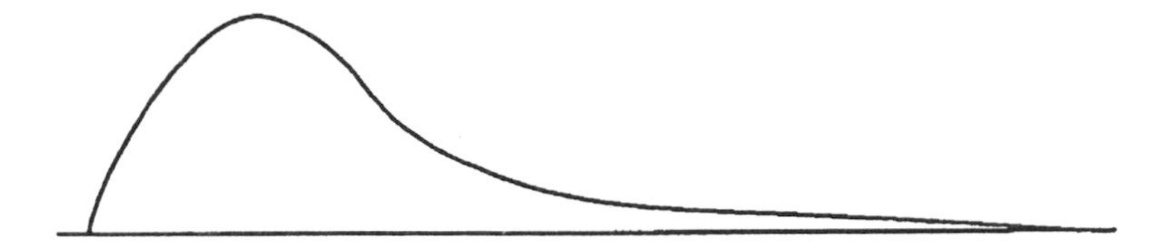

FIGURE 2.1. Skewed population.

trimmed mean. The median can be interpreted as an extreme case of the trimmed mean, where all the values have been deleted except for the middle observation (or the middle two observations).

Finally, we note that another measure, the **sample mode**, is sometimes considered as an estimate of centrality. The mode is the value that occurs most frequently in the sample. In Example 2.1, the mode is 4, since it is the only value that appears more than once. With small samples, using the mode to estimate the center of the population is not recommended. It is entirely possible that the most frequently occurring value is also the smallest (or largest) and therefore does not describe the center at all.

2.2 POINT ESTIMATES OF SPREAD

The estimation of center, or location, is a familiar problem. Most people have already computed a sample mean, median, trimmed mean, or mode, even if they did not use these terms to label their results. The estimation of spread in a population, on the other hand, is both more unusual and more difficult. The only measure of spread that is commonly employed outside the field of statistics is the **sample range**, the difference between the largest and smallest values in the sample. For Example 2.1, the range is $10 - 2 = 8$. Clearly the range is highly susceptible to erroneous or outlying values. In fact, the range measures spread using only the most extreme observations and ignores any information contained in the rest of the data. For this reason, the range is a poor estimate of spread in the population and is rarely used in statistics.

We can obtain a much better estimate of population spread if we recall the meaning of the population variance, σ^2. In Section 1.4, σ^2 was defined to be the long-run average squared distance from the mean. If the letter X is used to represent the random variable, σ^2 is the long-run average value of $(X-\mu)^2$. (Example 1.2 in Chapter 1 dealt with the radius of iron disks and used the letter R to represent the random variable. There we defined the population variance to be the long-run average of $(R-\mu)^2$. The name of the variable is unimportant; squared distance from the mean is the crucial concept.) Since the population variance is the *long-run* average squared distance from the mean, a natural way of estimating σ^2

is to calculate the average squared distance from the mean *in our sample*. That is, we might want to compute the average value of $(X-\mu)^2$ in our sample by adding this value for each observation and dividing by the number of observations. This would yield the formula

$$\frac{\Sigma(X-\mu)^2}{n}.$$

Unfortunately, we cannot actually use this approach, since μ is unknown. We circumvent this difficulty by replacing μ with an estimate, usually $\overline{X}$. Our numerator now is $\Sigma(X-\overline{X})^2$. When making this substitution, however, we should also replace n in the denominator with n - 1. (The explanation for this substitution will be discussed in Section 2.7.) By this reasoning we arrive at the **sample variance**, denoted by the symbol S^2, with the formula

$$S^2 = \frac{\Sigma(X-\overline{X})^2}{n-1}. \tag{2.2.1}$$

To demonstrate computation of the sample variance, consider the six observations in Example 2.1. The sample mean was found above to be $\overline{X} = 6$. To find S^2, we subtract 6 from each observation, square these differences, sum, and divide by n - 1. These calculations are displayed in Table 2.1. Here, $\Sigma(X - \overline{X})^2 = 50$ and

$$S^2 = \frac{\Sigma(X-\overline{X})^2}{n-1} = \frac{50}{6-1} = \frac{50}{5} = 10.$$

The sample variance, S^2, is the usual estimate of the population variance, σ^2.

We noted in Section 1.4 that the population variance is expressed in the square of the original units of measurement. To obtain a measure of spread in the original units, we defined the population standard deviation, σ, to be the square root of σ^2. Logically, to estimate σ we need only take the square root of our estimate of σ^2. The sample standard deviation, denoted by S, is the square root of S^2, and serves as the estimate of σ. For Example 2.1, $S = \sqrt{S^2} = \sqrt{10} = 3.162$. By the same logic used in Section 1.4, we can see that S^2 is in the square of the original units, while S is back in the original units.

While S^2 and S are the usual estimates of population spread, it is occasionally useful to be able to characterize *spread relative to the magnitude of the numbers*.

TABLE 2.1 Calculations for Sample Variance

X	$X - \overline{X}$	$(X - \overline{X})^2$
7	1	1
4	-2	4
10	4	16
2	-4	16
9	3	9
4	-2	4
		50

To demonstrate this concept, suppose we drew a sample of size 5 from a population and observed the values 11, 23, 1, 15, and 10. There appears to be a great deal of spread in this sample; the largest value is 23 times as big as the smallest. Now consider a second population from which we drew a sample of size 5 and observed the values 1011, 1023, 1001, 1015, and 1010. The spread here is not nearly so severe, since the ratio of largest to smallest (1023/1001) is just slightly bigger than 1. Clearly, variability is not as great a concern in the second population as in the first. However, both samples yield identical values of S, namely S = 8.

Looking carefully at the second sample, we see that each observation is exactly 1000 more than the corresponding observation in the first sample. If we plot both samples, the plot for the second sample looks exactly like the plot for the first, except shifted 1000 units to the right. The two plots have the same spread and, thus, the same value of S. Yet, relative to the size of the numbers, the spread in the first sample is far greater than in the second. To measure *spread relative to the size of the numbers,* we use the **coefficient of variation**, given by 100 times $(S/\overline{X})$. For the first sample, the coefficient of variation is 100(8/12) = 66.667, whereas the second sample yields a coefficient of variation of 100(8/1012) = 0.791. The difference in these values reflects the fact that spread relative to the size of the numbers is far greater in the first sample than in the second.

The coefficient of variation is a dimensionless quantity. Since both S and $\overline{X}$ are in the original units of measurement, the units cancel in the fraction. Thus, we can use the coefficient of variation to compare spread relative to the magnitude of the numbers, even when two samples are measured on different scales.

Earlier we noted that extreme values, or outliers, have a strong influence on estimates of center, such as the sample mean. Their effect on measures of spread is even more dramatic. Example 2.1 yielded a sample variance of S^2 = 10. If the observation with value of 10 were erroneously recorded as 100, the resulting sample variance would be 1504. For a highly skewed population, extreme values are likely to be present in a sample. In this case, the sample variance can paint a very misleading picture of the spread of typical observations. To describe spread in these situations, we use the **interquartile range**, which is the difference between the first and third quartiles. The first quartile is the number below which 25 percent of the sample values lie. The third quartile is the value below which 75 percent of the sample lies. The interquartile range, denoted by IQR, is the difference between these two, and reflects spread for the middle half of the data. While quartiles are easily understood, their computation often involves difficulties. For Example 2.1, we would need to divide six observations into four equal parts. There are several ways to approximate such a division, and different computer programs often yield slightly different answers for the interquartile range.

The probability distribution of a random variable, say X, describes the population with respect to this particular variable. All of the measures discussed above are designed to estimate the spread of the distribution of X. Frequently, however, we are also interested in probability distributions of other random variables. In Section 1.6 we discussed the distribution of $\overline{X}$, the sample mean for a random sample. We found that $\overline{X}$ has a variance given by $\sigma^2_{\overline{X}} = \sigma^2/n$, where σ^2 is the variance of X and n is the sample size. By taking the square root, we found the standard deviation of $\overline{X}$ to be $\sigma_{\overline{X}} = \sigma/\sqrt{n}$. To estimate $\sigma_{\overline{X}}$, we need only to replace σ with its usual estimate, S. This yields $S/\sqrt{n}$ as the estimated standard deviation of $\overline{X}$. For Example 2.1, the estimated standard deviation of $\overline{X}$ is

$$\frac{S}{\sqrt{n}} = \frac{3.162}{\sqrt{6}} = 1.291.$$

An estimated standard deviation is often referred to as a **standard error**. In Example 2.1, the standard error of the sample mean, denoted by SE($\overline{X}$), is 1.291. Note that we must be careful to indicate the variable for which we are estimating the standard deviation—in this case, $\overline{X}$. The expression "standard error" is sometimes used rather sloppily in statistics books and computer programs, without indicating the variable for which the standard deviation is being estimated.

2.3 INTRODUCTORY COMMENTS ABOUT COMPUTER PROGRAMS

The availability of modern computer programs has relieved us of most of the drudgery of performing statistics. On the other hand, this freedom carries with it the responsibility that we use such programs carefully. Too often, people ignore their own common sense in favor of results generated on a computer. This occurs even in cases where the wrong computations are requested from the computer, or where the right computations are requested but the results are misinterpreted or misrepresented. We should only request computer programs to perform analyses that we already understand on a logical basis, and should only draw conclusions from outputs we fully comprehend. The computer is an awesome tool for performing calculations, but it is not a substitute for thinking.

This book demonstrates examples using three of the most frequently employed computer packages, namely SAS, SPSS, and MINITAB. The SAS package is by far the most popular program for performing statistical analyses. It is probably the most powerful program in common use, in that it can analyze data from extremely complex experiments. SPSS, while possibly not quite as general as SAS, allows more options for data display and presentation. The third program, MINITAB, is designed as much for teaching statistics as for practical applications. MINITAB is broad enough to perform most of the standard analyses, while being basic enough to handle situations that are sometimes unrealistically simple. An example of the latter type is computing probabilities when the population variance is known, which is almost never the case in real applications.

Programs appearing in this book were run on release 6.06 of SAS, release 4.1 of SPSS, and release 7.1 of MINITAB. Future releases of SAS and SPSS are unlikely to vary greatly from current versions, as both programs have been in use for many years. Future releases of MINITAB, a much newer program, may include procedures unavailable in the current version.

All three of these programs are available in both mainframe (i.e., large computer) and microcomputer versions. Each of them is also available in both batch and interactive modes. **Batch mode** refers to typing a whole list of instructions, submitting them as input to a computer, and receiving output when the whole batch of instructions has been performed. **Interactive mode** allows the user to enter one request at a time, receive the results, and move on to the next request. The user and computer "take turns," as if the user were conducting a conversation with the program. For reasons related to both cost of operation and design of the programs, most users employ SAS and SPSS in batch mode and MINITAB in interactive mode. In keeping with this common usage, all our examples on SAS and SPSS are presented in batch mode, and all our examples on MINITAB are performed in interactive mode.

In general, computer programs produce far more results than requested. If asked for a sample mean, for example, most programs also calculate and print the sample variance and standard deviation. As a result, virtually every output contains some irrelevant information or at least some results that are only relevant in special circumstances. Part of learning to use a computer package intelligently is the ability to separate the useful output from the superfluous. In our examples we generally do not explain every number shown on each output. Our purpose is to demonstrate the use of computer packages for performing statistical analyses, not to explain every peculiarity of each program. Only the meaningful and relevant numbers on computer outputs are discussed.

By contrast, *all* the SAS, SPSS, or MINITAB commands needed to input and run each example are presented and discussed. The only input statements not shown in this text are the one or two lines necessary to identify you as a valid user of the computer, and to request access to a particular statistical package. Since these lines differ at every computer installation, you will have to learn the specific statements needed for your computer. In our demonstrations, we include as input only the commands needed to obtain the statistical analyses discussed in this book. Optional statements designed to produce additional information and fancy printouts are avoided.

Demonstrations in this book present and discuss the input and output for each of these three programs. If you plan to use only one of these programs, you may want to read only the corresponding discussion. If your installation has a statistical program other than these three, you should try the data in our examples on your program and compare the results to those shown here. All of our examples are presented with the data "in line"—that is, the data to be analyzed are contained in the input statements. Each program also provides options for reading data that have been previously stored in a file. For complete information on the program you are using, consult the manual.

2.4 POINT ESTIMATION IN COMPUTER PROGRAMS

All three of the programs—SAS, SPSS, and MINITAB—produce a number of point estimates for center and spread. For Example 2.1, we present and discuss in this section the input and output for each of these three programs.

SAS Input

```
OPTIONS LS = 80;
TITLE 'EXAMPLE 2.1 - SALES CONTRACTS';
DATA;
INPUT SALES;
LINES;
 7
 4
10
 2
 9
 4
PROC UNIVARIATE;
```

Discussion of SAS Input

Every SAS statement must end in a semicolon. The semicolon serves the same function as a period at the end of a sentence in English. If a statement extends for more than a single line, the program continues to search successive lines until it finds a semicolon. On the other hand, more than one statement may be entered on a line by inserting a semicolon after each statement. For clarity, we typically show one statement per line, but this is not necessary in applications.

The first line of input asks for the output to be formatted into lines of 80 characters each, as would be appropriate for viewing on most CRT (cathode ray tube) terminals. Since most users employ such terminals, we start every SAS job with this request. The TITLE command is optional; if included, each page of the output is identified with this title. The DATA statement informs the program that a dataset is forthcoming. Each variable for which values appear in the dataset must be named on the INPUT line; in our example, only one variable, SALES, is entered for each observation. Variable names can be from one to eight characters and should begin with a letter. The LINES statement immediately precedes the actual data lines. The data to be analyzed should be entered with one observation per line; data lines are the only input lines that do not end in semicolons. In the final command we request that SAS use its procedure (i.e., subroutine) named UNIVARIATE to analyze our data.

SAS Output

```
                 EXAMPLE 2.1 - SALES CONTRACTS                    1
                      Univariate Procedure

Variable = SALES
                            Moments
N                  6           Sum Wgts            6
Mean               6           Sum                36
Std Dev     3.162278           Variance           10
Skewness    0.113842           Kurtosis       -1.868
USS              266           CSS                50
CV          52.70463           Std Mean     1.290994
T:Mean=0     4.64758           Prob>|T|       0.0056
Num ^= 0           6           Num > 0             6
M(Sign)            3           Prob>|M|       0.0313
Sgn Rank        10.5           Prob>|S|       0.0313

                      Quantiles (Def = 5)
100% Max          10           99%                10
 75% Q3            9           95%                10
```

```
 50% Med              5.5        90%                10
 25% Q1                 4        10%                 2
  0% Min                2         5%                 2
                                  1%                 2
Range                   8
Q3-Q1                   5
Mode                    4

                        Extremes
          Lowest      Obs      Highest      Obs
              2(        4)          4(        2)
              4(        6)          4(        6)
              4(        2)          7(        1)
              7(        1)          9(        5)
              9(        5)         10(        3)
```

Discussion of SAS Output

The 1 in the upper right corner identifies this as the first page of output. SAS always numbers the output pages. From top to bottom, values in the first block of output that we have discussed are: number of observations (N), sample mean (Mean), sample standard deviation (Std Dev), sample variance (Variance), coefficient of variation (CV), and standard error of the mean (Std Mean). The use and interpretation of some of the other values in this block will be discussed in Chapter 3.

Under the heading Quantiles, we find the largest value (100% Max), third quartile (75% Q3), sample median (50% Med), first quartile (25% Q1), and smallest value (0% Min). To the right of these figures are other percentiles, and below are the sample range (Range), interquartile range (Q3-Q1), and sample mode (Mode). At the bottom of the page, SAS prints the five lowest and five highest values, identified by their observation numbers.

SPSS Input

```
SET WIDTH = 80
TITLE 'EXAMPLE 2.1 - SALES CONTRACTS'
DATA LIST FREE / SALES
EXAMINE VARIABLES = SALES / PLOT = NONE
BEGIN DATA
 7
 4
10
```

```
 2
 9
 4
END DATA
FINISH
```

Discussion of SPSS Input

SPSS statements must begin in the first column of a line, and can be continued for as many lines as needed by indenting successive lines at least one space. The SET WIDTH command asks for the output to be formatted into lines of 80 characters each, as would be appropriate for viewing on most CRT (cathode ray tube) terminals. Since most users employ such terminals, we start every SPSS job with this request.

The TITLE command is optional; if included, each page of the output is identified with this title. On the DATA LIST statement, the keyword FREE is used to indicate that numeric values in the input may be any number of digits,with or without decimal points. The names of all variables are listed after the slash; in our example, only one variable, SALES, is entered for each observation. Variable names can be up to eight characters long, and should begin with a letter.

The next line of input asks SPSS to use its subroutine called EXAMINE to analyze the data. All variables to be described should be listed after the keyword VARIABLES. The slash on this line is followed by an optional request, in this case to suppress plots that would otherwise be printed automatically. (These plots will be demonstrated in Section 2.13, after we have introduced them and explained their purpose in Section 2.12.) The actual data are submitted one line per observation and are delineated with BEGIN DATA and END DATA statements. *Note that the data follow the first request for an analysis.* If we wanted to do further analyses with these same data, additional requests would be inserted between END DATA and the FINISH statement.

SPSS Output

```
                   EXAMPLE 2.1 - SALES CONTRACTS                Page 1
  SALES

Valid cases:     6.0     Missing cases:   .0     Percent missing:   .0

Mean      6.0000    Std Err     1.2910    Min     2.0000    Skewness     .1138
Median    5.5000    Variance 10.0000    Max    10.0000    S E Skew     .8452
5% Trim   6.0000    Std Dev     3.1623    Range  8.0000    Kurtosis  -1.8680
                                          IQR     5.7500    S E Kurt   1.7408
```

Discussion of SPSS Output

SPSS numbers the output pages, the first couple of which contain information related to the computer operating system. We have taken the liberty of renumbering the pages to eliminate this information. From left to right, the values on this output that we have discussed are: number of observations (Valid cases), sample mean (Mean), sample median (Median), 5% trimmed mean (5% Trim), standard error of the mean (Std Err), sample variance (Variance), sample standard deviation (Std Dev), sample range (Range), and interquartile range (IQR).

MINITAB Transcript

```
MTB > NAME C1 = 'SALES'
MTB > SET 'SALES'
DATA > 7 4 10 2 9 4
DATA > END

MTB > DESCRIBE 'SALES'

               N      MEAN    MEDIAN    TRMEAN    STDEV   SEMEAN
SALES          6      6.00      5.50      6.00     3.16     1.29

             MIN       MAX        Q1        Q3
SALES       2.00     10.00      3.50      9.25

MTB > STOP
```

Discussion of MINITAB Transcript

Since MINITAB is interactive, separate files do not exist for input and output. Instead, the transcript of a conversation between MINITAB and the user is a single list of commands and responses, such as that shown here. The symbol > and the keyword to its left are MINITAB's prompt for user input. Any entry to the right of the > symbol is specified by the user. In all of our MINITAB examples, the user's input is underlined for clarity. The prompt MTB > indicates that MINITAB is ready for a new request from the user. A different prompt, such as DATA >, is used when MINITAB needs additional information to perform an earlier request.

MINITAB keeps track of the data by recording it on a rectangular worksheet. The columns of the worksheet are labeled C1, C2, and so on. Each column of the worksheet contains the values of a different variable. Each row of the worksheet corresponds to an observation, with the values of all the variables for this observation listed in the appropriate columns. Summary values descriptive of all the data may be stored in a separate list of constants, identified by K1, K2, and so on.

In the first line of this transcript, column C1 on the worksheet is identified to be a variable named SALES. Variables not specifically named by the user can be referenced by their column number, but it is recommended that you name each variable, so you will not have to remember its column number. A variable name can be up to eight characters long, and should begin with a letter. The SET statement on the second line of the transcript informs MINITAB that we are ready to enter the values of the variable named SALES for each observation. MINITAB responds with a prompt for the data, and continues to expect more data until we tell it we are finished with the keyword END. Note that MINITAB now is ready for a new command and thus responds with the prompt MTB >.

The DESCRIBE command on the fifth line produced all the output printed below it. The values in this output that we have discussed are: number of observations (N), sample mean (MEAN), sample median (MEDIAN), 5% trimmed mean (TRMEAN), sample standard deviation (STDEV), and standard error of the mean (SEMEAN). The sample range is not listed, but can be obtained as MAX - MIN = 10 - 2 = 8. Similarly, the interquartile range is not printed, but can be computed as the difference between the third and first quartiles: IQR = Q3 - Q1 = 9.25 - 3.50 = 5.75. In the final line of the transcript, the STOP statement ends the MINITAB session and returns control to the operating system of the computer.

2.5 INTERVAL ESTIMATION

Statistics computed from a sample are used to estimate parameters of a population. For example, the sample mean, $\overline{X}$, is an estimate of the population mean, μ. This estimate has several desirable properties: it is logical, it uses all the information in the sample, and its probability distribution is centered at the desired value (that is, the mean of the $\overline{X}$ distribution, $\mu_{\overline{X}}$, is the same as the population mean, μ). However, it also has one undesirable property: it is *correct* with probability zero! Since $\overline{X}$ is a continuous random variable, the probability that it takes any specific value is zero. In particular, the probability that it equals μ *exactly* is zero. While common sense tells us that $\overline{X}$ is a reasonable estimate of μ, it is unlikely that it is precisely equal to μ. The same problem is encountered with all the usual point estimates, since they are all continuous variables. S^2 is equal to σ^2 with probability zero, and S is equal to σ with probability zero. The only way to obtain a nonzero probability for a continuous variable is to look at a region, or interval of values. This process was demonstrated in Section 1.6, where we found the probability of a range of values for $\overline{X}$.

A point estimate is a single "best guess" at the value of a parameter. This estimate always has probability zero of being correct. Instead of trying to specify the value of a population parameter exactly, we may wish to consider an interval around the parameter. In Example 2.1, for instance, we found $\overline{X} = 6$, so we might think that μ is between, say, 5 and 7. There is a nonzero probability that this interval includes μ. Ideally, we would like to find the end points of the interval in such a way that this probability is large, say 95 percent. If these end points are designated by a and b, the interval from a to b is referred to as a 95% **confidence interval** for μ. In the next section we show how to use the information contained in a random sample to find the values a and b.

In finding a confidence interval for the population mean, we consider two cases: when the population standard deviation (σ) is known and when it is unknown. The first case clearly is unrealistic. No practical situations exist in which the population mean must be estimated, but the population standard deviation is already known. Nonetheless, we use this simpler case to develop the logic of confidence intervals. Then we consider the realistic case of unknown standard deviation, and learn how to incorporate an estimate of σ into our result. Finally, we formulate a confidence interval for the population variance.

2.6 CONFIDENCE INTERVAL FOR THE MEAN WHEN STANDARD DEVIATION IS KNOWN

To find the 95% confidence interval for μ, we need some of the facts we learned in the first chapter. For a variable with a standard normal distribution, commonly denoted by Z, the area between the values -2 and +2 is about 95 percent—that is, the probability that $-2 < Z < +2$ is approximately 0.95. We also learned that any variable with a normal distribution can be converted to a standard normal variable by subtracting its mean and dividing by its standard deviation. The sample mean, $\overline{X}$, is normally distributed with mean $\mu_{\overline{X}} = \mu$, and standard deviation, $\sigma_{\overline{X}} = \sigma/\sqrt{n}$. Standardizing $\overline{X}$, we obtain the standard normal variable

$$Z = \frac{\overline{X} - \mu_{\overline{X}}}{\sigma_{\overline{X}}}.$$

Substituting $\mu_{\overline{X}} = \mu$ and $\sigma_{\overline{X}} = \sigma/\sqrt{n}$, this standard normal variable is generally written as

$$Z = \frac{\overline{X} - \mu}{\sigma/\sqrt{n}}. \qquad (2.6.1)$$

Since $-2 < Z < +2$ with probability 0.95, we can replace Z in this inequality with the expression shown in (2.6.1), to find that

$$-2 < \frac{\overline{X} - \mu}{\sigma/\sqrt{n}} < +2$$

with probability 0.95. With a bit of algebra, this expression can be rearranged to yield

$$\overline{X} - 2\frac{\sigma}{\sqrt{n}} < \mu < \overline{X} + 2\frac{\sigma}{\sqrt{n}},$$

without affecting the probability. Taking advantage of the similarity of the end points, the interval for μ can be written as

$$\overline{X} \pm 2\frac{\sigma}{\sqrt{n}},$$

where the lower bound for μ is obtained using the minus sign, and the upper bound is computed with the plus sign.

A minor refinement is generally made in this expression for the 95% confidence interval to achieve slightly greater precision. We stated in Section 1.5 that *about* 95 percent of the area lies between the Z values of -2 and +2. When we referred to the normal table in the appendix, we found that the area between -2 and +2 is actually 0.9544. To more precisely capture 95 percent of the area in the center of the normal distribution, we should slightly modify the Z values of ±2. For 95% center area, the total of the two tail areas must be 5 percent, with 0.025 in each tail. Since this is a tail area, we search the body of Table 1 for this number and read outward to the margins to find the Z value of 1.96. Thus, ±1.96 should be used in place of ±2, yielding the interval

$$\overline{X} \pm 1.96\frac{\sigma}{\sqrt{n}}. \tag{2.6.2}$$

In summary, if we draw a random sample of size n from a population with mean μ and standard deviation σ, the bounds shown in (2.6.2) will include the population mean, μ, with probability 0.95.

Computation and interpretation of this confidence interval are demonstrated in connection with Example 2.2. The first step in finding the desired confidence interval is to compute the sample mean, $\overline{X}$. The sample size in this example is n = 15, and the sum of observations is $\Sigma X = 291$, yielding $\overline{X} = 291/15 = 19.4$. We are told that the population standard deviation, σ, is 6.3. Substituting these values into the formula for the 95% confidence interval, we find the bounds

$$\overline{X} \pm 1.96\frac{\sigma}{\sqrt{n}} = 19.4 \pm 1.96\frac{6.3}{\sqrt{15}} = 19.4 \pm 3.188.$$

The lower bound is 19.4 - 3.188 = 16.212, and the upper bound is 19.4 + 3.188 = 22.588. We are 95 percent certain that the interval from 16.212 to 22.588 contains the population mean, μ.

To interpret this interval, we note that the only random variable involved is the sample mean, $\overline{X}$. If we were to draw another sample of size 15,we would probably find a slightly different value for $\overline{X}$ and, therefore, slightly different lower and upper bounds. Repeating this process with many different samples would result in many different intervals. In the long run, 95 percent of these intervals would include the true mean.

Sample size plays a role in determining the *width* of the confidence interval. Suppose we had used a sample size of 30 instead of 15. The value of $\overline{X}$ would probably have been different, but not by very much. As we learned in Section 1.6, the probability distribution for $\overline{X}$ is centered at μ. A larger sample makes this

---

Example 2.2

In a study commissioned by the seafood industry, 15 lobsters were randomly selected from recent catches along a certain region of the Maine shore line. The lobsters were weighed to the nearest ounce, with results:

26	15	12
14	24	31
18	21	19
13	29	16
22	10	21

Suppose it is known that the population standard deviation is 6.3 ounces. Find the 95% confidence interval for μ, the mean weight of all lobsters in this region of the Maine shore line.

distribution narrower, but does not shift its center—that is, $\overline{X}$ is a good estimate of μ regardless of the sample size. The only major change expected in our confidence interval would be the value of n. Assuming that $\overline{X}$ did not change much, we would now find the interval

$$\overline{X} \pm 1.96\frac{\sigma}{\sqrt{n}} = 19.4 \pm 1.96\frac{6.3}{\sqrt{30}} = 19.4 \pm 2.254\ .$$

The resulting bounds for μ would have been 17.146 and 21.654. Note that this interval is narrower than the interval obtained above with a sample size of only 15 (bounds of 16.212 and 22.588). Common sense tells us that more information should enable us to do a better job of estimation. And indeed, a larger sample size results in a *tighter* 95% confidence interval.

In some circumstances, 95% confidence still allows too much chance for error. We may need an interval in which we can have, say, 99% confidence. The only change needed is in the Z values of ±1.96. These values include 95 percent of the area under the standard normal curve. To include 99 percent of the area in the center, each tail area must be 0.005. The closest value to 0.0050 in the body of Table 1 is 0.0049, corresponding to the Z value 2.58. (A normal table with more digits of accuracy would confirm that 2.58 should be used instead of 2.57.) The resulting 99% confidence interval is

$$\overline{X} \pm 2.58\frac{\sigma}{\sqrt{n}} = 19.4 \pm 2.58\frac{6.3}{\sqrt{15}} = 19.4 \pm 4.197,$$

with bounds 15.203 and 23.597. This interval is somewhat wider than the 95% confidence interval, which has bounds of 16.212 and 22.588. Logically, to be more sure of trapping the true value of μ, we need to allow more space. However, we can never be 100% certain that our interval includes the true mean. The normal distribution approaches the axis as the Z values become more extreme, but it never quite hits the axis. We can never trap 100 percent of the area. Statistics gives us answers that are *probably* correct. If you need to be 100% certain, statistics cannot help; you must measure the entire population.

2.7 THE t DISTRIBUTION

If the population standard deviation (σ) is known, the procedure described above can be used to obtain a confidence interval for the population mean. In practical applications, however, σ is almost never known. In practice, σ is replaced with its estimate, S. Instead of using

$$Z = \frac{\overline{X} - \mu}{\sigma/\sqrt{n}},$$

we now form the ratio

$$\frac{\overline{X} - \mu}{S/\sqrt{n}}. \tag{2.7.1}$$

Since this quantity is slightly different than Z, its probability distribution is also slightly different. The ratio in (2.7.1) has a distribution known as the **t distribution**.

The normal distribution is actually a family of distributions, all sharing the same general properties, but differing in their centers (values of μ) and in their spreads (values of σ). In the same sense, the t is actually a family of distributions. While the normal family is indexed by the two parameters, μ and σ, the family of t distributions is indexed by a single parameter, called **degrees of freedom**. For the t ratio in (2.7.1), the degrees of freedom, denoted by df, is given by df = n - 1, where n is the sample size. The term "degrees of freedom" is used in many statistical procedures, and an understanding of this expression will be most helpful.

When we take a random sample from a population, each observation is free to vary independently. For instance, if we draw a sample of size 10, knowing the first nine values does not tell us the tenth. If the random variable is named X, each of the 10 values of X is free to vary independently of the others.

If we modify the observations, however, some of this independence may be lost. For example, if we subtract the sample mean from each observation, and examine the values of $X-\overline{X}$, we find that they are not totally independent. In fact, the values of $X-\overline{X}$ always sum to 0. When we demonstrated the computation of the sample variance (S^2) in Section 2.2, our display included a column labeled $X-\overline{X}$. The numbers in this column sum to 0. This column sums to 0 for *any* sample, since $X - \overline{X}$ is positive for observations higher than the average, and negative for observations lower than average. By the definition of average, the positives and negatives must cancel, leaving a sum of 0. Thus, knowing the first n - 1 values of $X-\overline{X}$, we could determine the last. Only n - 1 of these quantities are free to vary. As a result, any computation based on the $X-\overline{X}$ values has n - 1 degrees of freedom. The sample variance, which has the sum of the squares of these values in its numerator, has df = n - 1.

We will see many other variance estimates for special situations in future chapters. Every variance estimate is a ratio with a sum of squares in the numerator and degrees of freedom in the denominator. This explains why we use n - 1 instead of n in the denominator of S^2; the numerator of S^2 is the sum of squares of $X-\overline{X}$, and these quantities have n - 1 degrees of freedom. The sample standard deviation, S, is the square root of the sample variance. It is based on the same information as S^2 and therefore also has n - 1 degrees of freedom.

The t distribution is employed in answering many of the questions we ask about a population. But the expression given in (2.7.1) is not the only variable that has a t distribution. In each application, the numerator of the t ratio is the difference between a random variable and the parameter it estimates. In the current setting, the numerator is $\overline{X}-\mu$, where the random variable $\overline{X}$ estimates the parameter μ. The denominator of the t ratio is the estimated standard deviation (i.e., standard error) of the random variable in the numerator. In our case, the random variable in the numerator is $\overline{X}$, with *true* standard deviation $\sigma_{\overline{X}} = \sigma/\sqrt{n}$. The *estimated* standard deviation of $\overline{X}$, its standard error, is $S/\sqrt{n}$, which appears in the denominator of the t. The degrees of freedom for a t ratio is the same as the degrees of freedom for the estimated standard deviation in its denominator. Thus, in the present case, the ratio in (2.7.1) has a t distribution with n − 1 degrees of freedom.

On a more intuitive level, we can think of degrees of freedom as a form of currency. In the everyday world, the fruits of our labors are rewarded with dollars. We spend these dollars in the marketplace for food, shelter, clothing, and other necessities and luxuries of life. In a broad sense, statistics deals with the marketplace of knowledge. Labor, in this framework, corresponds to the task of gathering information. More precisely, our job is to draw a random sample from a population. Each observation we obtain results in a degree of freedom, much like every few minutes of work at a job results in a dollar earned. In statistics, we spend our degrees of freedom to estimate parameters, to increase the probability of reaching correct decisions, or to form models of the way the world behaves. In short, degrees of freedom buy knowledge. While we admit that degrees of freedom is a strange name for the coin of the realm, the analogy is appropriate. One of the purposes of this book is to teach travelers in the world of statistics to spend their degrees of freedom wisely.

The t ratio is very similar to Z, the standard normal ratio. Not surprisingly, the t probability distribution is quite similar to the standard normal distribution. The t distribution is continuous, has a mean of 0, and is symmetric. It is unimodal, with the single lump occurring at the mean, and approaches the axis at both ends. The t distribution has a variance slightly greater than one (which is the variance of the standard normal distribution). Therefore, the t distribution is slightly more spread out than the Z distribution.

Figure 2.2 shows a t distribution (dashed curve) in relation to the standard

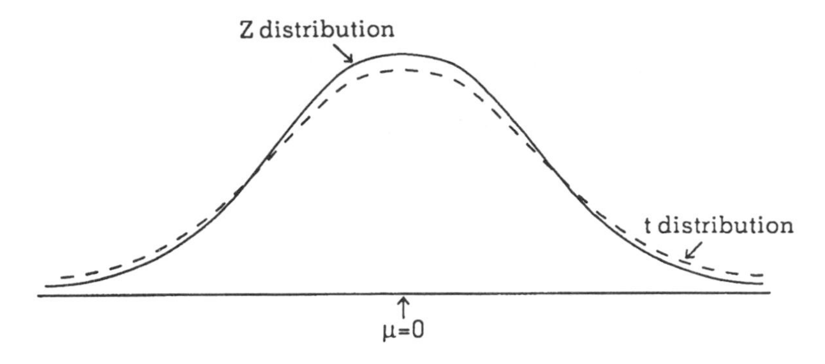

FIGURE 2.2. Comparison of t and Z distributions.

normal distribution (solid curve). For different values of df, the t distribution may be slightly wider or slightly narrower than pictured here. As the value of df increases, the variance of the t distribution approaches one, and the t curve approaches the Z curve. Common sense also tells us that the Z curve is the limiting shape of the t curve. The only difference between a t ratio and a Z ratio is in the denominator; the t contains an estimate of the standard deviation, whereas Z has the true standard deviation. For a large sample (and therefore large df), the estimated standard deviation should be very close to the true standard deviation, and the t ratio should be almost identical to the Z ratio.

2.8 CONFIDENCE INTERVAL FOR THE MEAN WHEN STANDARD DEVIATION IS UNKNOWN

In Example 2.2, the standard deviation was presumed known. This allowed us to use the Z distribution to form a confidence interval for the mean. Let us reconsider this example, but now suppose the standard deviation is *unknown.* When σ is unknown, we use the t distribution to obtain a confidence interval for μ.

The process of constructing a confidence interval is similar to that used in Section 2.6 when σ was known. Now, however, we must use the t distribution to find the desired cutoff values. Table 2 in the Appendix lists cutoff points on the t axis for several upper tail areas and various values of df. To obtain a 95% confidence interval, each tail must have area 0.025. This is one of the values shown along the top margin of the table. In Example 2.2, the sample size is 15, and therefore df = n - 1 = 14. In the row for 14 degrees of freedom, we find the 0.025 cutoff to be 2.145. Since the t distribution is symmetric about 0, the area below -2.145 would also be 0.025. Thus, 95 percent of the area is contained between the values -2.145 and 2.145. The t ratio given in (2.7.1) lies between these two values with probability 0.95—that is, the probability is 0.95 that

$$-2.145 < \frac{\overline{X}-\mu}{S/\sqrt{n}} < +2.145. \tag{2.8.1}$$

Without affecting the probability, this statement can be rearranged to yield

$$\overline{X} - 2.145\frac{S}{\sqrt{n}} < \mu < \overline{X} + 2.145\frac{S}{\sqrt{n}}.$$

Based on the similarity of the end points in this inequality, the 95% confidence interval for μ can be written as

$$\overline{X} \pm 2.145\frac{S}{\sqrt{n}}. \tag{2.8.2}$$

This interval is similar to the 95% confidence interval based on the normal distribution, given in (2.6.2). The only differences between the two intervals are the tabled value (1.96 for the normal, 2.145 for the t) and the multiplier ($\sigma/\sqrt{n}$ for the normal, $S/\sqrt{n}$ for the t).

In Example 2.2, we found the sample mean to be $\overline{X} = 19.4$. The sample variance is computed from formula (2.2.1), and results in $S^2 = 39.257$. Taking the square root yields the sample standard deviation, $S = \sqrt{39.257} = 6.266$. Substituting $\overline{X}$, S, and n into (2.8.2), we obtain

$$\overline{X} \pm 2.145\frac{S}{\sqrt{n}} = 19.4 \pm 2.145\,\frac{6.266}{\sqrt{15}} = 19.4 \pm 3.470.$$

The 95% confidence interval for μ has bounds of 15.930 and 22.870.

The confidence interval for μ based on the t distribution behaves much the same as the confidence interval based on the normal distribution. If the sample size is increased, the interval becomes narrower; more information yields a more precise estimate of the mean. If the level of confidence is increased, the interval becomes wider; to be more sure of trapping μ, we need to allow more room. We can never be 100% certain of trapping μ, since the t distribution, like the normal, never quite hits the axis.

For Example 2.2, the 95% confidence interval for μ based on the Z distribution has bounds of 16.212 and 22.588. The interval based on the t distribution has bounds of 15.930 and 22.870. Note that the interval based on the t distribution is *wider* than the one obtained with the Z distribution. The logic of this increased width follows from examination of Figure 2.2. To include 95 percent of the area under the curve, we need to use bounds further from the center for the t curve than for the Z curve; the t bounds are ± 2.145, while the Z bounds are ± 1.96. The quantity that is added to or subtracted from $\overline{X}$ to obtain the confidence interval is greater for the t than for the z. On a rational level, not knowing the standard deviation costs us some precision in our estimate.

2.9 THE CHI-SQUARE DISTRIBUTION

The population variance, σ^2, measures spread in the population. In Section 2.2 we introduced the usual point estimate for the population variance, namely the sample variance, S^2. Since S^2, like $\overline{X}$, is a continuous random variable, the probability that it equals σ^2 *exactly* is zero. In the next section, we discuss a confidence interval for σ^2—that is, a set of bounds that include the true value of σ^2 with a specified probability. To find these bounds, we must first examine the probability distribution of S^2.

To investigate the distribution for S^2, consider the experiment of drawing a random sample from the population. Each time we perform this experiment, we get a single value for the sample variance, S^2. Therefore, S^2 is a random variable associated with this experiment. If we perform this experiment repeatedly (i.e., draw lots of random samples), we can plot the obtained S^2 values to see which values are more likely and which are less likely. Our plot will approximate the probability distribution for S^2.

Since the sample variance is computed from a sum of *squares*, it can never be negative. Its distribution is bounded at the lower end by 0. S^2 is equal to 0 only if all the observations in the sample are identical and, therefore, the same as the average. That is, S^2 is 0 when the sample has no spread. However, there is no upper bound for S^2 since the values of the observations can be arbitrarily far apart. Because it is bounded at only one end, the distribution of S^2 is asymmetric.

Typically, instead of plotting the distribution of S^2, we work with a standardized version of S^2 known as the **χ^2 (chi-square) distribution**. The sample variance is converted to the χ^2 axis through the relationship

$$\chi^2 = \frac{(n-1)S^2}{\sigma^2}. \qquad (2.9.1)$$

Earlier, we standardized $\overline{X}$ to obtain a Z statistic. The standardization for the sample variance is rather different, but then so is the shape of its curve. Like S^2, the value of χ^2 can never be negative and has no upper bound. Figure 2.3 shows the shape of the χ^2 distribution.

Actually, χ^2 is a family of distributions, indexed by a single parameter—degrees of freedom. Since S^2 has n - 1 degrees of freedom, so does the χ^2 computed from it. For different values of df, the χ^2 distribution may be more or less spread out than shown in the figure, but it is always bounded below by zero and always approaches the axis at the right end.

2.10 CONFIDENCE INTERVAL FOR THE VARIANCE

To obtain the 95% confidence interval for σ^2, we must find bounds that include 95 percent of the area in the center of the χ^2 distribution. For an asymmetric distribution, it is difficult to visualize a center area, but we can still put 5 percent in the two tails, with area 0.025 in each tail. Since the distribution is asymmetric, lower tail and upper tail cutoffs must be tabled separately. Table 3a in the Appendix lists lower tail cutoffs, while Table 3b lists upper tail cutoffs.

In Example 2.2, the sample is size 15, and thus df = n - 1 = 14. From Table 3a we find the 0.025 lower tail cutoff to be 5.629, and from Table 3b the 0.025 upper cutoff is 26.12. The χ^2 statistic given in (2.9.1) lies between these two bounds with probability 0.95. That is, the probability is 95 percent that

$$5.629 < \frac{(n-1)S^2}{\sigma^2} < 26.12.$$

Rearranging this expression, we obtain the inequality

$$\frac{(n-1)S^2}{26.12} < \sigma^2 < \frac{(n-1)S^2}{5.629}.$$

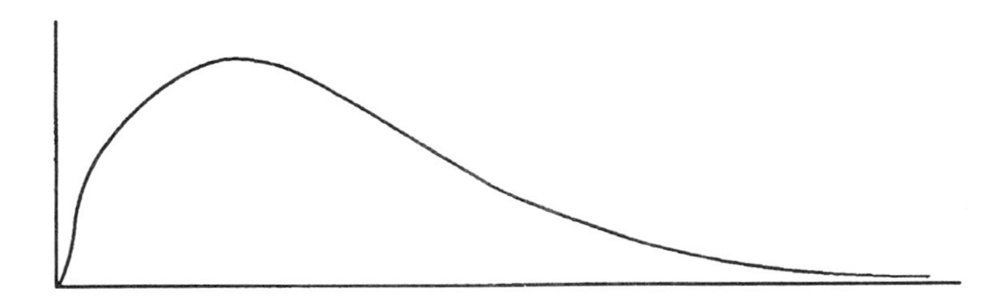

FIGURE 2.3. χ^2 distribution.

For the data of Example 2.2, we found $S^2 = 39.257$, yielding the 95% confidence interval

$$\frac{(15-1)39.257}{26.12} < \sigma^2 < \frac{(15-1)39.257}{5.629}$$

or

$$21.041 < \sigma^2 < 97.637.$$

Clearly, this interval does not provide a very precise estimate of population variance; it is extremely wide. In part, this is because spread is a harder concept to measure than center is. The confidence interval for the mean is much tighter than the interval for the variance. In general, when trying to estimate variance, we should use larger sample sizes than when estimating center. However, this is not the only reason the interval appears wide. Recall that variance is measured in the *square* of the original units of measurement. Thus, in part, the interval is wide because it is in the square of the original units. Since all quantities in this interval are positive, we can take the square root throughout to obtain the 95% confidence interval for the standard deviation, which is in the original units. For our example, we obtain the interval

$$\sqrt{21.041} < \sqrt{\sigma^2} < \sqrt{97.637},$$

or

$$4.587 < \sigma < 9.881.$$

The confidence interval for the population variance is not centered at the value of its point estimate—that is, $S^2 = 39.257$ is not halfway between the bounds 21.041 and 97.637. Similarly, the confidence interval for the standard deviation is not centered at the value of S. These off-center intervals result because the χ^2 distribution is asymmetric. We do not encounter this problem with confidence intervals for the mean, since the Z and t distributions are both symmetric. Confidence intervals for μ are always centered at the value of the point estimate, $\overline{X}$.

2.11 INTERVAL ESTIMATION IN COMPUTER PROGRAMS

In the last few sections, we have discussed confidence intervals for the mean when the standard deviation is known and when it is unknown, and the confidence interval for the variance. Neither SAS nor SPSS will compute these intervals. They do, however, calculate the sample mean, the standard error of the mean, and the sample variance, from which confidence intervals can be constructed. The commands to obtain these values are discussed in Section 2.4. MINITAB, on the other hand, will produce all three of these intervals. The Z and t confidence intervals for the mean are produced with single commands. The confidence interval for the variance is not built into MINITAB, but can be computed using special functions. The transcript shown here illustrates how to obtain these confidence intervals for the data of Example 2.2.

MINITAB Transcript

```
MTB > NAME C1 = 'WEIGHT'
MTB > SET 'WEIGHT'
DATA > 26 14 18 13 22 15 24 21 29 10 12 31 19 16 21
DATA > END

MTB > ZINTERVAL SIGMA = 6.3 'WEIGHT'

THE ASSUMED SIGMA = 6.30

            N     MEAN     STDEV     SE MEAN     95.0 PERCENT C.I.
WEIGHT     15    19.40      6.27        1.63       ( 16.21, 22.59)

MTB > TINTERVAL 'WEIGHT'

            N     MEAN     STDEV     SE MEAN     95.0 PERCENT C.I.
WEIGHT     15    19.40      6.27        1.62       ( 15.93, 22.87)

MTB > LET K1 = N ('WEIGHT') -1
MTB > LET K2 = STDEV ('WEIGHT') **2
MTB > INVCDF 0.975 K3;
SUBC > CHISQUARE K1.
MTB > LET K3 = K1*K2/K3
MTB > INVCDF 0.025 K4;
SUBC > CHISQUARE K1.
MTB > LET K4 = K1*K2/K4
MTB > PRINT K3 K4
K3        21.0423
K4        97.6419
MTB > STOP
```

Discussion of MINITAB Transcript

As with Example 2.1, user inputs are shown to the right of the > symbol, and are underlined for clarity. In the first four lines of the preceding MINITAB transcript, column C1 on the worksheet is identified to be a variable named WEIGHT, and the values for this variable are entered. The ZINTERVAL command on the fifth line requests the 95% confidence interval for the mean of WEIGHT when σ is known to be 6.3. Output from this command includes various descriptive statistics and the lower and upper bounds of the requested confidence interval. The TINTERVAL command produces comparable results when the standard deviation is unknown.

Since MINITAB does not have a command to produce a confidence interval for the variance, we must take advantage of its flexibility to perform user-specified actions. The next two lines of input store the degrees of freedom in constant K1 and the sample variance in K2.

Some MINITAB commands can be followed by one or more subcommands, specifying additional information needed to carry out the command. In such cases, the command line ends in a semicolon to indicate that subcommands follow, and each subcommand except the last ends in a semicolon. The last subcommand ends in a period, designating the end of information associated with the original command.

The INVCDF command requests a cutoff point on the axis of a probability distribution. In this case, the cutoff should have area 0.975 to the left, and the value of the cutoff is stored in K3. However, we still need to specify the shape of the distribution, and the subsequent subcommand identifies the χ^2 distribution whose degrees of freedom are stored in K1. Although this value is not displayed unless requested, the appropriate cutoff (26.12) is stored in K3. The following command multiplies the degrees of freedom times the sample variance, divides by this cutoff, and stores the result back in K3. At this point K3 holds the lower bound of the 95% confidence interval for the variance. The next three lines perform the same manipulation using the cutoff with area to the left equal to 0.025, with the result stored in K4. K4 now contains the upper bound of the confidence interval. Finally, we request that MINITAB print the values stored in K3 and K4.

2.12 ESTIMATING THE ENTIRE DISTRIBUTION

In preceding sections of this chapter, we have focused on two quantities, the mean and variance, which describe key features of the population. The mean defines the center of the probability distribution for the random variable of interest, and the variance measures its spread. Together, these two features tell us a great deal about what to expect when we observe values from the population. But they do not describe the shape of the distribution, and they do not provide information about outliers. In this section, we discuss methods of estimating the entire probability distribution and of identifying extreme values in a random sample. Example 2.3 is used to demonstrate these techniques.

The most common method for condensing an entire sample is a **frequency distribution**. In a frequency distribution, the range of sample values is divided into intervals, and the number of observations in each interval is displayed. For Example 2.3, a reasonable frequency distribution is shown in Table 2.2.

A frequency distribution is typically displayed graphically as a **histogram**, such as that shown in Figure 2.4. Although summarizing a sample with a frequency

---

Example 2.3

The National Science Foundation recently reviewed 22 completed projects, and determined for each project the percentage of goals outlined in the grant proposal that had actually been accomplished during the funding period. These percentages are shown below.

40	65	11	56
54	50	47	97
62	47	45	40
49	41	52	51
42	53	42	39
35	38		

TABLE 2.2 Frequency Distribution for Percentage of Grant Goals Accomplished

Interval	Frequency
90–99	1
80–89	0
70–79	0
60–69	2
50–59	6
40–49	9
30–39	3
20–29	0
10–19	1

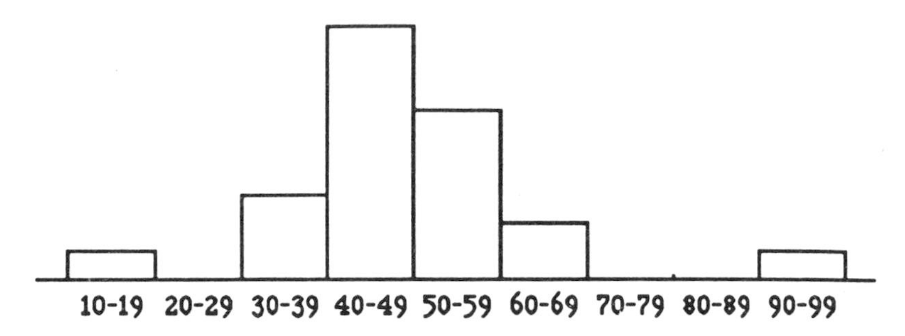

FIGURE 2.4. Histogram for percentage of grant goals accomplished.

distribution and histogram is simple and commonly understood, it does have certain drawbacks. One problem involves the arbitrariness of the intervals selected. A different picture of the sample might be created by using intervals of larger or smaller width, by centering the intervals differently (such as the interval 35–44), or by using open-ended intervals at the extremes (such as 70 and up). Clearly, some choices are undesirable, such as using only two intervals for an entire sample. But there are many reasonable sets of intervals for any sample. A second disadvantage is that we cannot determine any of the individual scores from a frequency distribution or histogram. All we know is the number of observations in each interval.

An alternative to these techniques is the **stem-and-leaf diagram**. In a stem-and-leaf diagram, each observation is displayed as a stem (the first digit of the observation) and a leaf (the second digit). The stems are shown to the left of a vertical line, with the ordered leaves for each stem listed to the right of the line. For the data of Example 2.3, the stem-and-leaf diagram is shown in Figure 2.5. Note that we can determine all of the original values from such a display. For instance, the fourth line, 6|25, represents the observations 62 and 65. To obtain the equivalent of a histogram, simply rotate the page 90 degrees.

The stem-and-leaf diagram is ideally suited for two-digit numbers. If observations have more than two digits, we can either round them to two digits or use leaves of more than one digit. To display the numbers 624 and 647, for example, we can round them to 620 and 650, and write 6|25, indicating at the bottom of the diagram that a multiplier of 10 is needed. Or we can include all the digits by using the form 6|24,47. For very small numbers, such as .062 and .065, we write 6|25, indicating at the bottom of the diagram that a multiplier of 1/1000 is needed.

For small samples, a stem-and-leaf diagram is preferable to a frequency distribution, since it shows the values of individual observations. With large samples, however, the leaves for a single stem may not fit on one line, and the diagram can become unwieldy. On the other hand, when working with very large samples, we

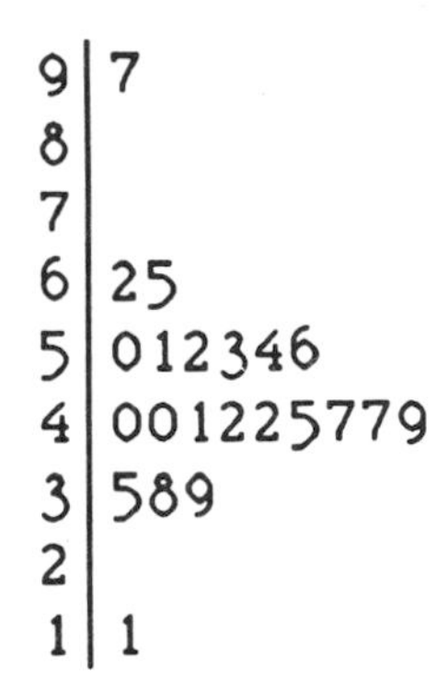

FIGURE 2.5. Stem-and-leaf diagram for percentage of grant goals accomplished.

may not need to know the values of individual observations. In this case, a frequency distribution would be adequate and might better summarize a large sample than a stem-and-leaf diagram.

While a stem-and-leaf diagram serves to provide information about the *shape* of the probability distribution, it is not ideal for showing *outliers*. In fact, outliers are often listed separately to avoid elongating the diagram. A **box plot** is a graphical technique designed to highlight outliers and display their relationship with the bulk of the sample. The box plot for the data of Example 2.3 is shown in Figure 2.6.

A box plot contains three components: the box, whiskers, and outliers. The box represents most of the data, and extends from the first quartile to the third quartile. Thus, the length of the box is the interquartile range (IQR). The median is drawn as a line, or waist, near the middle of the box. **Whiskers** extend from the box to the most extreme values that are still within 1.5 IQR of the box. Together, the box and whiskers span all the "reasonable" values in the sample. Observations that lie more than 1.5 IQR from the box are considered outliers. They are **mild outliers** if they lie within 3 IQR of the box and **extreme outliers** if they are more than 3 IQR from the box. Mild outliers are shown as open circles, and extreme outliers as asterisks. In this example, the value 11 is a mild outlier, while 97 is an extreme outlier.

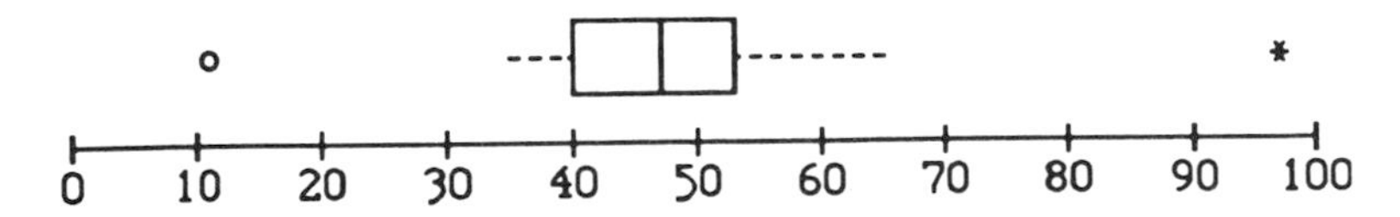

FIGURE 2.6. Box plot for percentage of grant goals accomplished.

Box plots can be drawn either horizontally or vertically. To compare two samples, we draw their box plots using the same axis, either one above the other, for horizontal box plots, or side by side, for vertical box plots.

2.13 STEM-AND-LEAF AND BOX PLOT IN COMPUTER PROGRAMS

All three of the programs—SAS, SPSS, and MINITAB—construct stem-and-leaf diagrams and box plots. These are demonstrated for the data of Example 2.3.

SAS Input

```
OPTIONS LS = 80;
TITLE 'EXAMPLE 2.3 - NSF GRANTS';
DATA;
INPUT PCTCOMP;
LINES;
40
54
62
49
42
35
65
50
47
41
53
38
11
47
45
52
42
56
97
40
51
39
PROC UNIVARIATE PLOT;
```

Discussion of SAS Input

Recall that every SAS statement, except the actual data lines, must end in a semicolon. For this problem, a single variable, named PCTCOMP, is entered for each of the 22 grant projects reviewed. The UNIVARIATE procedure, which provides general descriptive statistics, is used with the PLOT option to request the stem-and-leaf diagram and box plot.

SAS Output

```
                  EXAMPLE 2.3 - NSF GRANTS                        1
                    Univariate Procedure
Variable = PCTCOMP

                           Moments
    N                    22        Sum Wgts            22
    Mean                 48        Sum               1056
    Std Dev        15.46116        Variance      239.0476
    Skewness       0.989978        Kurtosis       5.20202
    USS               55708        CSS               5020
    CV             32.21076        Std Mean      3.296331
    T:Mean = 0     14.56164        Prob>|T|        0.0001
    Num ~= 0             22        Num > 0             22
    M(Sign)              11        Prob>|M|        0.0001
    Sgn Rank          126.5        Prob>|S|        0.0001

                     Quantiles (Def = 5)
    100% Max             97        99%                 97
     75% Q3              53        95%                 65
     50% Med             47        90%                 62
     25% Q1              40        10%                 38
      0% Min             11         5%                 35
                                    1%                 11
    Range                86
    Q3-Q1                13
    Mode                 40

                           Extremes
  Lowest            Obs        Highest           Obs
     11(            13)            54(             2)
     35(             6)            56(            18)
     38(            12)            62(             3)
```

```
    39(                 22)              65(                 7)
    40(                 20)              97(                19)

  Stem  Leaf                        #                     Boxplot
     9  7                           1                        *
     8
     7
     6  25                          2                        |
     5  012346                      6                     +-----+
     4  001225779                   9                     *--+--*
     3  589                         3                        |
     2
     1  1                           1                        0
        -+--+--+--+
     Multiply Stem.Leaf by 10**+1
```

Discussion of SAS Output

Descriptive information shown in the upper portions of this output is discussed in Section 2.4. Toward the bottom of the page are the stem-and-leaf diagram and box plot. The vertical box plot drawn by SAS leaves a great deal to be desired. In forcing it to fit the same scale as the stem-and-leaf, it is compressed beyond recognition. When the PLOT option is specified, SAS also draws a normal probability plot, helpful in determining if the population resembles a normal distribution. This plot has not been shown here, since it was also compressed beyond utility.

SPSS Input

```
SET WIDTH = 80
TITLE 'EXAMPLE 2.3 - NSF GRANTS'
DATA LIST FREE /  PCTCOMP
EXAMINE VARIABLES = PCTCOMP
BEGIN DATA
40
54
62
49
42
35
65
```

```
50
47
41
53
38
11
47
45
52
42
56
97
40
51
39
END DATA
FINISH
```

Discussion of SPSS Input

For this example, a single variable, named PCTCOMP, is read in free format. As in our previous example, subroutine EXAMINE is employed, but this time the default plots, a stem-and-leaf diagram and box plot, will be produced. Recall that SPSS requires the data, delineated with BEGIN DATA and END DATA statements, to follow the first request for analysis.

SPSS Output

```
                              EXAMPLE 2.3 - NSF GRANTS                  Page 1
   PCTCOMP
Valid cases:    22.0     Missing cases:    .0     Percent missing:    .0
Mean     48.0000    Std Err      3.2963    Min  11.0000   Skewness   .9900
Median   47.0000    Variance 239.0476    Max  97.0000   S E Skew   .4910
5% Trim 47.3737     Std Dev     15.4612    Rang 86.0000   Kurtosis 5.2020
                                          IQR  13.2500   S E Kurt   .9528


Frequency   Stem & Leaf
  1.00 Extremes    (11)
   .00        3  *
  3.00        3  .  589
  5.00        4  *  00122
  4.00        4  .  5779
  5.00        5  *  01234
```

```
  1.00        5  . 6
  1.00        6  * 2
  1.00        6  . 5
  1.00  Extremes    (97)
Stem width:     10.00
Each leaf:       1 case(s)
```

EXAMPLE 2.3 - NSF GRANTS Page 2

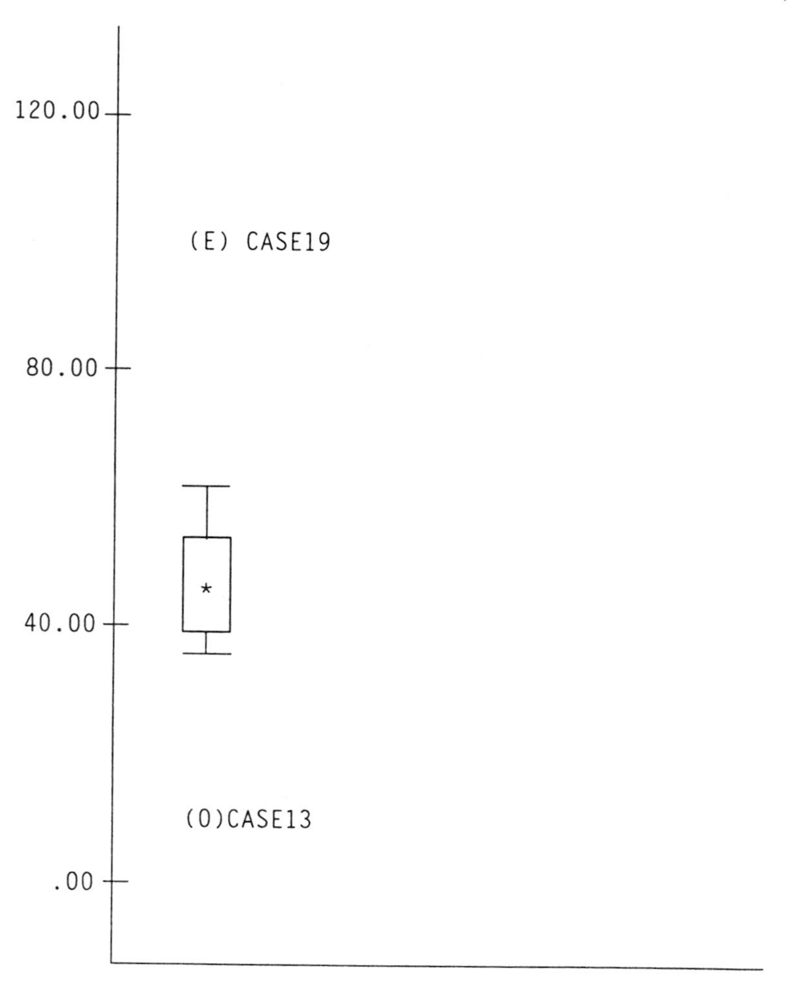


```
Variables       PCTCOMP
N of Cases        22.00
Symbol Key:         * - Median   (O) - Outlier   (E) - Extreme
```

Discussion of SPSS Output

Descriptive information shown in the top portion of page 1 is discussed in Section 2.4. The stem-and-leaf is shown upside down from usual displays, and the outliers (11 and 97) are indicated only as extreme values. In addition, each leading digit is broken into two stems, with an asterisk corresponding to leaves 0–4, and a period for leaves 5–9. On the second page of output, SPSS prints the box plot vertically, identifying each outlier with its observation number.

MINITAB Transcript

```
MTB > NAME C1 = 'PCTCOMP'
MTB > SET 'PCTCOMP'
DATA > 40 54 62 49 42 35 65 50 47 41 53 38 11 47 45 52
DATA > 42 56 97 40 51 39
DATA > END

MTB > STEM-AND-LEAF 'PCTCOMP'

Stem-and-leaf of PCTCOMP  N = 22
Leaf Unit = 1.0

   1    1 1
   1    2
   4    3 589
  (9)   4 001225779
   9    5 012346
   3    6 25
   1    7
   1    8
   1    9 7

MTB > BOXPLOT 'PCTCOMP'
```

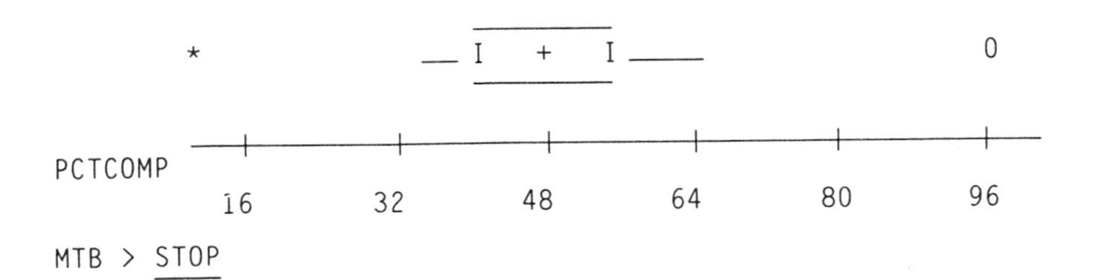


```
MTB > STOP
```

Discussion of MINITAB Transcript

As with previous examples, user inputs are shown to the right of the > symbol and are underlined for clarity. In the first five lines of the transcript above, column C1 on the worksheet is identified to be a variable named PCTCOMP, and the values for this variable are entered. The STEM-AND-LEAF command on the sixth line produces the stem-and-leaf diagram, although upside down from usual displays. To the left of each stem, MINITAB shows the cumulative number of observations, counting in from the extremes. The stem in which the median is located shows, in parentheses, the frequency for this stem alone. Reversing common usage, the box plot produced by MINITAB employs an asterisk for the mild outlier (11) and an open circle for the extreme outlier (97).

2.14 SUMMARY

This chapter has discussed point and interval estimation for the center and spread of a population. A point estimate is a single "best guess" at the value of a parameter, while an interval estimate is a range that includes the parameter with specified probability. Point estimates of center include the sample mean, sample median, trimmed mean, and sample mode. For estimating population spread, the sample variance or sample standard deviation is most commonly used, although the interquartile range may be advantageous when there are outliers. Spread relative to the size of the numbers is measured with the coefficient of variation.

Confidence intervals for the mean were examined when the standard deviation is known and when it is unknown. The standard normal distribution is used in the former case. In the latter case, the interval is based on the t distribution, which is similar to the standard normal distribution, but replaces the unknown standard deviation with an estimate. A confidence interval for the variance is obtained from the chi-square distribution, an asymmetric distribution that is a standardized form for the sample variance. Both the t and chi-square distributions are families of distributions, indexed by a single parameter called degrees of freedom.

An entire sample may be displayed in condensed form, using a frequency distribution, histogram, stem-and-leaf diagram, or box plot. Each of these describes the shape of the population, but the box plot is particularly useful in identifying outliers in the sample. SAS, SPSS, and MINITAB programs were used to illustrate the estimation techniques discussed in this chapter.

3

One-Sample Tests

Estimation is employed when there is no expectancy about the value of a population parameter prior to sampling. Sometimes, however, a claim is made about the value of a parameter, but we are not sure of its validity. Advertising provides some prime examples of such claims. "Rolaids consumes up to 47 times its weight in excess stomach acids." "The Crest group had 35 percent fewer cavities." What gives advertisers the right to make such claims? Are they believable? In the United States, the Food and Drug Administration requires manufacturers to support such statements with convincing evidence before allowing them to appear in advertisements. The procedure for evaluating evidence and determining whether or not it is convincing is the statistical methodology known as **hypothesis testing**. A hypothesis is a claim about the value of a population parameter. Statistics enables us to test that claim in light of the available evidence.

In this chapter wc introduce the logic of hypothesis testing and various applications for data gathered from a single random sample. Different procedures are used, depending on whether the claim involves a population mean, variance, or proportion. In most cases, the computations required are essentially the same as those used for construction of confidence intervals. When the hypothesis concerns the mean, different approaches are used, depending on whether the standard deviation is known or unknown, just as different methods are used for constructing confidence intervals for the mean in these two cases. Again, we recognize that the case of known standard deviation is unrealistic. In practice, when the population mean is not known for certain, neither is the standard deviation. Nonetheless, we treat this case first, in order to present the logic of hypothesis testing in the simplest situation.

3.1 TEST ON THE MEAN WHEN STANDARD DEVIATION IS KNOWN

A hypothesis makes a claim about the value of a parameter. We perform a **one-sided test** when we are only interested in departures from this claim in a single direction—for example, if the parameter is *smaller* than the stated value. Examples 3.1 through 3.3 are used to introduce the procedure for one-sided tests. We discuss the relative merits of one-sided and two-sided tests later, when two-sided tests are considered.

In Example 3.1, we wish to test whether the mean drying time with the additive is still $\mu = 170$. We call this the **null hypothesis**, and designate it by:

$$H_0: \mu = 170.$$

Here, μ refers to the mean drying time *with the additive*. We want to see if this unknown value is the same as the known mean of 170 without the additive. The null hypothesis states that there is no difference, hence the term "null."

We also specify what we are trying to show. In tnis case, the additive should reduce the average drying time. We call this the **alternative hypothesis**, and write:

$$H_1: \mu < 170.$$

The alternative hypothesis is generally associated with the claim we are trying to support.

In deciding between H_0 and H_1 we examine the evidence available, typically a random sample from the population. In Example 3.1, a sample of size $n = 25$ items painted with the paint containing the additive dried in an average time of $\overline{X} = 168$ minutes. We note that $\overline{X}$ is less than 170 minutes, the expected drying time without the additive. But is it ***enough less*** to convince us that the additive has reduced drying time?

To answer this question, let us suppose for now that the null hypothesis is true—that is, the mean drying time with the additive is still $\mu = 170$ minutes. How

---

Example 3.1

A particular brand of paint is known to have a mean drying time of $\mu = 170$ minutes, with standard deviation of $\sigma = 40$ minutes. In an attempt to improve the drying time, a new additive has been developed. Use of the additive in 25 test samples yielded an average drying time of $\overline{X} = 168$ minutes. Assuming that the standard deviation is unchanged, does the experimental evidence indicate that the additive reduces true average drying time?

likely is it that an average as low as 168 minutes could be obtained for a sample of 25 items? We learned how to answer such questions in Chapter 1, where it was found that $\overline{X}$ is normally distributed with $\mu_{\overline{X}} = \mu$ and $\sigma_{\overline{X}} = \sigma/\sqrt{n}$. This allows $\overline{X}$ to be standardized to

$$Z = \frac{\overline{X}-\mu_{\overline{X}}}{\sigma_{\overline{X}}} = \frac{\overline{X}-\mu}{\sigma/\sqrt{n}}. \tag{3.1.1}$$

In Example 3.1, the sample size is $n = 25$, and we are told that $\sigma = 40$. Assuming the null hypothesis is true ($\mu = 170$), the value of $\overline{X} = 168$ can be converted to the Z scale to obtain

$$Z = \frac{\overline{X}-\mu}{\sigma/\sqrt{n}} = \frac{168 - 170}{40/\sqrt{25}} = -0.25.$$

The probability that $\overline{X}$ is as low as 168 is equal to the probability that Z is as low as -0.25. By the symmetry of the normal distribution, the tail area below Z = -0.25 is the same as the tail area above Z = +0.25. From Table 1, we find this area to be 0.4013, as depicted in Figure 3.1. Therefore, when the null hypothesis is true, we would expect to find an average drying time of 168 minutes or less about 40 percent of the time. The observed value of $\overline{X}$ is *not* extremely unlikely. Accordingly, we accept the null hypothesis and conclude that the additive does not reduce drying time.

We only reject the null hypothesis when we have obtained convincing evidence that it is false. How convincing must the evidence be? This is clearly an arbitrary

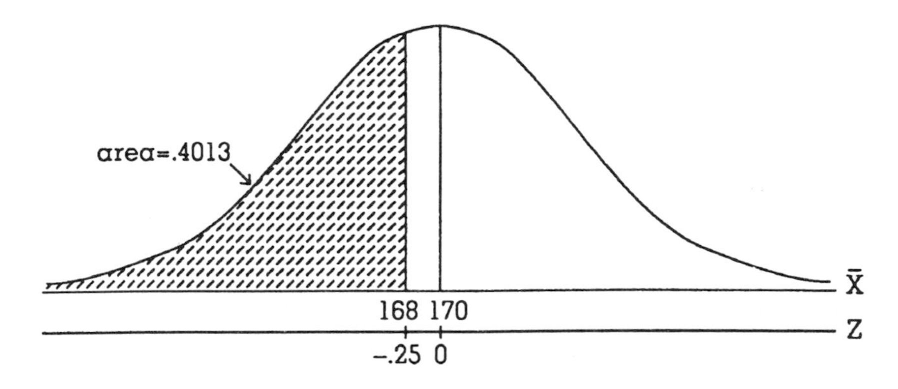

FIGURE 3.1. Probability of drying time less than 168 minutes.

and personal decision. People differ in their levels of skepticism and in how convincing the evidence for a claim must be. To achieve consistency, the scientific community has adopted a 5% standard. We only reject the null hypothesis if the probability of the obtained result is 5 percent or less. In Example 3.1, the probability of finding $\overline{X}$ less than 168 is 40 percent, so we accept the null hypothesis.

3.2 LOGIC OF HYPOTHESIS TESTING

The decision between the null and alternative hypotheses is not a "fair" decision, in that we are not choosing which of these is more likely. Instead, we set up the null hypothesis as a "straw man" that we would like to knock down. We are only allowed to knock it down if the evidence convinces us that it is false. This approach deliberately puts the burden of proof on the person who makes a claim, such as the paint manufacturer who offers a new, faster-drying paint. We only accept such claims when the evidence in favor of the claim is convincing. By using this strategy, we avoid wild goose chases in the pursuit of knowledge.

An analogy can be made between hypothesis testing and criminal law. The null hypothesis is accepted unless proven false, much as the defendant is presumed innocent until proven guilty in a criminal trial. However, a crucial difference in these situations exists in the definition of proof. In a trial,the judge instructs the jury that they must assume innocence unless they are convinced beyond a reasonable doubt of the defendant's guilt. But how big is a "reasonable doubt"? Is it the same for every juror? In statistics, we have the advantage of precisely defining our reasonable doubt by the 5% rule.

In Example 3.1, we are testing to see if the drying time with the additive is less than the known time of 170 minutes without the additive. Logically, if the average drying time in the sample is small enough, we should reject the null hypothesis and conclude that the additive has reduced the drying time. The decision is made, however, not in terms of the observed value of $\overline{X}$, but in terms of the tail area below this value. To find this area, it is necessary to convert $\overline{X}$ to the Z, or standardized, scale. Because the tail area is found from the standard normal distribution, this test is referred to as a **Z test**.

The obtained value of Z is called the **test statistic**, since this is the statistic used in making the test. In our example, the value of the test statistic was found to be $Z = -0.25$. In other testing situations, we may need a different standardization procedure. For instance, when the standard deviation is unknown, we will convert $\overline{X}$ to the t scale, and the obtained value of t will be the test statistic. In general, the data are standardized to a scale for which probabilities can be found from tables, and this standardized value is called the test statistic.

Listed below are five steps in performing a hypothesis test. These same steps can be used for any test, although the computations performed may differ, depending on the information available or the parameter being investigated.

1. State the null and alternative hypotheses.
2. Determine the rejection region.
3. Obtain the data.
4. Compute the test statistic.
5. Draw the conclusion.

These steps will be demonstrated in connection with Example 3.2.

1. State the null and alternative hypotheses.
   In this example, we wish to see if mean assembly line time has been reduced from the original value of 13 minutes. The null hypothesis states that there has been no change, and the alternative hypothesis indicates that the time has been reduced. To represent these statements, we write:

$$H_0: \mu = 13$$
$$H_1: \mu < 13.$$

2. Determine the rejection region.
   We will convert the observed value of $\overline{X}$ to the Z scale, and will reject H_0 if the tail area below Z is less than 5 percent. This region is depicted in Figure 3.2. To find the Z value with lower tail area of 5 percent, we use Table 1 in the Appendix. This table lists *upper* tail areas for the standard normal distribution, so we first seek the cutoff with upper tail area of 5 percent. Because .05 is a tail *area*, we search the *body* of the table for this number and read outward to the margins to obtain the Z value. The body of Table 1 lists two areas equally close to .05, namely .0505 and .0495. These correspond to Z values of 1.64 and 1.65, respectively. Ordinarily, we do not bother interpolating among tabled values. But since a tail area of 5 percent is used for most hypothesis tests, we make an

---

Example 3.2

Based on extensive experience, it is known that the average time to complete a certain assembly line job is 13 minutes, with a standard deviation of 0.6 minutes. An efficiency expert has suggested a modification of the process in an effort to reduce the time spent on this task. In a test of this modification, the average time was 12.7 minutes for 20 items. Has the alteration reduced the time for this job?

exception in this case. Splitting the difference between 1.64 and 1.65, we find the upper tail .05 cutoff point on the Z axis to be 1.645. By symmetry, the point cutting off 5 percent in the *lower* tail is -1.645.

We now state the **decision rule**: reject H_0 if Z is less than -1.645. The decision rule tells us what conclusion to draw, depending on the value of the test statistic. Specifically, it states the **rejection region** in terms of comparing the computed Z value with a cutoff on the Z axis. It is advantageous to formulate the decision rule in terms of a *standardized* variable, such as Z. If the decision rule were stated in terms of comparing $\overline{X}$ with a cutoff, we would have to calculate the $\overline{X}$ cutoff separately for each hypothesized mean. By phrasing the rule in terms of the Z scale, we can use the same Z cutoff of -1.645, regardless of the hypothesized mean on the $\overline{X}$ axis. The cutoff on the standardized scale, -1.645 in our example, is called the **critical value** for the test.

3. Obtain the data.

   For examples in this book, data are presented along with the problem statement. In practical applications, however, the problem should be formulated in terms of hypotheses and rejection region before gathering the data. The hypotheses represent the research question, which logically precedes obtaining and evaluating the evidence in a sample.

   In Example 3.2, we observed n = 20 items on the assembly line, and found the average time to be $\overline{X}$ = 12.7 minutes. The standard deviation is known to be σ = 0.6 minutes. (In realistic problems, where the standard deviation is unknown, we would estimate it in this step.)

4. Compute the test statistic.

   Substituting the hypothesized mean of μ = 13, and the values $\overline{X}$ = 12.7, σ = 0.6, and n = 20, we obtain

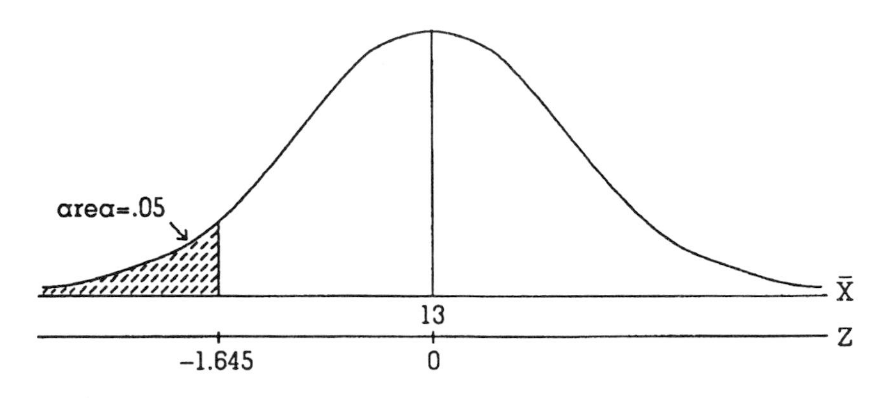

FIGURE 3.2. Left tail area of 5 percent for assembly line time.

$$Z = \frac{\overline{X}-\mu}{\sigma/\sqrt{n}} = \frac{12.7 - 13}{0.6/\sqrt{20}} = \frac{-0.3}{0.134} = -2.24.$$

5. Draw the conclusion.
   Since $Z = -2.24$ is less than -1.645, we reject H_0 and conclude that the efficiency expert's modification has indeed sped up the assembly line. A test that results in rejection of the null hypothesis is said to be **significant**.

In this example, the observed value of $\overline{X}$ is very unlikely when the true mean is 13, as given in the null hypothesis. In fact, when H_0 is true, we would obtain a value of $\overline{X}$ this small or smaller less than 5 percent of the time. Thus, we reject H_0 in favor of H_1. It is important to stress that we have *no proof* that H_0 is false, just some data we would be unlikely to obtain if H_0 were true. There are only two possible explanations for this occurrence: either H_0 is wrong, or we observed some very unusual data. Common sense tells us we should not make decisions by assuming we have been particularly lucky (or particularly unlucky). Instead, we draw the rational conclusion that the null hypothesis is wrong. In essence, all of statistics is based on the simple premise that "rare events never happen *to me*."

When we reject the null hypothesis, we are taking a risk because it could be correct. The 5% rule means that even when the null hypothesis is true, we will reject it about 5 percent of the time. This 5% risk level is referred to as the α **(alpha) level**, or **type one error rate**. It is the risk that we will erroneously reject the null hypothesis when it is true. If the consequences of making this type of error are extremely dire, we can be more conservative and use an α level of 1 percent. The only change needed in our testing procedure is the cutoff value. Instead of using the Z value -1.645, a 1% rule yields the more stringent cutoff of -2.33. The acceptable risk, or α level, must be specified in step 2 of the testing procedure, before looking at the data. This prevents us from deciding, after looking at the data, that we are willing to take a *6%* risk when the Z value is just barely in the acceptance region. We should determine *in advance* how convinced we must be to reject the null hypothesis. Then we look at the data to see if they satisfy this criterion.

3.3 FRAMING THE HYPOTHESES

Null and alternative hypotheses should be stated before examining the data. They reflect our interests in performing the experiment, not what the data suggest about the mean. The importance of this principle is highlighted in Example 3.3.

1. State the null and alternative hypotheses.
   To see if mean weight increases on this diet, we write:

Example 3.3

A strain of white mice used in laboratory research is known to have a mean weight of 44 grams, with a standard deviation of 6 grams. A special diet designed to increase weight was tested on ten mice. After six weeks on the diet, the following weights in grams were recorded: 46, 50, 40, 45, 32, 34, 41, 37, 48, 42. Assuming the diet does not effect the standard deviation, does it increase mean weight?

$$H_0: \mu = 44$$
$$H_1: \mu > 44.$$

2. Determine the rejection region.
   If mean weight on the diet is greater than 44, as stated in the alternative hypothesis, we would expect to find $\overline{X}$ higher than 44. Accordingly, we locate our 5% rejection region in the *upper* tail of the distribution, as shown in Figure 3.3. We found in section 3.2 that this Z cutoff is 1.645, so our decision rule is: reject H_0 if Z is greater than 1.645.
3. Obtain the data.
   For the 10 weights given in Example 3.3, the sample mean is $\overline{X} = 41.5$ grams. The standard deviation is known to be $\sigma = 6$ grams.
4. Compute the test statistic.
   The hypothesized value for μ is 44. This yields the test statistic

$$Z = \frac{\overline{X}-\mu}{\sigma/\sqrt{n}} = \frac{41.5 - 44}{6/\sqrt{10}} = -1.32.$$

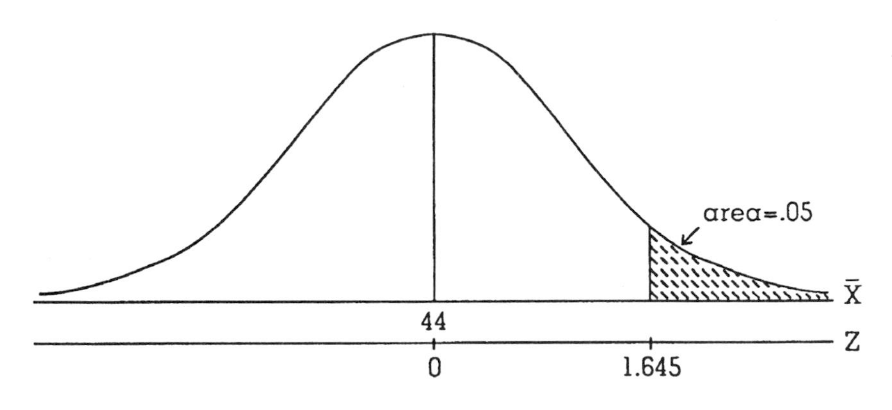

FIGURE 3.3. Right tail area of 5 percent for weight of mice.

5. Draw the conclusion.
   Because $Z = -1.32$ is not greater than 1.645, we accept H_0 and conclude that the diet does *not* increase the weight of this strain of white mice.

In this example, the mice appear to have lost weight instead of gaining it. It was evident that we would accept the null hypothesis when we calculated the sample mean in step 3 and found $\overline{X}$ less than 44. After computing $\overline{X}$, we might be tempted to change our mind, and test to see if the diet is *decreasing* weight. This is not acceptable for two reasons. First, if we are willing to put the 5% rejection region in either tail, depending on where the data fall, we are effectively operating with 5 percent in *each* tail. The total chance of a type one error is 10 percent, not 5. Second, we should not use the data to determine what we are going to test with that same data. Such circular reasoning is clearly unacceptable. If the present sample suggests a new hypothesis, it should be tested with a new sample. The order of the steps in hypothesis testing is important. We should state the hypotheses in step 1, before looking at the data. The hypotheses are a reflection of the research problem and should not be influenced by the outcome of the experiment.

We draw the curve according to the null hypothesis, with the distribution for $\overline{X}$ centered at the hypothesized value. The alternative hypothesis merely determines the proper tail for locating the rejection region. Since the data are examined in connection with the H_0 curve, the conclusion is stated in terms of accepting or rejecting H_0 not H_1. If the obtained value of $\overline{X}$ is near the center of the H_0 curve, we accept the null hypothesis; if it is extreme (in the direction of H_1), we reject the null hypothesis in favor of the alternative.

For a diet that purports to increase the weight of mice, we only reject the null hypothesis when $\overline{X}$ is convincingly larger than 44 grams. If $\overline{X}$ is near 44 or is substantially less than 44, we accept the null hypothesis. Thus, it would seem that we should have written the null hypothesis as H_0: $\mu \leq 44$, instead of H_0: $\mu = 44$. However, these two ways of writing the null hypothesis are handled identically, as long as the alternative hypothesis is H_1: $\mu > 44$.

Suppose we had stated the null hypothesis as H_0: $\mu \leq 44$. The $\overline{X}$ distribution could have been drawn centered at 44 or at any value less than 44. Figure 3.4 shows several possible locations for the H_0 curve. In each curve the 5% right tail cutoff point is indicated. Which cutoff point should be used to determine our rejection region? The answer is simply to act conservatively. We locate the rejection region to make it *as hard as possible* to reject H_0. This principle results in using the 5% cutoff point on the curve centered at 44. If $\overline{X}$ is above this cutoff point, it is automatically above the cutoff point for any other location of the H_0 curve. This strategy guarantees that the chance of a type one error will never be more than 5 percent. If we used any of the other cutoff points to determine our rejection region, and the true mean was actually 44, the chance of a type one error

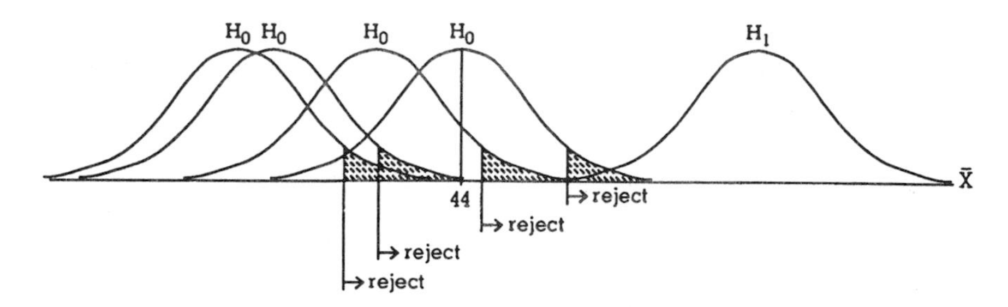

FIGURE 3.4. Right tail area of 5 percent for various locations of the null hypothesis curve.

would be considerably larger than 5 percent. In practice, the decision rule for the null hypothesis H_0: $\mu \leq 44$ is the same as that for H_0: $\mu = 44$, as long as the alternative hypothesis is H_1: $\mu > 44$.

Another way of arguing that H_0: $\mu = 44$ and H_0: $\mu \leq 44$ should be handled identically is to consider the worst case scenario. In effect, hypothesis testing attempts to decide whether the observed value of $\overline{X}$ came from the H_0 curve or the H_1 curve. This decision is hardest when the two curves *overlap as much as possible*. As seen in Figure 3.4, locating H_0 at $\mu = 44$ maximizes the overlap between H_0 and H_1. Therefore, adopting this worst case approach also leads us to locate the H_0 curve at 44. The philosophy of statistics is very conservative: even in the worst of cases, we only reject the null hypothesis when we are convinced that it is wrong. In this book we generally state the null hypothesis in terms of equality, such as H_0: $\mu = 44$, to emphasize the proper location of the H_0 curve. But the test procedure and conclusion for this hypothesis are identical to those obtained using H_0: $\mu \leq 44$.

In Example 3.3 we concluded that the diet does not increase average weight. We stress that this conclusion pertains to the population mean, μ, not to the sample mean, $\overline{X}$. To see if the sample mean is greater than 44, we need only compute it; hypothesis testing is not involved. The purpose of statistics is to draw an inference about the *population* based on what we see in the sample. The actual value of $\overline{X}$ is unimportant, except for the information it provides about the true population mean.

3.4 TYPE ONE AND TYPE TWO ERRORS

When conducting a hypothesis test, only two decisions are possible: we either accept H_0 or reject H_0. Whichever decision we reach may be correct or it may be in error, depending on whether H_0 or H_1 is actually true. When we accept H_0, we have made the correct decision if H_0 is really true, but are in error if H_1 is true. Likewise, when we reject H_0, we are correct if H_1 is true, but in error if H_0 is true. These two kinds of errors are differentiated with separate labels. **A type one error** is rejecting H_0 when H_0 is really true, and a **type two error** is accepting H_0 when H_1 is actually true. Table 3.1 summarizes the various possibilities for our decision and the truth.

In distinguishing two kinds of errors, statistical hypothesis testing is analogous with medical laboratory testing. An incorrect lab test may be either false positive or false negative. A false positive results when the lab test signals the presence of disease for a non-diseased patient, while a false negative is a reading of no disease for a diseased patient. In most cases, a false negative is the more serious error, because it results in a patient not getting needed treatment. Similarly, in statistics, a type one error is generally considered more serious than a type two error, because it results in believing an incorrect claim.

Statistical hypothesis testing is a reasonable decision procedure in the face of two types of unavoidable ignorance. The first ignorance is that *we will never know the truth.* All we ever know is our decision. Second, as a result of this, *we will never know whether our decision is correct or incorrect.* In Table 3.1 we will know which row we are in, but not which column. Since each row contains cells corresponding to both correct and incorrect results, we can never know whether or not we have made an error. Under these circumstances the best we can do is to employ a procedure for which the *probability* of an error is small. There are two kinds of errors and, thus, two error probabilities. The symbol α is used to represent the probability of a type one error, and β represents the probability of a type two error. That is, α is the probability that we reject the null hypothesis when it is true, and β is the probability that we accept the null hypothesis when it is false.

As discussed above, we set the α level (probability of a type one error), typically at 5 percent. When the null hypothesis is true, we allow a 5% risk of rejecting it anyway. By keeping α small, we can at least be sure that this is a small risk. But why should we tolerate *any* risk of making a type one error? Since α is within our control, why not set α to 0? The answer, upon reflection, is obvious. Refer to Figure 3.3, for example, in which the curve drawn is for the null hypothesis, and the shaded area is the rejection region. This area is the probability of rejecting the null hypothesis when it is true, that is, α. For a 5% tail area, the Z cutoff is 1.645. To

TABLE 3.1 Decision and Truth Posssibilities

		Truth	
		H_0 true	H_1 true
Decision	accept H_0	correct decision	type two error
	reject H_0	type one error	correct decision

reduce α to, say, 1 percent, we use a Z cutoff larger than 1.645. But a normal distribution never quite touches the axis. To reduce α to 0, we would have to move the Z cutoff infinitely far out. In effect, we would always be on the left side of the cutoff, and therefore would always accept the null hypothesis. The only way to avoid erroneously rejecting the null hypothesis is never to reject it. This is tantamount to a medical test that, in order to be certain of detecting every person who really has the disease, concludes that all patients are diseased.

Clearly, this is not a good decision procedure for either medicine or statistics. But to illustrate a point, let us follow this logic a bit further. If we adopted this approach, we would not even have to draw any data. We would already know we were going to accept the null hypothesis. Even if the null hypothesis were false, we would accept it. This is the definition of a type two error. The chance of accepting the null hypothesis when it is false would be 100 percent. To achieve a type one error rate of 0, we would have to tolerate a type two error rate of 100 percent. Type one and type two error rates do not always add up to 100 percent, but to some extent there is a trade-off between them. The smaller one error rate becomes, the larger the other becomes. However, this trade-off is not absolute. There are certain things we can do to minimize the type two error rate, even after setting the type one error rate at 5 percent. Before discussing these strategies, we introduce one additional term frequently used in hypothesis testing, namely power.

3.5 POWER

The **power** of a statistical test is the probability of rejecting the null hypothesis when it is false. In most experiments, we believe the null hypothesis is false and the alternative is correct. The chemist who develops the paint additive expects to produce a faster drying paint; the efficiency expert believes his or her modification will decrease assembly time; and the new diet for mice is designed to increase weight. Since we typically believe the null hypothesis is false, we would like a large chance of rejecting it (i.e., a large power). This word is well chosen in the sense that large power is desirable in both everyday and statistical definitions.

Power is the complement of β, the type two error probability. Power and β add to one. Both power and β pertain to decisions reached when the null hypothesis is false. Power is the probability of rejecting H_0, and β is the probability of accepting it. Since we must reach one or the other of these decisions when the null hypothesis is false, their probabilities add to one.

To examine the concepts of α, β, and power graphically, let us reconsider Example 3.3. This experiment concerns a diet designed to increase the weight of mice above the standard weight of 44 grams. We wrote the hypotheses:

$$H_0: \mu = 44$$
$$H_1: \mu > 44.$$

As defined above, α, β, and power are each probabilities—that is, areas under curves. However, they are not all areas under the same curve. As we will see, α is an area under the H_0 curve, while β and power are areas under the H_1 curve. In Figure 3.3, the probability distribution for $\overline{X}$ is drawn according to the null hypothesis and is centered at 44. The 5% rejection region is located in the upper tail of this distribution. This 5% area is α—the probability of rejecting H_0 when H_0 is true. In the preceding sentence, the word "probability" translates as "area," the phrase "rejecting H_0" implies "to the right of the line," and the phrase "when H_0 is true" corresponds to "under the H_0 curve." Thus, α is the area to the right of the line under the H_0 curve.

Figure 3.5 shows the area corresponding to α under the H_0 curve, and also displays the probability distribution for $\overline{X}$ according to the alternative hypothesis, centered at some value of μ above 44. The line determining the rejection region is extended upward until it hits the H_1 curve. The probability of a type two error, β, is the probability (area) of accepting H_0 (to the left of the line) when H_1 is true (under the H_1 curve). Therefore, β is the area to the left of the line under the H_1 curve. Power is the probability (area) of rejecting H_0 (to the right of the line) when H_1 is true (under the H_1 curve). Power is the area to the right of the line under the H_1 curve. Note that the area shown for power *includes* α. The two areas labeled β and power encompass the entire H_1 curve, and thus add up to 100 percent. As mentioned above, these two probabilities are complementary.

Figure 3.5 shows clearly the trade-off between α and β. If we reduce the chance of a type one error (α) by moving the cutoff line to the right, we increase the chance of a type two error (β). Attempting to achieve an α level of zero requires moving the cutoff infinitely far to the right, and makes β the whole area under the H_1 curve, or 100 percent.

However, we stated above that this trade-off is not absolute, that strategies exist for decreasing β, even after setting α at 5 percent. Since β and power are

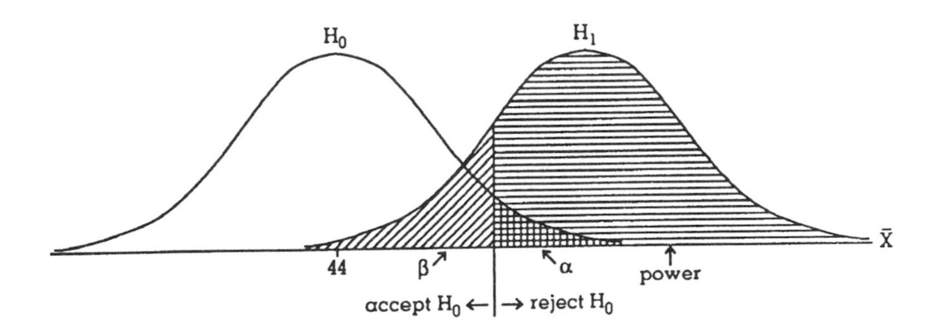

FIGURE 3.5. Regions for α, β, and power.

complementary, reducing β is the same as increasing power. Instead of phrasing our discussion in terms of decreasing β, we refer to methods for increasing power. The two practical methods for increasing power are decreasing variance and increasing sample size.

Figure 3.6 displays the effects of decreasing the variance. In this figure, the H_0 and H_1 curves are centered at the same values as in Figure 3.5, but both curves have been drawn narrower to reflect smaller variance. Because the H_0 curve is narrower, the line cutting off 5 percent of the area in the right tail of the curve has moved to the left. Since the line has moved to the left, the area to the right of this cutoff under the H_1 curve has increased. Power is larger. In addition to the H_0 curve, the H_1 curve is also narrower in Figure 3.6. More of the area under this curve is located to the right of any given cutoff. This also leads to increased power. Hypothesis testing attempts to determine whether the observed value of $\overline{X}$ came from the H_0 curve or the H_1 curve. This decision is much easier when the curves overlap less, as in Figure 3.6. One way to ensure that two curves overlap less is to draw them narrower (i.e., with lower variance).

But how can we decrease variance in practical applications? The answer is both simple and logical. We should run clean, tightly controlled experiments. If we are examining the effects of a new diet for laboratory mice, we should use a homogeneous group of test animals and treat them all alike. We should not use animals of varying ages, feed them according to different schedules, or subject them to other influences that could affect weight. Such factors will lead to large differences in weight among animals (i.e., to large variance). Similarly, to test a new paint formula, we should use samples of the same material, painted with coats of equal

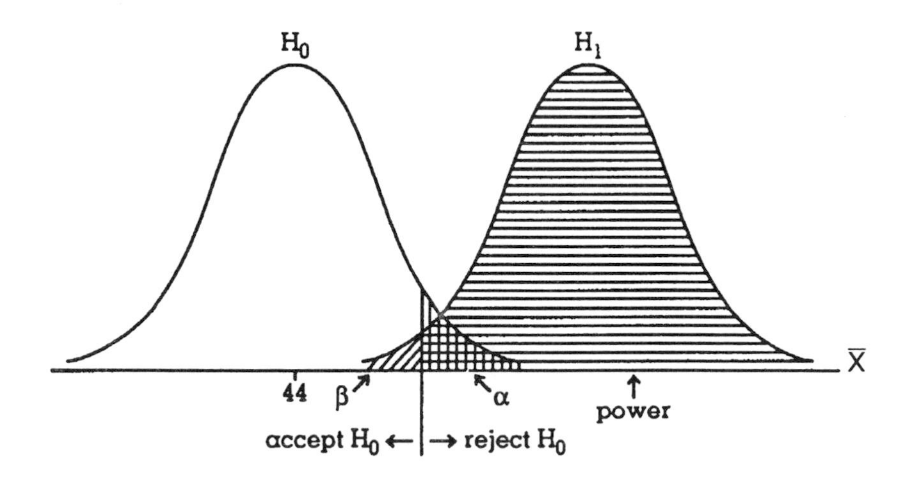

FIGURE 3.6. Increased power for decreased variance.

thickness and subjected to constant drying conditions. If we want to investigate drying times on different materials or under varying conditions, there are methods for designing appropriate experiments. These are discussed in detail in later chapters. The present lesson, the benefit of reducing variance, is pure common sense. Common sense dictates that good research requires careful planning and control. Now we see why these precautions are helpful from a statistical viewpoint. A well controlled experiment decreases variability and increases power. Power is the probability of reaching a correct decision when the alternative hypothesis is true. To a large extent, statistics is the science of reaching correct decisions.

The second method of increasing power is to increase sample size. At first glance, it may be difficult to see how sample size is involved in the curves we have been examining. The solution to this riddle lies in the label on the axis of our graphs. We are looking at probability distributions for $\overline{X}$. As we learned in Section 1.6, the variance of $\overline{X}$ is given by $\sigma^2_{\overline{X}} = \sigma^2/n$. This is the variance of each curve drawn in Figures 3.1 through 3.6. We can decrease this variance either by decreasing σ^2 (the variability among individual measurements) or by increasing n. Common sense suggests it is beneficial to increase sample size. If we gather more data, we should have a greater chance of reaching a correct decision. In general, statistics corresponds with and formalizes our common sense.

In the preceding paragraphs, we have looked at probability curves and seen graphically how decreasing σ and increasing n produce greater power. These effects can also be seen in terms of the test statistic,

$$Z = \frac{\overline{X}-\mu}{\sigma/\sqrt{n}}.$$

If σ is decreased, or n is increased, the ratio $\sigma/\sqrt{n}$ becomes smaller. Since this ratio is the denominator of Z, the value of Z becomes larger. A larger Z value is more likely to be beyond the cutoff needed to reject the null hypothesis. There is a greater chance that we will reject H_0 (i.e., greater power).

The probability of a type one error is controlled by the location of our rejection region, typically at 5 percent. Reducing variance and increasing sample size help decrease the chance of a type two error, or, equivalently, increase power. But can we *compute* power? Power is an area under the H_1 curve, and we know how to find areas under normal curves. The only difficulty is that we do not know precisely where the H_1 curve is located. All we know is that it lies somewhere to the right of the H_0 curve. In most practical applications, therefore, we cannot actually find the power of a statistical test. Nonetheless, it behooves us to design our research so as to maximize power. Reducing extraneous sources of variability and gathering more information are basic principles of good experimental technique.

3.6 SAMPLE SIZE CONSIDERATIONS

Researchers frequently want to know what sample size should be used in a decision-making situation. Unfortunately, to determine sample size, we must know the answers to several other questions. First, we must ask *why* we are trying to determine sample size. Typically, the answer is that the researcher would like to reject the null hypothesis. But reject H_0 with what probability (i.e., what power)? The larger the sample size is, the narrower are the curves and the greater is the power. If we know the desired power, we can work backwards to determine the sample size necessary to yield this power. However, we must also know the location of the H_1 curve—that is, the true value of μ. The true value of μ is, of course, generally unknown. In fact, the estimation of μ is typically one of the reasons for conducting the research in the first place.

Still, some ways of framing the sample size question do allow for answers. In Example 3.3, we might ask how big a sample size should be drawn to give us a 90% chance of rejecting H_0 when the diet increases average weight from 44 to 48 grams. This question can be answered, but it is so specific that it is unlikely to be a question of real interest. We have specified both the desired power and the amount by which the diet increased weight. In most practical applications, we cannot specify either of these quantities. In addition, we note that all of this discussion supposes the true standard deviation, σ, is known. In realistic situations, when σ is unknown, the answer to sample size questions cannot be found using the normal distribution. A different, more complex procedure is involved.

In conclusion, the sample size question typically cannot be answered exactly. Instead, common sense principles should govern our choice of sample size. We know that the larger the sample size, the greater the chance of reaching a correct decision. But there is clearly a point of diminishing returns. Increasing sample size from 10 to 20 observations will be much more beneficial than increasing it from 110 to 120 observations. This is because the change in $\sqrt{n}$ is much greater in the former case and, therefore, so is the degree to which the curves become narrower. We are also aware that obtaining data generally costs us something, whether it be money, time, or effort. Practical considerations often play the most critical role in determining how many observations we can afford to gather. As a general rule of thumb, a sample size of about 30 should be adequate for most applications. This will yield power of nearly 90 percent when the H_1 curve is located half a standard deviation above the H_0 curve.

The test characteristics discussed above—power and the probabilities of type one and type two errors—are relevant in all hypothesis testing situations. Regardless of the parameter being investigated or the information available, we set type one error rate at 5 percent and strive to maximize power. Decreasing variance and increasing sample size are key methods for achieving this goal in all statistical tests. We will return to these issues for other hypothesis testing settings in later chapters. Before leaving this section, however, we should emphasize one crucial fact. The observed value of $\overline{X}$ is not employed in determining the rejection region or the power of a statistical test. Power and the probabilities of type one and type two errors are characteristics of the *testing procedure*, not of the obtained data.

3.7 TWO-SIDED TEST

Every example presented so far has involved a one-sided alternative hypothesis. In each case, we were only interested in departures from the null hypothesis in a single direction. In some situations, however, we would like to determine if the true value of μ differs from the value specified in the null hypothesis *in either direction.* In a **two-sided test** we should reject H_0 if $\overline{X}$ is either much smaller or much larger than the hypothesized value of μ. In accordance with this decision strategy, we split the rejection region and put half of it in each tail of the distribution. Example 3.4 presents a typical problem of this sort.

1. State the null and alternative hypotheses.
   Letting μ represent the mean reading for span gas, we wish to test:

$$H_0: \mu = 70$$
$$H_1: \mu \neq 70.$$

2. Determine the rejection region.
   In this problem, we want to reject H_0 if $\overline{X}$ is far away from 70 in either direction. Splitting the 5% type one error probability, we put 2.5 percent in each tail, as shown in Figure 3.7. Since .025 is a tail area, we look for this number in the *body* of Table 1, and read outward to the margins to determine the Z value. The Z value for the upper cutoff is 1.96 and, by symmetry, the

---

Example 3.4

A spectrophotometer, used for measuring carbon monoxide concentration, is checked for accuracy by taking readings on a manufactured gas, called span gas. The carbon monoxide concentration in span gas is very precisely controlled at 70 parts per million, with a standard deviation of 5 parts per million. If the readings suggest that the mean differs from 70 parts per million, the spectrophotometer will have to be calibrated anew. On the basis of the six readings, 85, 77, 82, 68, 72, and 69 parts per million, is calibration necessary?

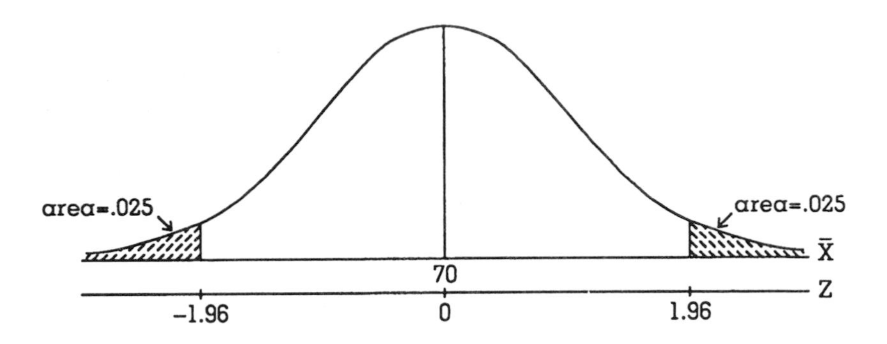

FIGURE 3.7. Two-sided rejection region for spectrophotometer.

lower cutoff is −1.96. Our decision rule is: reject H_0 if Z is greater than 1.96 or less than −1.96. Using absolute values, this can be written: reject H_0 if $|Z|$ is greater than 1.96.

3. Obtain the data.
   For our six observations the sample mean is $\overline{X} = 75.5$ parts per million. The standard deviation for span gas samples is known to be $\sigma = 5$ parts per million.
4. Compute the test statistic.
   Using the values of $\overline{X}$, σ, and n from step 3, and the hypothesized value of 70 for μ, we find

$$Z = \frac{\overline{X} - \mu}{\sigma/\sqrt{n}} = \frac{75.5 - 70}{5/\sqrt{6}} = 2.69.$$

5. Draw the conclusion.
   The observed Z value of 2.69 is greater than the upper critical value of 1.96. Consequently, we reject the null hypothesis (our result is significant), and we conclude that the spectrophotometer must be calibrated.

In this problem, we found an average of $\overline{X} = 75.5$, somewhat above the hypothesized mean value of 70. The obtained Z value of 2.69 was above the upper cutoff of the rejection region. Accordingly, we rejected the null hypothesis that μ is 70 in favor of the alternative that it is not 70. This leaves only two possibilities: either μ is greater than 70 or less than 70. Since $\overline{X}$ was higher than 70, common sense tells us that μ must be greater than 70 as well—that is, we can draw the *one-sided* conclusion that $\mu > 70$ from our two-sided test. In fact, we *should* draw this one-sided conclusion to aid in the calibration process. In general, we should state the conclusion of a statistical test as specifically as possible.

Some purists might disagree with drawing a one-sided conclusion from a two-sided test. They would argue that our conclusion must be limited to saying $\mu \neq 70$. We suggest, however, that this blind adherence to the phrasing of the alternative hypothesis is not rational. We found that $\overline{X} = 75.5$ was too high to believe that the distribution is centered at 70. If the distribution were centered *below* 70, the observed $\overline{X}$ value would be even further out in the tail and, therefore, even less likely. The chance of finding $\overline{X} = 75.5$ when μ is less than 70 is extremely small. We believe this extra risk is so small that it can safely be ignored. If the null hypothesis is rejected in a two-sided test, we *can* and *should* draw the appropriate one-sided conclusion.

The ability to draw a one-sided conclusion from a two-sided test raises an interesting question. Why should we ever do a one-sided test? In Example 3.3, for instance, we were interested in whether the diet increased weight of mice. We could have answered this question even with the two-sided alternative hypothesis H_1: $\mu \neq 44$. If our rejection were due to a large value of $\overline{X}$, we could conclude that the diet increased weight. In addition, if we rejected H_0 because of a small value of $\overline{X}$, we could conclude that the diet actually decreased weight. It would seem that we have something to gain and nothing to lose from always doing two-sided tests. But we are forgetting one important test characteristic, namely power.

Suppose, in Example 3.3, we decided to perform a two-sided test, even though we expected the diet to increase weight. Suppose, in addition, the diet does increase weight, so the H_1 curve is located to the right of 44. Have we lost anything by doing a two-sided test? Yes, we have lost power. Figure 3.8 shows the rejection regions for both the one-sided alternative hypothesis H_1: $\mu > 44$, and the two-sided alternative hypothesis H_1: $\mu \neq 44$. Power is the probability that we reject H_0 when H_1 is true—that is, the area to the right of the line under the H_1 curve. Note that this area is greater for the .05 cutoff, corresponding to the one-sided test, than for the .025 cutoff used in the two-sided test. The extra power we realize with the one-sided test is the region in Figure 3.8 with only horizontal shading. Simply sated, it is easier to obtain an $\overline{X}$ value above the .05 cutoff than above the .025 cutoff. If we found our $\overline{X}$ value *between* these two cutoffs, the one-sided test would reject H_0, while the two-sided test would have to accept H_0. In this case, we would surely regret having performed a two-sided test.

To be thorough, we should note that Figure 3.8 omits a small portion of the power of the two-sided test. If we extend the left tail of the H_1 curve, it approaches the axis,but never quite hits it. There is always a small area under the curve, even to the left of the lower .025 cutoff. Power is the probability that we reject H_0 when H_1 is true. For a two-sided test, power includes the area under the H_1 curve beyond the cutoffs, both above the upper .025 cutoff and below the lower .025 cutoff. However, the area under the H_1 curve below the lower .025 cutoff is virtually 0.

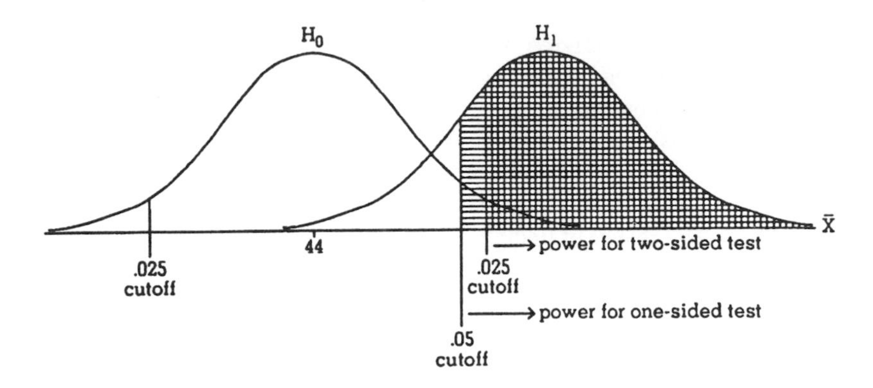

FIGURE 3.8. Loss of power in two-sided test.

Certainly, this area is trivial in comparison with the area between the upper .05 and .025 cutoffs. In short, this small region does not alter the fact that a one-sided test is more powerful than a two-sided test, as long as the departure from the null hypothesis is in the expected direction.

The decision to perform a one-sided test must be made *before looking at the data* or any statistics computed from the data. Suppose we are interested in departures from the hypothesized mean in either direction and, accordingly, plan a two-sided test. Before committing ourselves to the two-sided test, however, we cheat by computing the sample mean. If we see that $\overline{X}$ is above the hypothesized mean, we might be tempted to locate the 5% rejection region in the upper tail, as would be appropriate for a one-sided test. On the other hand, if we find that $\overline{X}$ is below the hypothesized mean, we might locate the 5 percent in the lower tail. For all intents and purposes, we are operating as if there is a 5% rejection region in *each* tail. This strategy allows a 10% chance of making a type one error. Recall the order of steps in hypothesis testing: *first* we state the null and alternative hypotheses and determine the rejection region, *then* we obtain the data. The data should not be used to determine what we are going to test with that same data.

The practical applications to be drawn from this discussion are simple and logical. If we are only interested in departures from the null hypothesis in a single direction, we should do the corresponding one-sided test. Putting all our rejection region in one tail makes the test more powerful if the expected departure occurs. We should only perform a two-sided test when it would be worth uncovering a departure from the hypothesized value in either direction. However, if we are able to reject the null hypothesis in a two-sided test, we should report the corresponding one-sided conclusion.

3.8 TEST ON THE MEAN WHEN STANDARD DEVIATION IS UNKNOWN

So far in this chapter we have discussed testing hypotheses about the mean when the standard deviation (σ) is known. In such cases, we compute a Z statistic and use the standard normal distribution to determine the rejection region. We now consider the more realistic problem of testing a hypothesis about the mean when the standard deviation is unknown. The groundwork for this situation was laid in Chapter 2. In Section 2.7, we learned that the ratio

$$t = \frac{\overline{X}-\mu}{S/\sqrt{n}} \tag{3.8.1}$$

follows the t distribution with n - 1 degrees of freedom. To test a hypothesis about the mean when the standard deviation is unknown, we use this statistic and perform a **t test**.

All the logic of hypothesis testing presented in this chapter is still appropriate. Only two minor changes are made in the case of unknown standard deviation: we use S instead of σ in the denominator of the test statistic, and we obtain the cutoff point from the t distribution instead of the normal. The rejection region may be located in the left tail, right tail, or split in both tails according to the alternative hypothesis. The t test is demonstrated for Example 3.5.

1. State the null and alternative hypotheses.
   The sprinkler system is meeting its design specification if activation time is 25

Example 3.5

A fire prevention sprinkler system is designed to activate within 25 seconds when the temperature reaches a certain threshold. In 20 trials at the threshold temperature, the following activation times were recorded:

21	23	33	29	25
39	34	19	24	30
24	35	31	41	27
29	39	18	37	21

Is the sprinkler system responding within the designed time limit?

seconds or less. Our concern is that the mean activation time may be more than 25 seconds. Thus, we wish to test:

$$H_0: \mu \le 25$$
$$H_1: \mu > 25.$$

As discussed above in connection with Example 3.3, we will treat the null hypothesis as if it were written H_0: $\mu = 25$.

2. Determine the rejection region.
   We locate the rejection region in the right tail since the alternative hypothesis would be likely to result in a large value for $\overline{X}$. Now, however, we are standardizing $\overline{X}$ to the t scale instead of the Z scale. Accordingly, we use the t distribution with $n - 1 = 20 - 1 = 19$ degrees of freedom. From Table 2 in the Appendix, we find the 5% cutoff to be 1.729. Our decision rule is to reject H_0 if t is greater than 1.729.
   Technically, Figure 3.9 slightly misrepresents the relationship between the $\overline{X}$ and t scales. When σ is known, $\overline{X}$ is standardized to the Z scale. Z, like $\overline{X}$, has a normal distribution, and it is appropriate to construct both Z and $\overline{X}$ scales under the same curve. However, when σ is unknown, $\overline{X}$ is standardized to the t statistic shown in (3.8.1). The t distribution is quite similar to the normal, but not identical, so it is technically incorrect to draw both t and $\overline{X}$ scales under the same curve. This slight inaccuracy is irrelevant, since the picture is only used as a conceptual tool. The t table properly accounts for the difference between t and normal distributions. Note that our t cutoff of 1.729 is slightly different from 1.645, the cutoff from the normal distribution.
3. Obtain the data.
   In this example, we have n = 20 observations and compute the sample mean to

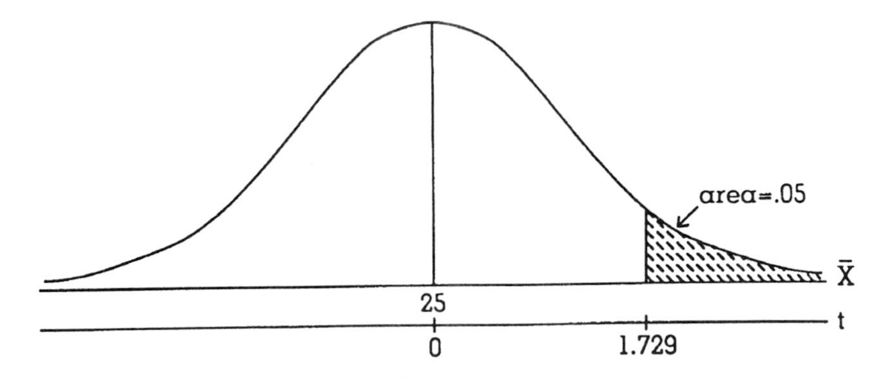

FIGURE 3.9. Right tail area of 5 percent for sprinkler activation time.

be $\overline{X} = 28.95$ seconds. Calculating the sample variance using (2.2.1) produces $S^2 = 49.734$. Taking the square root, we obtain the sample standard deviation $S = \sqrt{S^2} = 7.05$.

4. Compute the test statistic.
   The hypothesized value for μ is 25. We find the test statistic

$$t = \frac{\overline{X} - \mu}{S/\sqrt{n}} = \frac{28.95 - 25}{7.052/\sqrt{20}} = 2.505.$$

5. Draw the conclusion.
   The observed value $t = 2.505$ is greater than our cutoff of 1.729, so we reject the null hypothesis. We conclude that the sprinkler system is not activating within the 25-second design specification at the threshold temperature.

3.9 ACCEPTING THE NULL HYPOTHESIS

In Example 3.5, the average activation time of 28.95 seconds produced a significant result—that is, we were able to reject the null hypothesis. But suppose the sample mean had been slightly less than this figure and the observed t value fell just below the 1.729 critical point. If the average had been 27, for instance, we would have found $t = 1.268$. We would have accepted H_0 and concluded that the sprinkler system was operating adequately. With a sample mean of 27, we would have accepted the hypothesis that $\mu \leq 25$. Common sense tells us there is a problem with this conclusion.

The main difficulty in this example is that we would *like to accept* H_0. Accepting H_0 tells us that the sprinkler system is operating satisfactorily. Hypothesis testing, however, is designed for situations in which we *want to reject* H_0, and deliberately makes it difficult to reach this conclusion. It is very easy to accept the null hypothesis. If we use a small sample size and allow extraneous sources of variability to affect the research, power will be very low and we are quite likely to accept H_0. Since acceptance of H_0 can be virtually guaranteed by doing sloppy research, some statisticians never conclude they have accepted H_0. They either reject H_0 or *fail to reject* H_0. Simply stated, the procedures of hypothesis testing are not well suited to situations in which we would like to accept the null hypothesis.

Perhaps we could eliminate this problem by simply switching the null and alternative hypotheses, and writing:

$$H_0: \mu > 25$$
$$H_1: \mu \leq 25.$$

In hypothesis testing, we always assume that the null hypothesis is true unless shown to be false. By this choice of hypotheses, we would be assuming that the sprinkler system is faulty, unless it were proven adequate. This sounds like a reasonable and conservative strategy to adopt in testing a safety device. Unfortunately, this formulation of H_0 is inadequate, since it does not tell us precisely where to draw the H_0 curve. It says only that the curve must be drawn somewhere to the right of 25. We must know precisely where to draw the H_0 curve, in order to locate the cutoff for our rejection region.

We can remedy this problem by putting the "or equal to" part in H_0 and writing:

$$H_0: \mu \geq 25$$
$$H_1: \mu < 25.$$

At first glance, this approach seems to solve our problem. We would put our rejection region in the left tail of the distribution centered at $\mu = 25$, since the alternative hypothesis specifies that $\mu < 25$. With these hypotheses, we would only reject H_0 if the sample average were substantially less that 25 seconds. But what if the data show an average activation time of 25 seconds, or even slightly lower, but not low enough to reject H_0? We would be faced with evidence that falls within the design specifications (an average of 25 or slightly lower), yet would conclude that the sprinkler system is inadequate. For this reason we cannot eliminate our problem by simply switching the null and alternative hypotheses. We must frame them as originally planned.

We can, however, suggest a practical solution to this dilemma. The key lies in realizing that the usual type one error probability of 5 percent is inappropriate for situations in which we would like to accept H_0. We should use a larger α level, say 25 percent. The larger α level increases power, and makes it easier to reject H_0. We should also attempt to increase power by reducing extraneous sources of variability and using a large sample size. If the sprinkler system is not operating properly, we will have an excellent chance of detecting this problem. If we are *still* unable to reject the null hypothesis, even after taking all these precautions to ensure good power, we may be justified in concluding that the sprinkler system is adequate.

3.10 STATISTICAL SIGNIFICANCE AND PRACTICAL IMPORTANCE

In hypothesis testing we should be concerned with two separate questions. The first question is "How *sure* are we that H_0 is wrong?" The more convinced we are that the null hypothesis is wrong, the greater is the level of **statistical significance**. The second question is "How *wrong* is H_0?" A hypothesis may be only slightly incorrect, or the true value of μ may be very far from the hypothesized value. The greater the disparity between the true value of μ and the hypothesized value, the greater is the **practical importance** of our obtained result. These two questions address different aspects of the available evidence, and must be answered with different approaches. To demonstrate these two issues, consider Example 3.6.

Let us compare the chances of type one and type two errors for the two tests performed by Company A and Company B. The probability of a type one error is 5 percent for both A and B. The large difference in sample sizes is automatically accounted for by the difference in cutoff values. For A, the degrees of freedom are $n_A - 1 = 9$ and the critical t value is 1.833. For B, we have $n_B - 1 = 999$ degrees of freedom. Using the line on the t table for infinite degrees of freedom, we obtain the cutoff 1.645. (This is the 5% cutoff for the normal distribution. Recall from Section

---

Example 3.6

A large company is concerned with the rate at which it must replace drums in its copying machines. In the past few years drums have lasted, on the average, for only 30,000 copies. Two new types of copiers—let us label them A and B—are under consideration. Companies A and B have each been asked to provide evidence that their drums last longer than 30,000 copies.

Company A responded by testing the hypotheses:

$$H_0: \mu_A = 30{,}000$$
$$H_1: \mu_A > 30{,}000.$$

A sample of size $n_A = 10$ was employed and resulted in rejection of the null hypothesis using the 5% α level. The test was just barley significant, with the observed value of t falling only slightly beyond the 5% cutoff.

Likewise, Company B tested the hypotheses:

$$H_0: \mu_B = 30{,}000$$
$$H_1: \mu_B > 30{,}000.$$

Using a sample of size $n_B = 1000$, this test was also just barely significant at the 5% level. Comparison of the results showed that the two standard deviations, S_A and S_B, were nearly equal, so neither company appears to produce a more *consistent* copier.

Both companies have demonstrated that their drums average more than the 30,000 copies obtained from the machines currently in use. The two new copiers have roughly equal costs, so a decision is to be made on the basis of the statistical evidence presented above. Which copier should be selected?

2.7 that the t distribution approaches the standard normal as the degrees of freedom increase.) The two critical values, 1.833 and 1.645, are rather similar. Yet the difference between them is adequate to ensure that both A and B have the same chance of a type one error. Each cutoff value leaves exactly 5 percent in the upper tail of the t distribution with the appropriate degrees of freedom. Type one error rate clearly does not play a role in deciding between the two copiers.

Instead of comparing the probabilities of type two errors for these two tests, we phrase our discussion in terms of power. Power is the complement of type two error (power = $1 - \beta$). The much larger sample size used by B would seem to result in a more powerful test than for A. However, this is not the case. Our earlier discussion of power in Section 3.5 mentioned two factors affecting power, namely variance and sample size. A third factor is also involved—the difference between the locations of the H_0 and H_1 curves. Look back at Figure 3.5 and suppose that the H_1 curve was located even further to the right of the H_0 curve than pictured. This would result in greater power. In general, the greater the difference between the locations of the H_0 and H_1 curves, the greater is the power. We did not mention this location difference earlier because we were concerned with *practical* ways of increasing power, and the exact location of the H_1 curve is beyond the control of the researcher. But when *comparing* two statistical tests, we should be concerned with the difference between the H_0 and H_1 curves in the two tests.

In the next few paragraphs we show that the difference between the H_0 and H_1 curves is much greater for A than for B. The greater disparity between the curves for A compensates for the smaller sample size. As a result, the test for A is about as powerful as the test for B. Evidence of this rough equality in power is that both tests were just barely significant.

We can show that the difference between the H_0 and H_1 curves is greater for A than for B by comparing the observed t values. The t statistic for A is given by

$$t_A = \frac{\overline{X}_A - \mu}{S_A/\sqrt{n_A}} = \frac{\overline{X}_A - 30{,}000}{S_A/\sqrt{10}} .$$

Instead of dividing by $S_A/\sqrt{10}$ in this formula, we can invert the denominator and multiply. Thus, t_A can be written as

$$t_A = (\overline{X}_A - 30{,}000)\frac{\sqrt{10}}{S_A} .$$

In similar fashion, the t statistic for B is equivalent to

$$t_B = (\overline{X}_B - 30{,}000)\frac{\sqrt{1000}}{S_B}.$$

The two cutoff values (1.833 for A and 1.645 for B) are quite similar. Since both tests were just barely significant, t_A and t_B were approximately equal. Ignoring any slight difference, we equate t_A and t_B to find:

$$(\overline{X}_A - 30{,}000)\frac{\sqrt{10}}{S_A} = (\overline{X}_B - 30{,}000)\frac{\sqrt{1000}}{S_B}.$$

When both sides of an equation are multiplied by the same quantity, the results are still equal. Here we multiply by S_A, cancelling out this term on the left side of the equation. On the right, S_A approximately cancels out with S_B, since these values were nearly the same. Again, we ignore any slight difference and write:

$$(\overline{X}_A - 30{,}000)\sqrt{10} = (\overline{X}_B - 30{,}000)\sqrt{1000}.$$

Now divide on both sides by $\sqrt{10}$, cancelling this factor on the left side of the equation. On the right side, $(\overline{X}_B - 30{,}000)$ is now multiplied by

$$\frac{\sqrt{1000}}{\sqrt{10}} = \sqrt{\frac{1000}{10}} = \sqrt{100} = 10.$$

We therefore have the result:

$$(\overline{X}_A - 30{,}000) = (\overline{X}_B - 30{,}000)(10). \tag{3.10.1}$$

The left side of this equation, $(\overline{X}_A - 30{,}000)$, is an estimate of the *improvement* offered by copier A over the present machine. $\overline{X}_A$ estimates μ_A, the mean number of copies expected with copier A, and the present machine produces an average of 30,000 copies per drum. The difference between these values, $(\overline{X}_A - 30{,}000)$, estimates the additional copies we should expect to obtain from copier A. Similarly, the factor $(\overline{X}_B - 30{,}000)$ on the right side of this equation estimates the improvement for copier B over the present machine. Equation (3.10.1) says that the improvement for A is *ten times greater* than the improvement for B. This factor of ten far outweighs any slight differences ignored in earlier steps. Our choice is clearly to purchase copier A, because it offers a much longer expected drum life. For example, if copier B averages 31,000 copies before requiring drum replacement (an improvement of 1000 copies), copier A would average 40,000 copies (an improvement of 10,000 copies).

In Example 3.6, the null hypothesis curve is drawn at $\mu = 30{,}000$ for both tests. The alternative hypothesis says that μ is greater than 30,000 and, accordingly, the H_1 curve lies somewhat to the right of 30,000. The tests for both A and B rejected H_0, so we believe that H_1 is correct in each test. But the H_1 curves for A and B are *not equally far* above 30,000. For each copier, the difference $(\overline{X} - 30{,}000)$ is an estimate of how far above the H_0 curve the H_1 distribution lies. For A, this difference is ten times as much as for B.

Statistical significance addresses the question "How *sure* are we that H_0 is wrong?" The answer to this question is measured by how far out in the tail of the H_0 distribution the test statistic falls. The further out in the tail, the more convinced we are that H_0 is false. In our example, both t values were just barely beyond the 5% cutoff. Thus, we are equally sure that the null hypothesis is false in the two tests.

Practical importance, on the other hand, addresses the question "How *wrong* is H_0?" In the present situation, we have measured practical importance with the quantity $(\overline{X} - 30{,}000)$. The larger this quantity is, the greater the difference between the H_0 and H_1 curves, and the more wrong H_0 is. The practical importance for A was much greater than for B.

Sample size plays a role in determining statistical significance, but not practical importance. In our example, B used a sample size of 1000, but suppose a much smaller sample had been employed. The t value for B is

$$t_B = \frac{\overline{X}_B - \mu}{S_B/\sqrt{n_B}}.$$

What terms in this expression would change greatly if a smaller sample size had been used? The value of $\overline{X}_B$ is always an estimate of the true mean. With a smaller sample size, it would undoubtedly change, but there is no particular reason to expect $\overline{X}_B$ to be substantially smaller, nor substantially larger. Similarly, S_B estimates the true standard deviation, and would probably not change much with a smaller sample size. The hypothesized mean μ would still be 30,000. Only the value of n_B would change appreciably. A smaller value of n_B would result in a larger value of $S_B/\sqrt{n_B}$ and, therefore, a smaller value of t_B. Since we were just barely in the rejection region with our original sample, a smaller sample size would have resulted in accepting H_0. In this sense, B *had* to draw a large sample to reject

H_0. In order to reject the null hypothesis, B had to draw a much larger sample size than A, because the null hypothesis was more nearly true for B than for A.

This should not be construed as an argument against large sample sizes. Power is definitely greater with a larger sample size. But a large sample size carries with it a responsibility, because we can sometimes detect differences that are not worth finding. An excellent example of this is found in research on extrasensory perception (ESP). A simple ESP experiment might involve trying to guess the suit of a card drawn randomly from a deck. If there is no such phenomenon as ESP, the proportion of correct guesses should be 0.25 (the null hypothesis). If ESP exists, this proportion should be greater than 0.25 (the alternative hypothesis). Notice, however, how easy it is to obtain data. Each time a card is guessed we have another observation. Simple ESP experiments such as this have employed sample sizes of many thousands. Some of these experiments have produced observed values of t extremely far out in the tail. Such research produces very strong evidence that ESP exists. But how big is the effect? While some people are better guessers than others, the overall average proportion of correct guesses is typically a figure like 0.250001. The null hypothesis appears to be only a tiny bit wrong, since the proportion is just slightly above 0.25. With a huge sample size, however, we can be very sure that this tiny difference exists. This discussion is not intended to promote belief in ESP. Much of this research has been criticized for having inadequate controls. But even when flaws have not been uncovered in experimental procedure, the practical importance of the effect is negligible.

Statistical significance is and should be dependent on sample size. Even if the null hypothesis is only slightly wrong, the more evidence we have of this slight effect, the more sure we are that H_0 is wrong. But additional evidence does not make the null hypothesis *any more wrong*. Practical importance must be measured by some quantity that is independent of sample size. In our comparison of the two copiers, practical importance is measured by $(\overline{X} - 30{,}000) = (\overline{X} - \mu)$. This quantity should not change appreciably as sample size changes.

An even better measure of practical importance is

$$\frac{\overline{X}-\mu}{S},$$

since the dimensions cancel out and the ratio is **scale-free**. If we measure the lifetime of copier drums in *reams* of paper rather than individual sheets, the values of $\overline{X}$, μ, and S would all change, but the changes would cancel out in this ratio. A quantity is said to be scale-free if it remains the same when the data are measured on a different scale.

The issues of statistical significance and practical importance are relevant in all hypothesis tests, not just t tests. For any test, statistical significance is measured by how far out in the tail the test statistic falls. The further out in the tail, the more sure we are that H_0 is wrong. Most computer programs compute and print the actual tail area beyond the test statistic. This tail area is called the **p-value** or **significance level**. As long as the p-value is less than 5 percent, we can reject the null hypothesis. But the smaller the p-value, the more convincing the evidence against H_0 is. It is good practice to report the p-value whenever conducting a hypothesis test.

The p-value is an appropriate yardstick for assessing the degree of statistical significance in any hypothesis test. The appropriate measure of practical importance, however, depends on the kind of test performed. It must always reflect how wrong the null hypothesis is, be insensitive to sample size, and, preferably, be scale-free. We will continue to address these two sides of the coin—statistical significance and practical importance—throughout the remaining chapters of this book.

3.11 RELATIONSHIP BETWEEN HYPOTHESIS TESTING AND CONFIDENCE INTERVALS

In many situations, the same statistical tools are used for testing hypotheses as in the construction of confidence intervals. For instance, the t distribution is employed both in testing a hypothesis about μ when σ is unknown and in forming a confidence interval for μ when σ is unknown. It should come as no surprise that these two applications of the t distribution are intimately related.

In Section 2.8, we constructed a confidence interval for μ when σ was unknown. The example we used in that discussion had a sample size of 15, and therefore df $= n - 1 = 14$. The form of the 95% confidence interval was

$$\overline{X} \pm 2.145\frac{S}{\sqrt{n}}.$$

To demonstrate the relationship between confidence intervals and hypothesis testing, suppose we use the same sample size to test a hypothesis about μ. Specifically, consider a two-sided test at the 5% α level with 14 degrees of freedom. Putting 0.025 in each tail, we would reject H_0 if the observed t value were less than -2.145 or greater than 2.145. Equivalently, we would *accept* H_0 if $-2.145 < t < 2.145$. Substituting the formula given in (3.8.1) for t, we would accept H_0 if

$$-2.145 < \frac{\overline{X}-\mu}{S/\sqrt{n}} < 2.145,$$

where μ is the hypothesized value for the mean. With some algebra, we could rewrite this acceptance criterion as

$$\overline{X} - 2.145\frac{S}{\sqrt{n}} < \mu < \overline{X} + 2.145\frac{S}{\sqrt{n}}.$$

The bounds for μ in this acceptance region are

$$\overline{X} \pm 2.145\frac{S}{\sqrt{n}},$$

the same as the bounds for the confidence interval. In short, we would accept H_0 if the hypothesized value for μ lies in the confidence interval.

Using this relationship, we could avoid ever having to perform a hypothesis test. We could construct the confidence interval instead. If the interval contains the value we planned to specify for μ in the null hypothesis, we would accept H_0. In fact, for many purposes, a confidence interval may be more informative than a hypothesis test. The confidence interval includes all values for μ that would be accepted as hypothesized values. The 95% confidence interval is equivalent to a hypothesis test at the 5% α level. Similarly, a 99% confidence interval is equivalent to a hypothesis test at the 1% α level. If we wanted to perform a *one-sided* hypothesis test, it is possible to construct a corresponding one-sided confidence interval.

When the true standard deviation, σ, is known, we perform a Z test for a hypothesis about μ. This test is equivalent to the confidence interval for μ when σ is known, which is also based on the Z distribution. In general, for relatively simple hypothesis testing situations, such as one- and two-sample procedures, equivalent confidence intervals exist. However, in more complex experiments, this parallelism breaks down. The hypotheses we wish to test in complicated research plans do not correspond in any direct way to confidence intervals. Therefore, in subsequent chapters we continue our development of hypothesis testing separately from confidence interval construction.

3.12 ASSUMPTION OF NORMAL DATA

Technically, both the Z and t tests for hypotheses about the mean are valid only when the data are normally distributed. As we learned in Section 1.7, however, the distribution of $\overline{X}$ approaches the normal curve as sample size increases, even if the distribution of individual observations is not normal. For the Z test, in which the only random variable is $\overline{X}$, we can safely ignore the normality assumption.

The t ratio, on the other hand, depends on the data through both $\overline{X}$ and S. If the data come from a normal distribution, this ratio is known to have a t distribution. We can use the table of the t distribution to locate the 5% cutoff value. If the data do *not* come from a normal distribution, however, the t ratio we compute may have a distribution that is not the same as a true t curve. Theoretically, we want to locate our rejection region according to the *actual* distribution of this ratio. The 5% cutoff point from the t table can be used only if the actual distribution of our ratio is close to a true t distribution.

In general, increasing sample size tends to lessen the need for normality. For even moderate sample sizes, say 15 to 25 observations, the actual distribution of our computed t ratio is likely to be very close to a true t distribution, and we can use the t table to find our rejection region. With small sample sizes, however, it is prudent to plot the available data to see if they resemble the normal shape. If the data have the general features of a normal curve, symmetry with a single lump near the middle, the test is certainly valid. In fact, we need only be concerned about the normality assumption when the plot of the data appears substantially non-normal.

In the next section, we introduce a test for the mean that can be used for non-normal data. This test can also be used for normal data, but in the normal case the t test is more powerful. Since tests are available for both normal and non-normal data, it would be advantageous to have an objective method of deciding whether the data follow the normal shape. We would like to be able to *test* the null hypothesis that the data come from a normal distribution. Such tests exist, but they are infrequently used. The drawback to tests of normality is that their power is greatest at the wrong times. With large samples, the tests are very powerful and generally reject the null hypothesis of normality, even when the data are relatively close to the normal shape. Rejecting normality suggests that the t test is invalid. But with a large sample, close to normally distributed, the t test is definitely appropriate. On the other hand, if the sample size is very small, the tests for normality are weak. We are liable to accept the null hypothesis of normality even when the data depart rather markedly from the normal shape. Acceptance of normality suggests that the t test is okay. But with a very small sample, far from normal in shape, the t test is questionable. In short, testing for normality is not a solution to the problem of deciding whether it is safe to use the t test. The tests for normality are powerful only with large sample sizes, precisely when we need be *least* concerned with the normality assumption. Instead, we recommend plotting the data to see if they resemble the normal shape. When examining this plot, we should keep in mind that relatively greater departures from normality can be tolerated with larger sample sizes.

3.13 TESTS ON THE MEAN FOR NON-NORMAL DATA

Tests that do not assume normality are generally referred to as **distribution-free** or **nonparametric** tests. Virtually all such tests begin by ranking the data. Only the ranks are used in subsequent computations, not the original numeric values. Since the same ranks might be obtained from rather different samples, the actual distribution of the data is relatively important. However, all tests make *some* assumptions about the shape of the distribution. Thus, the phrase "distribution-free" is technically incorrect. In addition, these procedures test hypotheses about parameters, so the word "nonparametric" is also misleading. Nevertheless, "nonparametric" is the expression most often used to describe such rank tests, and we will use this term to refer to these methods.

The **signed rank test** is a nonparametric procedure that can be used to test hypotheses about the mean of a population. In theory, the test is only appropriate for testing the null hypothesis that the mean is 0. In practice, however, the test can be used with other hypothesized means by transforming the data prior to testing. As with Z and t tests, either one-sided or two-sided alternatives can be employed. We discuss the logic of the signed rank test in reference to Example 3.5.

The sprinkler activation times presented in Example 3.5 are plotted in Figure 3.10. These data appear relatively symmetric, with the points spread fairly evenly across the range of the sample. They do *not* tend to accumulate near the center, as would be expected for a sample from a normal distribution. In practice, the sample size is large enough (20 observations) that normality should not be an issue. In addition, the plot is not grossly non-normal. Considering the sample size and plot, the t test is certainly valid for these data. Nonetheless, since the plot departs at least slightly from the normal shape, we might consider employing the signed rank test.

In this example, we wish to test

$$H_0: \mu = 25$$
$$H_1: \mu > 25,$$

where μ refers to mean sprinkler activation time. Since the signed rank test requires that the hypothesized mean be 0, we begin by subtracting 25 from each observation. If the mean of the original data is 25, the mean of these transformed values is 0. The transformation is simply a standardization, like the Z scale, except that we change only the mean of the scale, not its variance. In terms of the transformed values, we wish to test

$$H_0: \mu = 0$$
$$H_1: \mu > 0.$$

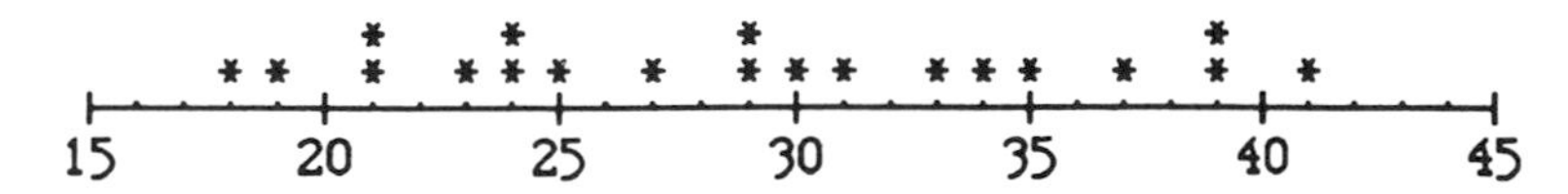

FIGURE 3.10. Sprinkler activation times.

If the null hypothesis is true, the average of the transformed values in our sample should be near 0. About half the values should be positive, and half negative. Furthermore, the positives and negatives should tend to be equally close to 0. We should not find all the positive values very near 0, and all the negative values far away. If we rank the observations in terms of their distance from 0, without regard to sign, we should find roughly equal sums of ranks for positives and negatives. The signed rank statistic is expressed in terms of the difference between these sums of ranks, the sum for positives, and the sum for negatives. When the null hypothesis is true, this difference should be near 0.

Now suppose, on the other hand, that the alternative hypothesis is true and the mean of the transformed data is greater than 0. We should find more positive than negative values in our sample, and the positives should tend to be further from 0 than the negatives. The sum of ranked distances from 0 should be greater for the positives than for the negatives. The difference between these sums should be greater than 0.

The distribution of the signed rank statistic can be shown to be approximately normal. To perform the test, this statistic is standardized to the Z scale, and the standard normal table is used to determine the rejection region. For Example 3.5, we put the 5% rejection region in the right tail and use the critical Z value of 1.645. Omitting the actual computations, the Z value for the signed rank test on these data is 2.13, and we reject the null hypothesis.

Recall that the t test performed in Section 3.8 also resulted in rejecting the null hypothesis for Example 3.5. These two tests will generally reach the same conclusion for large samples, as well as for small samples that are reasonably close to normal. When the data are normally distributed, the t test tends to be slightly more powerful than the signed rank test. This is because the t test employs the actual numeric values, while the signed rank test uses only their relative sizes (i.e., their ranks). Since the t test uses more of the information in the sample, it has a slightly greater chance of reaching the correct decision (i.e., more power). When dealing with small samples that depart substantially from the normal shape, the signed rank test is somewhat more powerful, and tends to reject H_0 more frequently than the t test.

Although they do not assume normality, nonparametric tests do make assumptions about the distribution of the data. The signed rank test requires that the data come from a continuous and symmetric distribution. Assuming that the distribution is continuous may cause difficulties. For a continuous distribution, the probability that two observations are exactly the same is technically 0. Ties should not occur. In practice, however, measurements are never infinitely precise, and rounding often produces ties. In most nonparametric methods, including the signed rank test, ties are handled by the **midranking procedure**. In this procedure, observations that are tied are each assigned the average of the ranks that would have been assigned had the values been slightly different, and therefore not tied. This strategy is reasonable, but the presence of ties alters the possible values that can be observed for the signed rank statistic and, therefore, its distribution. While a small to moderate number of ties can be tolerated without invalidating the approximate normality, the signed rank test should not be employed if ties are especially prevalent.

Asymmetry, a problem for the t test, is also a problem for the signed rank test. Another nonparametric procedure, called the **sign test**, should be used with asymmetric data. Technically, this is a test that the population median, not the mean, is 0. The sign test is based on the same transformed data as the signed rank test—that is, the hypothesized mean is subtracted from each observation, since the procedure is designed to test for a mean of 0. Instead of *ranking* the differences from 0, however, the sign test merely *counts* the number of positives and negatives. If the null hypothesis is correct, we should find about as many positives as negatives in our sample, and the difference between these two counts should be near 0. If the alternative hypothesis is correct in Example 3.5, we should find more positives than negatives, and the difference in counts should be larger than 0.

When the data are symmetric, as in Example 3.5, the median is essentially the same as the mean. The t, signed rank, and sign tests can all be used. The sign test is less powerful with symmetric data than the t and signed rank tests. The sign test uses less of the information in the data than the other two tests, and is less likely to reach the correct decision. When the data are substantially asymmetric, the sign test is generally more powerful than the t and signed rank tests.

3.14 P-VALUES IN COMPUTER PROGRAMS

The next section of this chapter uses Example 3.5 to demonstrate tests for the mean in computer programs. As will be seen in that section, most computer programs (including SAS and SPSS) do not allow the user to specify on input either an α level or the alternative hypothesis. However, the information they provide is sufficient to allow for any desired α level or alternative hypothesis. This section discusses how to use the information provided by computer programs to perform either one-sided or two-sided tests at any desired α level.

In Example 3.5, a sample of size 20 was used to test for a mean sprinkler activation time of 25 seconds. Suppose, for the time being, the alternative had been two-sided, corresponding to the hypotheses

$$H_0\colon \mu = 25$$
$$H_1\colon \mu \neq 25.$$

The t test uses $n - 1 = 19$ degrees of freedom. Putting 0.025 area in each tail, as would be appropriate for a 5% α level, the t table shows critical values of ±2.093. Earlier, we computed the t value for this sample to be 2.505, so we should reject the null hypothesis.

When this example is run on the computer, most programs report the computed t statistic and list the "two-tailed p-value" of 0.022. This is the tail area above t = 2.505 plus the corresponding tail area below t = −2.505, as pictured in Figure 3.11. By the symmetry of the t distribution, each tail area must be half this total, or 0.011. Since the tail area is less than 0.025, the computed t must lie beyond the 0.025 cutoff and, therefore, in the rejection region. We do not even have to know that the 0.025 cutoff is 2.093. The table of the t distribution is not needed when using a computer program. For a two-sided test, all we need is the two-tailed p-value. If this figure is less than 0.05, each tail separately is less than 0.025, and the observed t value is in the rejection region. In Example 3.5, since p = 0.022 is less than 0.05, we reject H_0 and report the p-value of 0.022.

If we wished to use a different α level for our test, say $\alpha = 0.01$, the two-tailed p-value must be less than 0.01, to reject the null hypothesis. Knowing the two-tailed p-value allows the user to employ any desired α level, simply by comparing the p-value with α. If p is less than α, we reject the null hypothesis.

Suppose, as in the original problem, we want to perform a one-sided test, such as

$$H_0: \mu = 25$$
$$H_1: \mu > 25.$$

In this case, we put our 5% rejection region all in the right tail of the t distribution. The computer, however, reports a two-tailed p-value automatically. This is not a problem, since we only need to halve the printed p-value to obtain the area in a

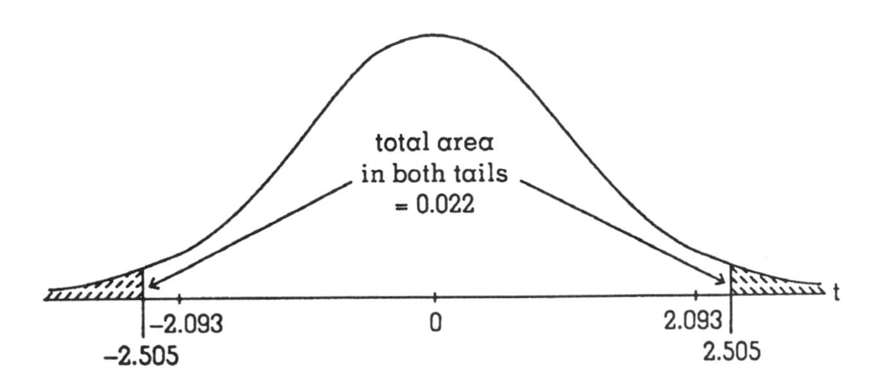

FIGURE 3.11. Two-tailed p-valve of 0.022.

single tail. For Example 3.5, the observed t value of 2.505 has upper tail area of 0.022/2 = 0.011. Since 0.011 is less than 0.05, we are in the rejection region. We should reject H_0, and report our p-value as 0.011. For a one-sided test, the one-tailed p-value is half of the two-tailed p-value printed automatically by most programs. Again, we can use any desired α level. If the one-tailed p-value is less than α, we reject the null hypothesis.

There is one additional check that must be made for a one-sided test. The p-value measures how far out in the tail the test statistic falls, but we must make sure that it falls in the *proper* tail. Suppose, in our example, we wished to test

$$H_0: \mu = 25$$
$$H_1: \mu < 25.$$

In this case, the 5% rejection region would be located in the *lower* tail of the t distribution. Since the observed t value was positive, we must surely accept H_0. The obtained one-sided tail area of 0.011 is misleading. It tells us that t = 2.505 is far out in the tail, but it doesn't tell us *which* tail. For a one-sided test, we must first check the sign of the observed t value to be sure it is in the predicted direction. If so, it is appropriate to compute the one-tailed p-value to see if it is *far enough* out in the tail to reject the null hypothesis.

For a two-sided test, we learned in Section 3.7 that a significant result leads to a one-sided conclusion. We conclude that the mean is less than the hypothesized value if t is negative, and greater than the hypothesized value if t is positive. Despite drawing a one-sided conclusion, however, we should *not* halve the two-tailed p-value printed by the computer. Doing so would be tantamount to performing a one-sided test. As we also learned in Section 3.7, the decision to perform a one-sided test must be made before looking at the data or any statistics depending on the data. When a two-sided test is significant, we can state a one-sided conclusion, but should report the two-tailed p-value. The one-sided conclusion reflects the *direction* of the obtained results. The two-tailed p-value acknowledges that we would have been equally convinced to reject the null hypothesis if t had been as extreme in the other direction.

In summary, computer programs generally assume the user wishes to perform a two-sided test. They automatically print the two-tailed p-value. To perform a two-sided test, we simply compare the printed two-tailed p-value with the desired α level. If p is less than α, we reject the null hypothesis and report the two-sided p-value as our obtained level of significance.

If we wish to perform a one-sided test, we should first check to make sure the results are in the predicted direction. If the sign of t is *inconsistent* with our alternative hypothesis, we must accept H_0. If the sign of t is *consistent* with the alternative hypothesis, we should halve the printed two-tailed p-value to obtain the one-tailed p-value. If the one-tailed p-value is less than α, we reject the null hypothesis and report the one-tailed p-value as our obtained level of significance.

3.15 TESTS ON THE MEAN IN COMPUTER PROGRAMS

We will not concern ourselves with the Z test, since in practical applications the standard deviation is never known. All three of the programs—SAS, SPSS, and MINITAB—perform the t test, signed rank test, and sign test. These are demonstrated for the data of Example 3.5.

SAS Input

```
OPTIONS LS = 80;
TITLE 'EXAMPLE 3.5 - SPRINKLER ACTIVATION TIME';
DATA;
INPUT TIME;
DELAY = TIME-25;
LINES;
21
39
24
29
23
34
35
39
33
19
31
18
29
24
41
37
25
30
27
21
PROC PRINT;
PROC UNIVARIATE;
VAR DELAY;
```

Discussion of SAS Input

As shown on the input line, the TIME of sprinkler activation is the only variable entered for each observation. In the t, signed rank, and sign tests, SAS tests for a mean of 0; there is no way to specify a different hypothesized value. To accommodate for this peculiarity, the line immediately following the input line defines a new variable, DELAY, as TIME minus 25. SAS automatically computes the new variable for each observation and adds it to the dataset, exactly as if we had input both TIME and DELAY. The PRINT procedure is requested so that we can see the dataset with both variables. Testing for a mean of 0 on DELAY is equivalent to testing for a mean of 25 on TIME.

The UNIVARIATE procedure, which was used to find point estimates and box plots in Chapter 2, also produces t, signed rank, and sign tests. The VAR statement tells PROC UNIVARIATE to describe only the variable DELAY, not other variables that may be in the dataset. SAS procedures frequently use additional information, such as VAR statements, following the PROC command. All such statements refer to the preceding PROC, until a new PROC is encountered.

SAS Output

```
            EXAMPLE 3.5 - SPRINKLER ACTIVATION TIME                 1
                    OBS        TIME        DELAY

                      1          21          -4
                      2          39          14
                      3          24          -1
                      4          29           4
                      5          23          -2
                      6          34           9
                      7          35          10
                      8          39          14
                      9          33           8
                     10          19          -6
                     11          31           6
                     12          18          -7
                     13          29           4
                     14          24          -1
                     15          41          16
                     16          37          12
                     17          25           0
                     18          30           5
                     19          27           2
                     20          21          -4
```

```
          EXAMPLE 3.5 - SPRINKLER ACTIVATION TIME                    2
                Univariate Procedure
Variable=DELAY
                          Moments

            N                  20       Sum Wgts          20
            Mean             3.95       Sum               79
            Std Dev      7.052249       Variance    49.73421
            Skewness     0.150517       Kurtosis    -1.11974
            USS              1257       CSS           944.95
            CV           178.5379       Std Mean    1.576931
            T:Mean=0     2.504866       Prob>|T|      0.0215
            Num ~= 0           19       Num > 0           12
            M(Sign)           2.5       Prob>|M|      0.3593
            Sgn Rank           53       Prob>|S|      0.0311

                      Quantiles (Def = 5)

            100% Max           16       99%               16
             75% Q3           9.5       95%               15
             50% Med            4       90%               14
             25% Q1          -1.5       10%               -5
              0% Min           -7        5%             -6.5
                                         1%               -7

            Range              23
            Q3-Q1              11
            Mode               -4

                      Extremes

         Lowest      Obs         Highest       Obs
             -7(      12)            10(         7)
             -6(      10)            12(        16)
             -4(      20)            14(         2)
             -4(       1)            14(         8)
             -2(       5)            16(        15)
```

Discussion of SAS Output

The first page of SAS output, a listing of the dataset, is produced by the PRINT procedure. On the second page, the seventh line under the heading Moments shows the t statistic for testing that the mean of DELAY is 0. Note that this value, 2.504866, is the same as the t value found when testing for a mean of 25 on the original variable, TIME. To the right of the t value is the two-tailed p-value, labeled

Prob > |T|. Since we want to perform a one-sided test with rejection region in the right tail, we first note that t is positive, and thus in the proper tail. Halving the two-tailed p-value, we obtain the one-tailed p-value of 0.0215/2 = 0.01075. This is less than 5 percent, so we reject the null hypothesis. The one-tailed p-value of 0.01075 is the obtained level of significance.

Three lines below this we see the signed rank statistic (before standardizing to the Z scale) and its two-tailed p-value of 0.0311. The signed rank statistic (53) is positive, in accordance with our alternative hypothesis. As with the t test, the one-tailed p-value (0.01555) is less than 5 percent, and the null hypothesis is rejected.

Note that the p-value for the t test (0.01075) is smaller than that for the signed rank test (0.01555). This reflects the greater power of the t test when the data are close to normally distributed, as in Example 3.5. If average sprinkler activation time had been somewhat closer to the hypothesized mean of 25 seconds, both p-values would have been larger. The t test might have been just barely significant (p slightly less than 5 percent), while the signed rank test resulted in accepting the null hypothesis (p slightly greater than 5 percent).

The sign test, shown directly above the signed rank test, has a two-tailed p-value of 0.3593, and therefore results in a one-tailed p-value of 0.17965. We are in the proper direction, since the sign statistic, labeled M(Sign), is positive, but we must accept the null hypothesis because the one-tailed p-value is greater than 5 percent. For symmetric data, as in this example, the sign test is generally less powerful than the t and signed rank tests. Here, it resulted in accepting the null hypothesis, whereas both the t and signed rank tests were significant.

SPSS Input

```
SET WIDTH = 80
TITLE 'EXAMPLE 3.5 - SPRINKLER ACTIVATION TIME'
DATA LIST FREE / TIME
COMPUTE CONSTANT = 25
LIST
BEGIN DATA
21
39
24
29
23
34
35
39
33
19
```

```
31
18
29
24
41
37
25
30
27
21
END DATA
T-TEST PAIRS = TIME CONSTANT
NPAR TESTS WILCOXON = TIME CONSTANT / SIGN = TIME CONSTANT
FINISH
```

Discussion of SPSS Input

For Example 3.5, a single variable, named TIME, is read in free format. In SPSS, the t, signed rank, and sign tests are all designed to test for a difference between means using paired data. We will discuss the comparison of two means in Chapter 4. For the present, we can force the program to perform a one-sample test by defining a second variable to have the value of our hypothesized mean. Since we want to test for a mean of 25, we use a COMPUTE command to define the variable CONSTANT to be 25 for each observation. SPSS computes the new variable for each observation and adds it to the dataset, exactly as if we had input both TIME and CONSTANT. The LIST command is included so that we can see the dataset with both variables.

Recall that SPSS requires the first command to be entered before the data, and all subsequent commands to follow the data. The T-TEST command specifies that TIME and CONSTANT are to be paired for each observation. The NPAR TESTS command requests the signed rank test (sometimes referred to as the WILCOXON signed rank test, after the name of its developer), and the SIGN test.

SPSS Output

```
       EXAMPLE 3.5 - SPRINKLER ACTIVATION TIME     Page 1
  TIME CONSTANT
  21.00   25.00
  39.00   25.00
  24.00   25.00
  29.00   25.00
  23.00   25.00
  34.00   25.00
```

```
  35.00   25.00
  39.00   25.00
  33.00   25.00
  19.00   25.00
  31.00   25.00
  18.00   25.00
  29.00   25.00
  24.00   25.00
  41.00   25.00
  37.00   25.00
  25.00   25.00
  30.00   25.00
  27.00   25.00
  21.00   25.00

Number of cases read: 20 Number of cases listed: 20
```

```
         EXAMPLE 3.5 - SPRINKLER ACTIVATION TIME            Page 2

                t-tests for paired samples

Variable     Number                  Standard      Standard
            of Cases        Mean     Deviation       Error
TIME           20         28.9500      7.052         1.577
CONSTANT       20         25.0000       .000          .000

(Difference) Standard  Standard |      2-tail|  t  Degrees of  2-tail
    Mean     Deviation   Error  |Corr. Prob. |Value Freedom     Prob.
   3.9500      7.052     1.577  | .         . | 2.50    19        .022
```

```
         EXAMPLE 3.5 - SPRINKLER ACTIVATION TIME     Page 3
         Wilcoxon Matched-Pairs Signed-Ranks Test
  TIME
with CONSTANT
    Mean Rank    Cases
      12.33        12   - Ranks (CONSTANT LT TIME)
       6.00         7   + Ranks (CONSTANT GT TIME)
                    1     Ties  (CONSTANT EQ TIME)
                   --
                   20     Total

      Z =   -2.1328             2-Tailed P =.0329
```

```
                              Sign Test

  TIME
with CONSTANT

Cases
 12   - Diffs (CONSTANT LT TIME)
  7   + Diffs (CONSTANT GT TIME)          (Binomial)
  1     Ties                              2-Tailed P =    .3593
 --
 20     Total
```

Discussion of SPSS Output

As in Chapter 2, we have taken the liberty of renumbering the SPSS output pages, to eliminate pages related to operating system information. The first page of output is produced by the LIST command. The variable CONSTANT, with the value of 25, is included for each observation.

The second page contains the results of the t test, including descriptive information for each variable (Number of Cases, Mean, Standard Deviation, and Standard Error of the mean). The line below the descriptive information includes a number of items irrelevant for our present purpose, but toward the right end of this line are printed the t value, its degrees of freedom, and two-tailed p-value (2-tail Prob). The t value of 2.50 agrees with our computations in Section 3.8. Since we want to perform a one-sided test with rejection region in the right tail, we first note that t is positive and thus in the proper tail. Halving the two-tailed p-value, we obtain the one-tailed p-value of 0.022/2 = 0.011. This is less than 5 percent, so we reject the null hypothesis. The one-tailed p-value of 0.011 is the obtained level of significance.

The third page of output reports on the signed rank and sign tests. For the signed rank test, the standardized test statistic Z = −2.1328 has a two-tailed p-value of 0.0329. The one-tailed p-value (0.01645) is less than 5 percent, but the fact that Z is negative suggests that we are in the wrong tail. Unfortunately, SPSS computes Z using an algorithm that *always* produces a negative Z value, thereby losing the directional information. We cannot tell which tail we are in from the Z value printed. The confusion can be cleared up by examining the rank information printed above the Z value. According to our alternative hypothesis, H_1: $\mu > 25$, we would expect more observations above 25 than below. Equivalently, we would expect to find the CONSTANT (25) *less than* the observation (TIME) for most of the data. In fact, this is what we obtained. The line corresponding to CONSTANT LT TIME shows more observations (12) and a greater average rank (12.33) than for CONSTANT GT TIME (7 observations, average rank of 6). Thus, the result is indeed in the proper direction for our alternative hypothesis. Since the one-tailed p-value of 0.01645 is less than 5 percent, we reject H_0.

Note that the p-value for the t test (0.011) is smaller than that for the signed rank test (0.01645). This reflects the greater power of the t test when the data are close to normally distributed, as in Example 3.5. If average sprinkler activation time had been somewhat closer to the hypothesized mean of 25 seconds, both p-values would have been larger. The t test might have been just barely significant (p slightly less than 5 percent), while the signed rank test resulted in accepting the null hypothesis (p slightly greater than 5 percent).

The sign test shows a two-tailed p-value of 0.3593 and, therefore, results in a one-tailed p-value of 0.17965. We are in the proper direction, since CONSTANT LT TIME occurs more often than CONSTANT GT TIME. But we must accept the null hypothesis because the p-value is greater than 5 percent. For symmetric data, as in this example, the sign test is generally less powerful than the t and signed rank tests. Here, it resulted in accepting the null hypothesis, whereas both the t and signed rank tests were significant.

MINITAB Transcript

```
MTB > NAME C1 = 'TIME'
MTB > SET 'TIME'
DATA > 21 39 24 29 23 34 35 39 33 19 31 18 29
DATA > 24 41 37 25 30 27 21
DATA > END

MTB > TTEST 25 'TIME';
SUBC > ALTERNATIVE 1.

TEST OF MU = 25.00 VS MU G.T. 25.00

          N      MEAN     STDEV   SE MEAN      T    P-VALUE
TIME     20     28.95      7.05      1.58   2.50      0.011

MTB > WTEST 25 'TIME';
SUBC > ALTERNATIVE 1.

TEST OF MEDIAN = 25.00 VERSUS MEDIAN G.T. 25.00
                 N FOR   WILCOXON              ESTIMATED
            N    TEST    STATISTIC  P-VALUE     MEDIAN
TIME       20     19       148.0     0.017      29.00

MTB > STEST 25 'TIME';
SUBC > ALTERNATIVE 1.
SIGN TEST OF MEDIAN = 25.00 VERSUS G.T. 25.00
```

```
              N   BELOW  EQUAL ABOVE  P-VALUE  MEDIAN
TIME         20       7      1    12   0.1796   29.00

MTB > STOP
```

Discussion of MINITAB Transcript

As with MINITAB examples in Chapter 2, user inputs are shown to the right of the > symbol and are underlined for clarity. In the first five lines of this transcript, column C1 on the worksheet is identified to be a variable named TIME, and the values for this variable are entered.

Remember that MINITAB commands that include subcommands end in a semicolon. Each subcommand except the last ends in a semicolon, and the last subcommand ends in a period, to designate the end of information for the original command. The TTEST command specifies the hypothesized mean of 25 for the variable TIME, and the ALTERNATIVE subcommand requests a one-sided test with rejection region in the upper tail. (The value -1 on the ALTERNATIVE subcommand requests a rejection region in the lower tail. A two-sided test is performed if the ALTERNATIVE subcommand is omitted.) MINITAB prints a number of descriptive statistics (N, MEAN, STDEV, SE MEAN), as well as the observed t value of 2.50, and the one-tailed p-value of 0.011, in accordance with our request for a one-sided test. Since 0.011 is less than 5 percent, we reject the null hypothesis. The one-tailed p-value of 0.011 is the obtained level of significance.

The WTEST command requests the signed rank test, sometimes referred to as the Wilcoxon signed rank test, after the name of its developer. Again, a one-sided test with rejection region in the upper tail is requested. The test is reported as a test on the median, but for a symmetric distribution, the mean and median are the same. The one-tailed p-value of 0.017 is less than 5 percent, so the signed rank test also rejects the null hypothesis.

Note that the p-value for the t test (0.011) is smaller than that for the signed rank test (0.017). This reflects the greater power of the t test when the data are close to normally distributed, as in Example 3.5. If average sprinkler activation time had been somewhat closer to the hypothesized mean of 25 seconds, both p-values would have been larger. The t test might have been just barely significant (p slightly less than 5 percent), while the signed rank test resulted in accepting the null hypothesis (p slightly greater than 5 percent).

The sign test is obtained using the STEST command and shows a one-tailed p-value of 0.1796. Since this is greater than 5 percent, we must accept the null hypothesis. For symmetric data, as in this example, the sign test is generally less powerful than the t and signed rank tests. Here, it resulted in accepting the null hypothesis, whereas both the t and signed rank tests were significant.

3.16 TEST ON THE VARIANCE

When we examine a sample, we are generally interested in the average and, accordingly, perform a test on the mean. Occasionally, however, important questions involve the population variance, σ^2. A test on the variance is based on the distribution of its estimate, the sample variance, S^2. In Section 2.9, we learned that the ratio

$$\frac{(n-1)\,S^2}{\sigma^2}$$

follows the χ^2 (chi-square) distribution with n - 1 degrees of freedom. To test a hypothesis about the variance, we replace σ^2 in this formula with the hypothesized value, just as we replaced μ in the t statistic with the hypothesized mean. This test is demonstrated for the situation of Example 3.7.

1. State the null and alternative hypotheses.
   The filaments are meeting the tolerance limit if the standard deviation of their diameters is less than or equal to 0.5 microns. If the standard deviation is greater than this value, corrective action must be taken. Accordingly, we write

$$H_0: \sigma \leq 0.5$$
$$H_1: \sigma > 0.5.$$

   We could write equivalent hypotheses in terms of the variance by squaring each side:

$$H_0: \sigma^2 \leq 0.25$$
$$H_1: \sigma^2 > 0.25.$$

Example 3.7

Filaments for a certain lamp must have the proper diameter in order for the lamp to burn properly. It is not enough, however, for the diameter to be correct *on average*. If the filaments are inconsistent, thick ones will not burn brightly enough, and thin ones will flare brilliantly and then snap. Careful study has established that a standard deviation greater than 0.5 microns cannot be tolerated. A random sample of eight filaments was measured and found to have a sample standard deviation of 0.86 microns. Are the filaments currently being produced satisfactory in this regard?

2. Determine the rejection region.
   The rejection region is located in the right tail of the chi-square distribution, since we expect a large value of S^2 under the alternative hypothesis. The degrees of freedom for this test are $n - 1 = 7$. From Table 3b, we find the upper 5% cutoff to be 14.07, as depicted in Figure 3.12.
3. Obtain the data.
   A sample of size $n = 8$ was found to have a sample standard deviation of $S = 0.86$. To compute the observed value of chi-square, we need the sample variance, $S^2 = (0.86)^2 = 0.7396$.
4. Compute the test statistic.
   The hypothesized value for σ^2 is 0.25. We find the test statistic

$$\chi^2 = \frac{(n-1)\,S^2}{\sigma^2} = \frac{(8-1)(0.7396)}{0.25} = 20.71.$$

5. Draw the conclusion.
   The observed value $\chi^2 = 20.71$ is greater than our cutoff of 14.07, so we reject the null hypothesis. The filament diameters show too much variability.

In this problem, as in Example 3.5 on sprinkler activation time, we really would like to *accept* the null hypothesis. As discussed in Section 3.9, such situations should be handled by using an α level of 25 percent and a large sample size to guarantee adequate power. If we had decided to accept H_0 in the present problem, the conclusion that the filaments are satisfactory would be unconvincing. With a sample size of only eight, and a 5% α level, the power of the test is rather low. However, since we were able to reject H_0 under these circumstances, the variability in filament diameters is probably far outside the tolerance range.

While it is possible to use the chi-square statistic for a two-sided alternative, or to locate the rejection region in the lower tail, such tests are rare. In most situations involving the variance, the major concern is with *too much* variability. Consistency of production is one of the main issues in the current emphasis on quality control.

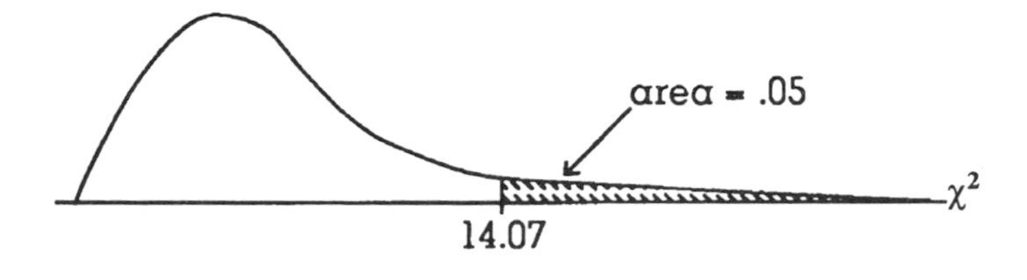

FIGURE 3.12. Right tail area of 5 percent for χ^2 with 7 degrees of freedom.

3.17 TEST ON THE VARIANCE IN COMPUTER PROGRAMS

Neither SAS nor SPSS will perform a test on the variance. These programs can be used to find the sample variance, as illustrated in Section 2.4, and then the test statistic can be computed by hand. In MINITAB, the test on the variance is not a single command, but the observed chi-square value can be calculated using the LET command, and the p-value can be obtained using a distributional function. The transcript shown here illustrates this process.

MINITAB Transcript

```
MTB > NAME C1 = 'DIAMETER'
MTB > SET 'DIAMETER'
DATA > 54.57 55.12 56.55 56.04 57.24 55.99 56.69 55.80
DATA > END
MTB > LET K1 = N ('DIAMETER') - 1
MTB > LET K2 = K1 * STDEV ('DIAMETER') **2 / 0.25
MTB > CDF K2 K3;
SUBC > CHISQUARE K1.
MTB > LET K3 = 1 - K3
MTB > PRINT K1 K2 K3
K1      7.00000
K2      20.7087
K3      0.00421721
MTB > STOP
```

Discussion of MINITAB Transcript

As presented above, Example 3.7 gives the sample standard deviation for eight observations, but not the raw data. In the first four lines of the MINITAB transcript, the actual data are entered as the values of a variable named DIAMETER. In the next line, the degrees of freedom are computed and stored in K1. The following line calculates the chi-square value and puts the result in K2. The CDF command and its CHISQUARE subcommand calculate the area below this chi-square value and store the result in K3. The p-value is the tail area *above* our chi-square value, obtained by subtracting the lower tail area from one. The PRINT command shows the values of df (K1), the observed chi-square (K2), and p-value (K3). Since the p-value of 0.00421721 is less than 5 percent, we reject the null hypothesis and conclude that the filament diameters show too much variability.

3.18 TEST ON A PROPORTION

Section 1.2 introduced a classification of variables into types referred to as numeric, categorical, and ordered categories. So far in Chapter 3, all the procedures discussed have been designed to answer questions about numeric variables. We now consider the most common issue with categorical data, the proportion of observations that fall in a specified category. In practice, a categorical variable may have any number of categories. But since we are only concerned with a single category, it is sufficient to redefine the categorical variable to have only two categories: the desired category, and *not* the desired category.

We are concerned here with data gathered from an experiment that satisfies four conditions:

1. The experiment consists of a fixed number of trials or observations, denoted by n.
2. Each trial results in one of two possible outcomes, typically called "success" and "failure."
3. The probability of success, denoted by π, is the same for each trial. (This use of the symbol π does not refer to the irrational number 3.14159... Here, π stands for a probability between 0 and 1.)
4. The trials are independent—that is, the result of one trial, success or failure, does not influence the result of any other trial.

In such an experiment, the number of successes is said to have a **binomial distribution** with parameters π and n.

In Section 1.7, we considered the experiment of flipping a coin several times and observing the number of heads. For a fixed number of coin flips, the number of heads is a binomial variable. Each trial (coin flip) results in either success (heads) or failure (tails). The probability of success is the same for each trial (π = probability of heads = ½). The trials are independent, since the outcome of one coin flip cannot influence any other flip. Exact probabilities for the binomial distribution can be found by enumeration, as demonstrated in Section 1.7. However, as we also saw in that section, the binomial distribution tends to approach the normal distribution as the number of trials increases. For most practical applications, it is sufficient to use the normal approximation to the binomial distribution to obtain our probabilities.

Typically, we work with the observed *proportion* of successes, P, found by dividing the number of successes by the number of trials. This variable is approximately normally distributed with mean $\mu_P = \pi$, and standard deviation $\sigma_P = \sqrt{\pi(1-\pi)/n}$. We convert P to a standard normal variable in the usual fashion, by subtracting its mean and dividing by its standard deviation:

$$Z = \frac{P - \mu_P}{\sigma_P} = \frac{P - \pi}{\sqrt{\pi(1 - \pi)/n}}. \tag{3.18.1}$$

This conversion allows us to test hypotheses about the value of the binomial parameter π. We simply substitute the hypothesized probability of success for π, and use the normal table to obtain critical values for this Z statistic. Example 3.8 demonstrates this procedure.

1. State the null and alternative hypotheses.
   Gasoline-powered lawn mowers typically start on the first pull 40 percent of the time. We want to see if the probability of starting on the first pull is greater than this figure for EZSTART mowers. We wish to test

$$H_0: \pi = 0.4$$
$$H_1: \pi > 0.4.$$

2. Determine the rejection region.
   Putting 5 percent in the right tail of the standard normal distribution, we will reject the null hypothesis if Z is greater than 1.645.
3. Obtain the data.
   For $n = 50$ EZSTART mowers, the sample proportion starting on the first pull is $P = 28/50 = 0.56$.
4. Compute the test statistic.
   The hypothesized value for π is 0.4. Since (3.18.1), we find the test statistic

$$Z = \frac{P - \pi}{\sqrt{\pi(1 - \pi)/n}} = \frac{0.56 - 0.4}{\sqrt{0.4(1 - 0.4)/50}} = 2.31.$$

5. Draw the conclusion.
   The observed value $Z = 2.31$ is greater than our cutoff of 1.645, so we reject the null hypothesis and conclude that EZSTART mowers are indeed appropriately named.

Example 3.8

Most gasoline-powered lawn mowers start on the first pull of the starter cord about 40 percent of the time. To see if the EZSTART mower is better than average, 50 EZSTART mowers were tested and 28 started on the first pull. At the 5% level, are EZSTART mowers aptly named?

3.19 TEST ON A PROPORTION IN COMPUTER PROGRAMS

SAS does not perform the test for a single proportion, but SPSS and MINITAB do. For each of these programs, the data of Example 3.8 were entered, using the value 1 for each mower that started on the first pull (success), and 2 for each mower that did not start (failure).

While the normal distribution can be used to approximate binomial probabilities, more precise results are obtained by using a continuity correction to compute the Z value. The correction term accounts for the fact that a continuous distribution (normal) is being used to approximate a discrete distribution (binomial). Several different forms for the continuity correction exist and, consequently, computer programs sometimes disagree slightly in computed p-values.

SPSS Input

```
SET WIDTH = 80
TITLE 'EXAMPLE 3.8 - EZSTART MOWERS'
DATA LIST FREE / MOWER START
NPAR TESTS BINOMIAL (0.4) = START (1,2)
BEGIN DATA
 1   1
 2   1
 3   2
 4   1
 5   1
 6   1
 7   2
 8   2
 9   2
10   2
11   1
12   2
13   1
14   2
15   2
16   1
17   1
18   1
19   2
```

```
20    1
21    1
22    2
23    2
24    1
25    1
26    1
27    1
28    1
29    2
30    2
31    1
32    1
33    2
34    2
35    2
36    2
37    1
38    2
39    1
40    1
41    1
42    1
43    1
44    1
45    2
46    2
47    2
48    1
49    2
50    1
END DATA
FINISH
```

Discussion of SPSS Input

For each mower, the observation number (MOWER) and the value of the variable START (either 1 or 2) are entered. The NPAR TESTS command requests the BINOMIAL test for the hypothesized value of $\pi = 0.4$, and lists the values of START corresponding to success and failure.

SPSS Output

```
                    EXAMPLE 3.8 - EZSTART MOWERS              Page 1
                            Binomial Test
 START

 Cases                Test Prop. = .4000
  28   = 1.00         Obs. Prop. = .5600
  22   = 2.00
                      Z Approximation
  50   Total          1-Tailed P = .0152
```

Discussion of SPSS Output

The observed proportion of successes (P = 0.56) is greater than the hypothesized value (π = 0.4) and, therefore, is in the direction specified by our alternative hypothesis. SPSS prints a one-tailed p-value of 0.0152. Since this is less than 5 percent, we reject the null hypothesis and conclude that EZSTART mowers start on the first pull more often than other mowers.

MINITAB Transcript

```
MTB > NAME C1 = 'START'
MTB > SET 'START'
DATA > 1 1 2 1 1 1 2 2 2 2 1 2 1 2 2 1 1 1 2 1
DATA > 1 2 2 1 1 1 1 1 2 2 1 1 2 2 2 2 1 2 1 1
DATA > 1 1 1 1 2 2 2 1 2 1
DATA > END

MTB > TALLY 'START'

       START COUNT
         1     28
         2     22
         N =   50

MTB > CDF 27 K1;
SUBC > BINOMIAL 50 0.4.
MTB > LET K1 = 1-K1
MTB > PRINT K1
K1    0.0160348
MTB > STOP
```

Discussion of MINITAB Transcript

The 50 values of START (either 1 or 2) are entered in the first six lines of this transcript. The TALLY command tells us that 28 mowers started on the first pull (START value of 1). According to our alternative hypothesis, we wish to put the 5% rejection region in the right tail. Thus, we need to find the probability that P is greater than or equal to 28 under the null hypothesis. If this tail area is less than 5 percent, we will reject H_0.

CDF stands for cumulative distribution function and accumulates the probability up to and including the specified value. The probability that $P \geq 28$ is the complement of the probability that $P < 28$ or, equivalently, $P \leq 27$. Therefore, we wish to accumulate the probability up to and including 27 in a binomial distribution with 50 trials and probability of success 0.4. The CDF command and BINOMIAL subcommand store this result in K1. The probability that $P \leq 27$ is now in K1, but we want to find the complement probability that $P \geq 28$, so we subtract K1 from 1. The desired one-tailed p-value is printed as 0.0160348. Since this is less than 5 percent, we reject the null hypothesis and conclude that EZSTART mowers start on the first pull more often than other mowers.

3.20 SUMMARY

In this chapter, we have introduced the concepts of hypothesis testing, as well as a number of the more common applications for a single sample. Hypothesis testing is a technique for making decisions that are probably correct, and putting limits on the chance of reaching an erroneous decision. The null hypothesis states that there has been no change in the value of a parameter or that a treatment has had no effect. The alternative hypothesis, which is usually what we want to show, states that a specific change has occurred or that the treatment has had the desired effect. The decision between these hypotheses is made on the basis of a test statistic, although different statistics are used, depending on the parameter we are investigating. The null hypothesis determines the probability distribution of this test statistic, while the alternative hypothesis determines the location of a region for rejecting the null. Hypothesis tests can be either one-sided or two-sided, depending on whether the direction of change is predicted. In many cases, hypothesis testing is equivalent to construction of confidence intervals, since the acceptance region for the hypothesis test is identical to the confidence interval for the parameter under test.

A type one error is rejecting the null hypothesis when it is true,while a type two error is accepting the null hypothesis when it is false. The probability of a type one error is controlled by the researcher, typically at 5 percent. Power is the probability of rejecting the null hypothesis when it is false, which is the complement of type two error rate. Power can be increased by decreasing the variation in observed measurements or by increasing sample size. Generally, however, it is not possible to determine power exactly. Similarly, it is not possible to precisely determine the sample size that should be used in a hypothesis test.

Accepting the null hypothesis is an unconvincing conclusion, since this result can be obtained by conducting poor research (large variability, small sample size). In situations where we would like to accept the null hypothesis, we can partially overcome this difficulty by taking steps to ensure large power and by using a type one error rate of 25 percent.

Statistical significance addresses the issue of how certain we are that the null hypothesis is wrong, while practical importance refers to the degree to which the null hypothesis is wrong. Statistical significance can be measured by the p-value, the tail area beyond the observed value of the test statistic. For two-sided tests, the p-value also includes the corresponding area in the opposite tail. The smaller the p-value, the more sure we are that the null hypothesis is wrong. Larger sample sizes include extra information that may help convince us that the null hypothesis is wrong and, consequently, tend to produce smaller p-values. Practical importance must be measured by a statistic that is insensitive to sample size and preferably scale-free.

In this chapter, several tests have been presented for hypotheses that deal with the population mean. For normally distributed data, the Z test is used when the variance is known, and the t test is used when it is unknown. Tests designed for non-normal data are referred to as nonparametric tests. For non-normal but symmetric data, the signed rank test can be used to test hypotheses about the mean. For asymmetric data, the sign test is recommended. In general, the more information we use from the sample, the more powerful the testing procedure is. Therefore, when possible, we should use the t test instead of a nonparametric procedure. The t test uses the actual values of the observations, while nonparametric tests typically use only rank information.

To test hypotheses about the population variance, the chi-square test is employed. For categorical data, we test a hypothesis about the population proportion in one of the categories using a test based on the binomial distribution. SAS, SPSS, and MINITAB programs were used to demonstrate the t test, signed rank test, and sign test on the mean, the chi-square test on the variance, and the binomial test on a proportion.

4

Two-Sample Tests

The previous chapter presented several procedures for testing hypotheses based on a single sample. In each case, we required that the sample be drawn at random in order to be representative of the whole population. Information from the sample is used to compare a population parameter to a specified value. In many applications, however, no standard value exists for comparison. Instead, we wish to compare the values of two parameters (e.g., means of two populations) when neither of them is known. To provide information about both parameters, random samples must be drawn from each population.

In this chapter, our research questions deal with comparing two parameters and, consequently, slightly different statistics are computed from those used in Chapter 3. But, conceptually, hypothesis testing for two samples is identical to the one-sample case. Either one-sided or two-sided tests can be constructed, depending on the situation. We still draw the curve to represent the null hypothesis and locate the rejection region according to the alternative. Type one error rate is controlled, again at 5 percent. Power is still influenced by sample size and variance. In short, all of the concepts of the previous chapter apply, with only minor alterations, to the two-sample case.

We present tests in this chapter for comparing two means, two variances, and two proportions. For the comparison of means, two different sampling paradigms are considered: samples drawn independently from the two populations, and samples that are paired or matched. We present both parametric tests (which assume normality) and nonparametric tests. We begin with the independent sample situation, and start by considering the simplest case, when the variances are known.

4.1 COMPARISON OF MEANS WHEN VARIANCES ARE KNOWN

Before presenting the procedure for comparing two means, let us briefly review the one-sample case. For a hypothesis about the population mean, μ, we base our test on its estimate, $\overline{X}$. As we learned earlier, $\overline{X}$ is normally distributed with mean $\mu_{\overline{x}} = \mu$, and variance $\sigma^2_{\overline{x}} = \sigma^2/n$. To standardize $\overline{X}$ to the Z scale, we subtract its mean and divide by its standard deviation (the square root of its variance). The test statistic is obtained by replacing μ in this Z statistic with the hypothesized mean.

We now wish to compare the means of two populations to see if they are equal. Letting μ_1 and μ_2 represent the two population means, we write the null hypothesis

$$H_0: \mu_1 = \mu_2,$$

Depending on the experimental situation, the alternative hypothesis is one of the following:

$$H_1: \mu_1 < \mu_2,$$

$$H_1: \mu_1 > \mu_2 \text{, or}$$

$$H_1: \mu_1 \neq \mu_2.$$

Similar to the one-sample case, the rejection region is located in the left tail for the first of these alternative hypotheses, in the right tail for the second, and split half in each tail for the third.

For this section and through Section 4.6, we suppose that information to test the null hypothesis is obtained by drawing independent samples from the two populations—that is, the observations sampled from one population do not influence which observations are obtained from the other. At present, we consider the simple, but unrealistic case, when the population variances, σ_1^2 and σ_2^2, are known. To allow for the possibility of unequal sample sizes, we use n_1 to designate the first sample size and n_2 the second. If the sample sizes are equal, we simply have $n_1 = n_2$.

Let $\overline{X}_1$ and $\overline{X}_2$ represent the two sample means. In each sample separately, the sample mean is normally distributed, as in the one-sample case. Thus, $\overline{X}_1$ is normal with mean μ_1 and variance σ_1^2/n_1, and $\overline{X}_2$ is normal with mean μ_2 and variance σ_2^2/n_2.

To see if the *population* means are equal, we should examine the difference between the *sample* means, $(\overline{X}_1 - \overline{X}_2)$. It can be shown that this difference is also normally distributed. The mean of the variable $(\overline{X}_1 - \overline{X}_2)$ is $(\mu_1 - \mu_2)$. Logically, since $\overline{X}_1$ estimates μ_1, and $\overline{X}_2$ estimates μ_2, their difference $(\overline{X}_1 - \overline{X}_2)$ estimates $(\mu_1 - \mu_2)$. The variance of $(\overline{X}_1 - \overline{X}_2)$ is the *sum* of the separate variances of $\overline{X}_1$ and $\overline{X}_2$, since both samples contribute variability to the experiment. Therefore, the variance of $(\overline{X}_1 - \overline{X}_2)$ is

$$\frac{\sigma_1^2}{n_1} + \frac{\sigma_2^2}{n_2}.$$

As with any other normal variable, $(\overline{X}_1 - \overline{X}_2)$ is standardized to the Z scale by subtracting its mean and dividing by the square root of its variance:

$$Z = \frac{(\overline{X}_1 - \overline{X}_2) - (\mu_1 - \mu_2)}{\sqrt{\frac{\sigma_1^2}{n_1} + \frac{\sigma_2^2}{n_2}}}.$$

In the one-sample case, we obtain the test statistic by replacing μ in the Z ratio with its hypothesized value. In the two-sample case, our null hypothesis is that the population means are equal, or, equivalently, $(\mu_1 - \mu_2)$ is 0. Making this substitution in the previous equation, we arrive at the test statistic

$$Z = \frac{(\overline{X}_1 - \overline{X}_2)}{\sqrt{\frac{\sigma_1^2}{n_1} + \frac{\sigma_2^2}{n_2}}}. \quad (4.1.1)$$

If the variances are known, this Z can be compared with the appropriate cutoff point from the normal distribution. In practice, however, the variances are never known. Nonetheless, the Z ratio in equation (4.1.1) will be useful in finding a corresponding t statistic for the realistic case of unknown variances.

4.2 COMPARISON OF MEANS WHEN VARIANCES ARE UNKNOWN BUT BELIEVED EQUAL

In the one-sample case when the standard deviation is unknown, we replace it with its estimate—the sample standard deviation. The resulting ratio no longer has a Z distribution, but a t distribution with n - 1 degrees of freedom. Similarly, in the two-sample case when the variances are unknown, we replace them with estimates to obtain a t statistic. Two separate t tests exist, however, depending on whether we believe the population variances are equal or unequal. It may seem unrealistic to speculate that the population variances are equal, especially when we are testing to see if the means are equal. Nevertheless, we begin with this assumption in developing our t test in this section. The test we present is valid even when the equal variance condition is rather seriously in error and is widely applied in practice. We will return to the issue of validity after presenting the test statistic and an example.

When the variances are equal, we can use a common symbol, σ^2, to replace σ_1^2 and σ_2^2 in the Z statistic of (4.1.1). This term can be factored out in the denominator, so that Z becomes

$$Z = \frac{(\overline{X}_1 - \overline{X}_2)}{\sqrt{\sigma^2\left(\frac{1}{n_1} + \frac{1}{n_2}\right)}}. \tag{4.2.1}$$

Since the population variances are assumed to be equal, we can estimate the unknown σ^2 with either of the sample variances, S_1^2 or S_2^2. Rather than picking one or the other of these estimates, we should combine the information in the two terms by averaging them. An even better idea is to take a **weighted average,** to give more emphasis to the estimate from the larger sample size.

Weighted averages are used frequently in statistics. In general, a weighted average of two terms is obtained by multiplying each term by a weight, and then dividing by the sum of the weights. Using w_1 and w_2 to represent the weights, a weighted average of two-sample variances has the form

$$\frac{w_1 S_1^2 + w_2 S_2^2}{w_1 + w_2}.$$

If the weights are equal, say both have the value w, then this becomes

$$\frac{w(S_1^2 + S_2^2)}{2w} = \frac{S_1^2 + S_2^2}{2},$$

that is, the regular average. The weighted average is the more general concept. The regular average is the special case of a weighted average in which the weights are equal.

In statistics, we generally use degrees of freedom as weights. Since each sample variance has (n - 1)degrees of freedom, we form a weighted average known as the **pooled variance estimate**, denoted by S_p^2:

$$S_p^2 = \frac{(n_1 - 1)S_1^2 + (n_2 - 1)S_2^2}{(n_1 - 1) + (n_2 - 1)}. \tag{4.2.2}$$

This estimate is used in place of the unknown σ^2 in (4.2.1) to obtain the t statistic

$$t = \frac{(\overline{X}_1 - \overline{X}_2)}{\sqrt{S_p^2\left(\frac{1}{n_1} + \frac{1}{n_2}\right)}}. \tag{4.2.3}$$

A test based on this ratio is called a **pooled t test,** to reflect its use of the pooled estimate of the variance.

As discussed in Section 2.7, the degrees of freedom for any statistic are the number of independent pieces of information from the sample used in its computation. Logically, if two samples are independent, we can add degrees of freedom. In the present situation, each sample variance has (n - 1) degrees of freedom. The pooled variance estimate uses information from both samples and has $(n_1 - 1) + (n_2 - 1) = (n_1 + n_2 - 2)$ degrees of freedom. The pooled t employs the pooled variance estimate in its denominator, and therefore has $(n_1 + n_2 - 2)$ degrees of

---

Example 4.1

A building contractor wanted to compare the efficiencies of two types of heat pumps. Fourteen identical houses under construction in a subdivision were used in the experiment. For each house, a coin was flipped to determine which type of heat pump was installed. Heads resulted in installing a type 1 heat pump, and tails in a type 2 heat pump. Based on separate coin flips for each house, type 1 heat pumps were installed in eight of the houses, and type 2 heat pumps in the other six. Efficiency was monitored during a one-month period while interior finishing was being completed. Based on the efficiency figures reported below, do the two types of heat pumps differ in efficiency?

type 1 heat pump	type 2 heat pump
72	82
78	74
73	81
69	71
75	85
74	73
69	
75	

freedom. Example 4.1 is used to demonstrate the pooled t test, and will also be used for other two-sample testing procedures presented in later sections.

1. State the null and alternative hypotheses.
   We would be interested in learning if *either* heat pump is more efficient than the other, so a two-sided test is in order:

$$H_0: \mu_1 = \mu_2$$
$$H_1: \mu_1 \neq \mu_2.$$

2. Determine the rejection region.
   The degrees of freedom for this test are $(n_1 + n_2 - 2) = (8 + 6 - 2) = 12$. For a two-sided test, we split the 5% rejection region, and put 0.025 in each tail. From Table 2 in the Appendix, we find the cutoff 2.179. We will reject H_0 if t is greater than 2.179 or less than -2.179. Equivalently, we will reject H_0 if $|t|$ is greater than 2.179.
3. Obtain the data.
   For each sample separately, we compute the sample mean and sample variance. These summary statistics are shown in Table 4.1.
4. Compute the test statistic.
   Before computing t, we must find the pooled variance estimate

$$S_p^2 = \frac{(n_1 - 1)S_1^2 + (n_2 - 1)S_2^2}{(n_1 - 1) + (n_2 - 1)} = \frac{(8 - 1)(9.55) + (6 - 1)(32.67)}{(8 - 1) + (6 - 1)} = 19.18.$$

   Now we use (4.2.3) to obtain

$$t = \frac{(\overline{X} - \overline{X}_2)}{\sqrt{S_p^2\left(\frac{1}{n_1} + \frac{1}{n_2}\right)}} = \frac{73.13 - 77.67}{\sqrt{19.18\left(\frac{1}{8} + \frac{1}{6}\right)}} = -1.92.$$

5. Draw the conclusion.
   The observed value $t = -1.92$ is just inside the critical values of ±2.179, so we must accept the null hypothesis. With small sample sizes, of course, power is

TABLE 4.1 Summary Statistics for Heat Pump Efficiencies

type 1 heat pump	type 2 heat pump
$n_1 = 8$	$n_2 = 6$
$\overline{X}_1 = 73.13$	$\overline{X}_2 = 77.67$
$S_1^2 = 9.55$	$S_2^2 = 32.67$

relatively low. Slightly more information might have produced a significant result. Based on the present experiment, the evidence is insufficient to conclude that the two types of heat pumps differ in efficiency.

In theory, this test requires independent random samples from two normal distributions with equal variances. As in the case of the one-sample t test, the normality requirement can be virtually ignored with moderate to large sample sizes. The assumption of equal variances, frequently referred to as the **homogeneity of variance** condition, seems to pose a more substantial problem. In practice, we can never know if the population variances are equal. But we can get some information about the population variances by examining the sample variances. If the sample variances are reasonably close, the pooled t test is clearly valid. In our example, the variances (9.55 and 32.67) are decidedly different. However, even when the variances are rather unequal, the test has been shown to be valid if the sample sizes are large, or if they are small and nearly equal. The sample sizes in our example, eight and six, are not large, but are close enough to justify using the pooled t test, despite the disparity in sample variances. Further discussion of the homogeneity of variance assumption is given in the next section and in Section 4.11.

4.3 COMPARISON OF MEANS WHEN VARIANCES ARE UNKNOWN AND BELIEVED UNEQUAL

The pooled t test can be used if the sample variances are reasonably close, if the sample sizes are large, or if the sample sizes are small and nearly equal. But, occasionally, all these conditions fail, and an alternative procedure is required. Unfortunately, an exact t test is impossible under these circumstances. However, an approximate t test can be found by following a common sense strategy—we simply estimate each variance separately. In the Z statistic shown in (4.1.1), we replace each population variance with its sample estimate, obtaining

$$t = \frac{(\overline{X}_1 - \overline{X}_2)}{\sqrt{\dfrac{S_1^2}{n_1} + \dfrac{S_2^2}{n_2}}}. \qquad (4.3.1)$$

The **unequal variance t test** employs the statistic above, but involves a complicated formula for the approximate degrees of freedom. In essence, the degrees of freedom are computed so as to maximize the similarity between the distribution of (4.3.1) and a true t distribution. We will not present the formula for degrees of freedom in this book, since it does not shed any light on the logical use of this statistic.

For Example 4.1, the approximate degrees of freedom are 7.19. Since this is not a whole number, we use the line for df = 7 in Table 2 to obtain the critical values of ±2.365. We will reject H_0 if $|t| > 2.365$. Substituting the data into (4.3.1) yields the value

$$t = \frac{(\overline{X}_1 - \overline{X}_2)}{\sqrt{\frac{S_1^2}{n_1} + \frac{S_2^2}{n_2}}} = \frac{73.13 - 77.67}{\sqrt{\frac{9.55}{8} + \frac{32.67}{6}}} = -1.76.$$

As with the pooled t test, we must accept the null hypothesis.

Since (4.3.1) can be used with *unequal* variances, we might wonder whether it could also be used with *equal* variances. The answer is yes, but in such cases the pooled t test is more powerful. Typically, the two tests produce rather similar results. In our example, the observed values of t are quite close (pooled t = −1.92, unequal variance t = −1.76), as are the critical values (±2.179 for pooled t, ±2.365 for unequal variance t). But note that the pooled t value is slightly farther from 0, and its critical values are slightly closer to 0, than for the unequal variance t test. As a result, the pooled t is closer to its rejection region than is the unequal variance t. In our example, both tests are nonsignificant. But with slightly different data, the pooled t test might reject the null hypothesis, while the unequal variance t test does not. When both tests are valid, the pooled t test is slightly more powerful and should certainly be employed. The unequal variance t test should only be used when the pooled t is invalid—that is, when the sample sizes are small and rather different, and the sample variances are far from equal.

When possible, it is beneficial to use equal sample sizes for comparing two means. In Example 4.1, the contractor could have divided his 14 houses into two groups of 7 houses for each heat pump. In addition to ensuring the validity of the pooled t test, equal sample sizes maximize power. The pooled t test would have been slightly more powerful with seven observations in each group than with eight in one and six in the other.

4.4 COMPARISON OF MEANS FOR NON-NORMAL DATA

For testing on a single mean with non-normal data, Chapter 3 presents two nonparametric procedures—the signed rank and sign tests. The signed rank test uses ranks of the data, while the sign test merely counts the number of observations above and below the hypothesized mean. Similarly, in the case of two samples, two nonparametric approaches exist. One uses ranks and the other employs only counts.

In the **rank sum test**, the two samples are combined and the total sample is ranked. Sums of ranks are then computed separately for each sample. The test statistic is a function of the difference between these two sums of ranks. (An alternative, but equivalent, test, known as the **U test**, is a slightly different function of this difference in rank sums.) When the null hypothesis is true and the means are equal, the observations in each sample should be dispersed relatively evenly throughout the combined ranking. We should not find most of the data from one sample ranked above most of the data from the other sample. The difference between the sums of ranks for the two samples should be small. If the alternative hypothesis is true and the means are unequal, data from the population with the higher mean should tend to be ranked above data from the other population. The difference in sums of ranks should be substantially farther from zero.

If the sample sizes are greatly unequal, we might expect the difference between sums of ranks to be large even when the null hypothesis is true. The larger sample size will have a greater sum of ranks by virtue of summing over more observations. The test statistic accounts for any disparity in sample sizes, essentially by comparing *average* ranks instead of sums of ranks.

The distribution of the test statistic approaches a normal distribution as both sample sizes increase. To perform the test, this statistic is standardized to the Z scale, and the standard normal table is used to determine the rejection region. In Example 4.1, a two-sided test is desired to compare the efficiencies of the two types of heat pumps. Splitting the 5% rejection region and putting half in each tail of the Z distribution, we find the critical values of ± 1.96. Omitting the actual computations, the Z value for the rank sum test on these data is 1.297, and the null hypothesis is accepted. The evidence is insufficient to conclude that these two types of heat pumps differ in efficiency.

Recall that the pooled t test reached the same conclusion for this example. When the data are reasonably close to normal, the two tests will generally result in the same decision. Under exact normality, the t test is slightly more powerful than the rank sum test, since it uses more of the information in the sample. As in the one sample case, the t test uses actual numeric values, while the nonparametric test employs only relative ranks of the observations. If the distribution departs markedly from normality, the rank sum test is substantially more powerful than the t test.

The rank sum test assumes that the data represent independent samples from two continuous distributions that differ only in their locations. As for the pooled t test, the variance in the two populations is assumed to be the same. Fortunately, like the t test, the rank sum test is relatively robust with respect to violations of the equal variance assumption—that is, the test is approximately correct even when the variances of the two populations differ. Note that symmetry is not assumed for the rank sum test. It can be used for skewed data, as long as the direction of skewness is the same in the two samples. When employed with asymmetric data, the test should be interpreted as a comparison of population medians rather than means.

As in the one sample case, the assumption of a continuous distribution means that ties should not occur. In practice, a small to moderate number of ties can be handled by the midranking procedure. Observations that are tied are each assigned the average of the ranks that would have been assigned had the values been slightly different. If a large number of observations are involved in ties, the validity of the rank sum test may be questionable.

The **median test** is an alternative nonparametric procedure for comparing two populations that does not assume identically shaped distributions. The two samples are combined, and the overall median is determined. For each sample, the proportion of observations above the overall median is determined. If the null hypothesis is correct, these proportions should be roughly the same, and their difference near 0. When the alternative hypothesis is true, the values in one sample should tend to be higher than those in the other, and the proportion of observations above the median should be different in the two groups. With large samples, the difference between these proportions can be standardized to either the Z scale or chi-square scale, depending on the form of standardization employed. Some computer programs present the Z scale standardization, while others print the approximate chi-square test.

Unlike the rank sum test, the median test can be used even if the two distributions have different shapes. However, it still assumes continuous data so that large numbers of ties are problematic. The median test only *counts* the number of observations in each group above and below the overall median. It does not depend on *how far* above or below the overall median the observations are. Thus, it does not use the relative positions, or ranks, like the rank sum test. Since it uses less information from the data, the median test is generally less powerful than the rank sum test. The median test should only be used when the two samples have decidedly different shapes, in which case the rank sum test is invalid.

4.5 COMPARISON OF MEANS IN COMPUTER PROGRAMS

Four tests have been discussed for comparing means: pooled t, unequal variance t, rank sum, and median test. All four tests are available in each of the three programs, SAS, SPSS, and MINITAB. For each program, Example 4.1 is analyzed, using each of the four tests.

SAS Input

```
OPTIONS LS = 80;
TITLE 'EXAMPLE 4.1 - HEAT PUMP EFFICIENCIES';
DATA;
INPUT TYPE EFFIC;
LINES;
1  72
```

```
1  78
1  73
1  69
1  75
1  74
1  69
1  75
2  82
2  74
2  81
2  71
2  85
2  73
PROC TTEST;
CLASS TYPE;
VAR EFFIC;
PROC NPAR1WAY WILCOXON MEDIAN;
CLASS TYPE;
VAR EFFIC;
```

Discussion of SAS Input

For each observation, two variables are entered: TYPE (type 1 or type 2 heat pump) and EFFIC (efficiency). In our input, all the observations for the first heat pump are listed before those for the second heat pump, but this need not be the case. Most SAS procedures allow the data to be entered in any order.

The TTEST procedure automatically performs both pooled t and unequal variance t tests. The CLASS statement informs this subroutine that TYPE is the classification variable, determining which observations are in each sample. The VAR statement indicates that the t test is to be performed on the variable EFFIC. Recall that SAS procedures frequently use additional information, such as CLASS and VAR statements, following the PROC command. All such statements refer to the preceding PROC, until a new PROC is encountered.

The second procedure, NPAR1WAY, performs both the rank sum and median tests. The first option requested (WILCOXON) asks for the rank sum test (developed by Wilcoxon), and the second option (MEDIAN) for the median test. Again, both CLASS and VAR statements are used in connection with this procedure.

SAS Output

```
                    EXAMPLE 4.1 - HEAT PUMP EFFICIENCIES                  1
                                TTEST PROCEDURE
Variable: EFFIC
```

TYPE	N	Mean	Std Dev	Std Error	Minimum	Maximum
1	8	73.12500000	3.09088522	1.09279295	69.00000000	78.00000000
2	6	77.66666667	5.71547607	2.33333333	71.00000000	85.00000000

Variances	T	DF	Prob>|T|
Unequal	-1.7627	7.2	0.1203
Equal	-1.9200	12.0	0.0789

For H0: Variances are equal, F' = 3.42 DF = (5,7) Prob>F' = 0.1399

EXAMPLE 4.1 - HEAT PUMP EFFICIENCIES 2

N P A R 1 W A Y P R O C E D U R E

Wilcoxon Scores (Rank Sums) for Variable EFFIC
Classified by Variable TYPE

TYPE	N	Sum of Scores	Expected Under H0	Std Dev Under H0	Mean Score
1	8	50.0	60.0	7.71184333	6.25000000
2	6	55.0	45.0	7.71184333	9.16666667

Average Scores were used for Ties
Wilcoxon 2-Sample Test (Normal Approximation)
(with Continuity Correction of.5)
S= 55.0000 Z= 1.23187 Prob >|Z| = 0.2180

T-Test approx. Significance = 0.2398

Kruskal-Wallis Test (Chi-Square Approximation)
CHISQ= 1.6814 DF= 1 Prob > CHISQ= 0.1947

EXAMPLE 4.1 - HEAT PUMP EFFICIENCIES 3

N P A R 1 W A Y P R O C E D U R E

Median Scores (Number of Points above Median)
for Variable EFFIC
Classified by Variable TYPE

TYPE	N	Sum of Scores	Expected Under H0	Std Dev Under H0	Mean Score
1	8	3.0	3.42857143	0.950914622	0.375000000
2	6	3.0	2.57142857	0.950914622	0.500000000

```
        Average Scores were used for Ties
      Median 2-Sample Test (Normal Approximation)
  S= 3.00000    Z= 0.450694    Prob >|Z| = 0.6522

     Median 1-Way Analysis (Chi-Square Approximation)
 CHISQ= 0.20313    DF= 1     Prob > CHISQ= 0.6522
```

Discussion of SAS Output

The first page of SAS output is produced by the TTEST procedure and is divided into three separate blocks of information. The first block provides descriptive statistics for each group, including the sample size, sample mean, sample standard deviation, standard error of the mean, and smallest and largest observations.

The second block shows the results of both pooled and unequal variance t tests. The second line of this block (Variances listed as Equal) pertains to the pooled t test and shows the t value of -1.92 and its degrees of freedom as 12. These results are the same as our hand computations in Section 4.2. The two-tailed p-value of 0.0789 (PROB > |T|) is greater than 5 percent, so we accept the null hypothesis in our two-sided test. For the unequal variance t test, SAS shows the observed t value (-1.7627) and its approximate degrees of freedom (7.2). Again, the t value parallels our hand computations. SAS can compute tail areas for t distributions with fractional degrees of freedom, and prints the two-tailed p-value of 0.1203. The p-value for the unequal variance t test is slightly larger than that for the pooled t test, reflecting the greater power of the pooled t test when it is valid.

The last block of information produced by PROC TTEST is a single line for testing equality of variances, and will be discussed in Section 4.12.

Page 2 of the output is produced by the NPAR1WAY procedure, and gives results of the rank sum (WILCOXON) test. Although the rank sum statistic approaches normality for large sample sizes, other approximate tests exist for relatively small sample sizes, as in our example. Three different approximate tests are shown by SAS. The Z value printed (1.23187) differs slightly from the Z value (1.297) mentioned in Section 4.4, since SAS employs a continuity correction. For this approximation, the two-tailed p-value of 0.2180 is greater than 5 percent, leading to acceptance of the null hypothesis. Comparable p-values are printed for the other two approximate tests. The p-value for the rank sum test is larger than for either of the t tests, since the rank sum test has lower power when the data are reasonably close to normally distributed.

On page 3, SAS displays results of the median test, including the approximate Z value of 0.450694 and its associated two-tailed p-value of 0.6522. (The chi-square approximation results in the same p-value.) This p-value is substantially larger than for any of the previous tests. The median test uses the least information from the data and should be employed only when the requirements of the other tests are seriously violated.

SPSS Input

```
SET WIDTH = 80
TITLE 'EXAMPLE 4.1 - HEAT PUMP EFFICIENCIES'
DATA LIST FREE / TYPE EFFIC
T-TEST GROUPS = TYPE (1,2) / VARIABLES = EFFIC
BEGIN DATA
1 72
1 78
1 73
1 69
1 75
1 74
1 69
1 75
2 82
2 74
2 81
2 71
2 85
2 73
END DATA
NPAR TESTS M-W = EFFIC BY TYPE (1,2) / MEDIAN = EFFIC BY TYPE (1,2)
FINISH
```

Discussion of SPSS Input

For each observation, two variables are entered: TYPE (type 1 or type 2 heat pump) and EFFIC (efficiency). Recall that SPSS requires the first command to be entered before the data, and all subsequent commands to follow the data. In our input, all the observations for the first heat pump are listed before those for the second heat pump, but this need not be the case. Most SPSS procedures allow the data to be entered in any order.

The T-TEST command automatically performs both pooled t and unequal variance t tests. The groups to be compared are specified as TYPE 1 and TYPE 2, and the variable to be used is EFFIC. After the data, the NPAR TESTS command is used to request both the U test (M-W stands for the Mann-Whitney U test, which is equivalent to the rank sum test) and the MEDIAN test. Each test is performed on the variable EFFIC for the two groups, TYPE 1 and TYPE 2.

SPSS Output

```
                EXAMPLE 4.1 - HEAT PUMP EFFICIENCIES              Page 1
              t-tests for independent samples of TYPE

GROUP 1 - TYPE EQ       1.00
GROUP 2 - TYPE EQ       2.00

Variable        Number                    Standard      Standard
                of cases      Mean        Deviation      Error
- - - - - - - - - - - - - - - - - - - - - - - - - - - - - - - - - -
EFFIC
      GROUP 1    8         73.1250          3.091          1.093
      GROUP 2    6         77.6667          5.715          2.333

                | Pooled Variance Estimate |Separate Variance Estimate
                |                          |
  F    2-tail  |   t    Degrees of  2-tail|  t    Degrees of  2-tail
Value   Prob.  | Value    Freedom    Prob.| Value    Freedom    Prob.
- - - - - - - -|- - - - - - - - - - - - - |- - - - - - - - - - - - - -
 3.42   .140   | -1.92      12        .079 | -1.76      7.19      .120
```

```
             EXAMPLE 4.1 - HEAT PUMP EFFICIENCIES            Page 2
         Mann-Whitney U - Wilcoxon Rank Sum W Test
   EFFIC
by TYPE

     Mean Rank      Cases
       6.25            8  TYPE = 1.00
       9.17            6  TYPE = 2.00
                      --
                      14  Total

                              Exact       Corrected for ties
       U        W          2-tailed P        Z       2-Tailed P
     14.0     55.0           .2284       -1.2967       .1947

                                 Median Test
   EFFIC
by TYPE

                              TYPE
                          1.00        2.00
                      +----------+----------+
          GT median   |    3     |     3    |
EFFIC                 +----------+----------+
          LE median   |    5     |     3    |
                      +----------+----------+
```

```
Cases          Median          Exact probability
 14            74.000                1.0000
```

Discussion of SPSS Output

As in previous chapters, we have taken the liberty of renumbering the SPSS output pages, to eliminate pages related to operating system information. The first page of output is produced by the T-TEST command and is divided into two blocks of information. The top block shows descriptive information for each group, including the sample size, sample mean, sample standard deviation, and standard error of the mean.

The second block begins with an F test for comparing variances, which will be discussed in Section 4.12. In the center are the results of the pooled t test, and to the right, the results of the unequal variance t test. For the pooled t test, the output shows the t value of -1.92 and degrees of freedom of 12, as computed in Section 4.2. The two-tailed p-value of 0.079 is greater than 5 percent, so we accept the null hypothesis in our two-sided test. For the unequal variance t test, the t value is -1.76 and its approximate degrees of freedom are 7.19. Again, the t value parallels our hand computations. SPSS can compute tail areas for t distributions with fractional degrees of freedom, and prints the two-tailed p-value of 0.120. The p-value for the unequal variance t test is slightly larger than that for the pooled t test, reflecting the greater power of the pooled t test when it is valid.

Page 2 of the output is produced by the NPAR TESTS command, and gives results of both the U test (equivalent to the rank sum test) and median test. For the U test, the output shows the Z value of -1.2967 and two-tailed p-value of 0.1947. The two-tailed p-value is greater than 5 percent, leading to acceptance of the null hypothesis. The p-value for the rank sum test is larger than for either of the t tests, since the rank sum test has lower power when the data are reasonably close to normally distributed.

For the median test, the sample size is small enough so that SPSS performs an exact test instead of using the normal approximation. The two-tailed p-value of 1.0000 is substantially larger than for any of the previous tests. The median test uses the least information from the data and should be employed only when the requirements of the other tests are seriously violated.

MINITAB Transcript

```
MTB > NAME C1 = 'TYPE'
MTB > NAME C2 = 'EFFIC'
```

```
MTB > READ 'TYPE' 'EFFIC'.
DATA > 1 72
DATA > 1 78
DATA > 1 73
DATA > 1 69
DATA > 1 75
DATA > 1 74
DATA > 1 69
DATA > 1 75
DATA > 2 82
DATA > 2 74
DATA > 2 81
DATA > 2 71
DATA > 2 85
DATA > 2 73
DATA > END
       14 ROWS READ

MTB > TWOT 'EFFIC' 'TYPE';
SUBC > POOLED.

TWOSAMPLE T FOR EFFIC
TYPE N    MEAN    STDEV    SE MEAN
1    8   73.12     3.09        1.1
2    6   77.67     5.72        2.3

95 PCT CI FOR MU 1 - MU 2: (-9.7, 0.6)
TTEST MU 1 = MU 2 (VS NE): T=-1.92 P=0.079 DF=12.0

MTB > TWOT 'EFFIC' 'TYPE'

TWOSAMPLE T FOR EFFIC
TYPE N    MEAN    STDEV    SE MEAN
1    8    73.12   3.09        1.1
2    6    77.67   5.72        2.3

95 PCT CI FOR MU 1 - MU 2: (-10.6, 1.6)
TTEST MU 1 = MU 2 (VS NE): T=-1.76 P=0.12 DF=7.2
MTB > NAME C3 = 'TYPE1EFF'
MTB > NAME C4 = 'TYPE2EFF'
MTB > UNSTACK 'EFFIC' 'TYPE1EFF' 'TYPE2EFF';
```

```
SUBC > SUBSCRIPTS 'TYPE'.
MTB > PRINT C1 C2 C3 C4

 ROW   TYPE    EFFIC    TYPE1EFF  TYPE2EFF
   1     1       72        72        82
   2     1       78        78        74
   3     1       73        73        81
   4     1       69        69        71
   5     1       75        75        85
   6     1       74        74        73
   7     1       69        69
   8     1       75        75
   9     2       82
  10     2       74
  11     2       81
  12     2       71
  13     2       85
  14     2       73

MTB > MANN-WHITNEY 'TYPE1EFF' 'TYPE2EFF'

Mann-Whitney Confidence Interval and Test

 TYPE1EFF   N =   8     MEDIAN =  73.500
 TYPE2EFF   N =   6     MEDIAN =  77.500
POINT ESTIMATE FOR ETA1-ETA2 IS  -4.5011
95.5 PCT C.I. FOR ETA1-ETA2 IS   (-11.0, 2.0)
W = 50.0
 TEST OF ETA1 = ETA2 VS. ETA1 N.E. ETA2 IS SIGNIFICANT AT 0.2200

 CANNOT REJECT AT ALPHA = 0.05

MTB > MOOD 'EFFIC' 'TYPE'

Mood median test of effic

Chisquare = 0.22    df = 1       p = 0.640

                                         Individual 95.0% CI's

type N<= N> Median Q3-Q1--+---------+--------+---------+----
 1   5   3   73.5   5.3 (-------+--)
 2   3   3   77.5  10.2       (----------+-----------)
                        --+---------+--------+---------+----
                          70.0      75.0     80.0      85.0
```

```
Overall median = 74.0
A 95.0% C.I. for median(1) - median(2): (-11.0,2.3)

MTB > STOP
```

Discussion of MINITAB Transcript

As with MINITAB examples in previous chapters, user inputs are shown to the right of the > symbol, and are underlined for clarity. The first two lines define columns C1 and C2 on the worksheet to be variables named TYPE and EFFIC, respectively. When each observation includes values for more than one variable, a READ command is used to enter the data. MINITAB reads each line of input as a separate observation. In our transcript, all the observations for the first heat pump are entered before those for the second heat pump, but this need not be the case. MINITAB allows the data to be input in any order.

Remember that MINITAB commands that include subcommands end in a semicolon. Each subcommand except the last ends in a semicolon, and the last subcommand ends in a period to designate the end of information for the original command. The TWOT command and its POOLED subcommand request a (two-sample) pooled t test on the variable EFFIC. The second variable listed, TYPE, is the classification variable determining the two samples to be compared.

In response, MINITAB first prints descriptive information for each sample, including the sample size, sample mean, sample standard deviation, and standard error of the mean. The 95% confidence interval for $(\mu_1-\mu_2)$ is shown as $-9.7 < (\mu_1-\mu_2) < 0.6$. Note that this interval includes 0, so it is reasonable that $(\mu_1-\mu_2) = 0$, or equivalently, $\mu_1 = \mu_2$. This confidence interval is equivalent to the hypothesis test for equality of means. On the next line, MINITAB prints the results of the hypothesis test, including the t value (-1.92), the two-tailed p-value (0.079), and degrees of freedom (12). The t value and degrees of freedom are the same as computed by hand in Section 4.2. Since the two-tailed p-value is greater than 5 percent, we accept the null hypothesis in our two-sided test.

The next command again requests a two-sample t test, but now since the POOLED subcommand is omitted, the unequal variance t test is performed. The t value of -1.76 parallels our hand computations. The approximate degrees of freedom for this test are shown as 7.2. MINITAB can compute tail areas for t distributions with fractional degrees of freedom and prints the two-tailed p-value of 0.12. The p-value for the unequal variance t test is slightly larger than that for the pooled t test, reflecting the greater power of the pooled t test when it is valid.

To perform the U test (equivalent to the rank sum test), MINITAB requires that the data be structured differently, with each sample in a separate column of the worksheet. Although the data could be reentered, this transcript demonstrates

MINITAB's ability to reorganize the data when needed. Columns C3 and C4 are labeled TYPE1EFF and TYPE2EFF, to represent the separate samples. The UNSTACK command and associated SUBSCRIPTS subcommand put the value of EFFIC for each observation into either column C3 or C4, according to the TYPE of heat pump. The PRINT command and its output are included to display the entire worksheet, including the two new columns.

The MANN-WHITNEY command asks for the Mann-Whitney U test (equivalent to the rank sum test), to compare the samples in TYPE1EFF and TYPE2EFF. As mentioned in Section 4.4, this test should be interpreted as a comparison of population medians when the data are skewed. (If the data are not skewed, the median is the same as the mean.) MINITAB uses the term ETA to represent the median, and allows for the possibility of skewed data by phrasing its test in terms of ETA—that is, the null hypothesis is stated as ETA1 = ETA2, instead of $\mu_1 = \mu_2$. The 95% confidence interval for (ETA1-ETA2) is shown as $-11.0 <$ (ETA1-ETA2) < 2.0. As discussed above in connection with the pooled t test, it is reasonable to conclude that ETA1 = ETA2, since this interval includes 0. MINITAB prints W, the signed rank statistic, but not its approximate standardized Z value. The two-tailed p-value of 0.2200 is greater than 5 percent, leading to acceptance of the null hypothesis. The p-value for the rank sum test is larger than for either of the t tests, since the rank sum test has lower power when the data are reasonably close to normally distributed.

The median test (developed by Mood) is requested with the MOOD command. MINITAB uses a chi-square approximation and displays the two-tailed p-value of 0.640. This p-value is substantially larger than for any of the previous tests. The median test uses the least information from the data and should be employed only when the requirements of the other tests are seriously violated.

4.6 INFERRING CAUSALITY

Intelligent use of statistics involves more than using the correct computer program. It begins with sensible design of the experiment, includes proper choice of analytic technique, and concludes with logical interpretation of results. Example 4.2 highlights the importance of good experimental design in the ability to draw rational conclusions.

Clearly, based on the results of the study described in this example, we can draw no conclusion about the relative effectiveness of the two antiperspirants. Chicago is colder than Atlanta, and the obtained results may be entirely due to differences in temperature between the two cities, not differences in effectiveness between the antiperspirants. City and antiperspirant are completely **confounded** in this experiment, since we cannot separately measure their effects.

We hasten to add that the statistical test performed on these data was perfectly

Example 4.2

Two antiperspirant manufacturers, in fierce competition, each tested their products on local residents. The Chicago-based firm randomly sampled 100 local residents, provided each person with a can of its antiperspirant, and obtained a laboratory measurement of the amount of underarm mositure after two weeks. Similarly, the Atlanta-based firm randomly sampled 100 local residents, provided each with its anitperspirant, and measured moisture after two weeks. In a two-sided pooled t test, the null hypothesis was soundly rejected. The average wetness for those using the antiperspirant of the Chicago firm was much less than that for the Atlanta firm. What can we conclude about the relative effectiveness of the two antiperspirants?

---

valid. Two independent random samples were employed, and the large sample sizes certainly allow for use of the pooled t test. It is legitimate to conclude that mean wetness is less for Chicago residents on the Chicago antiperspirant than for Atlanta residents on the Atlanta antiperspirant. But we are unable to determine *why* the difference occurs. Specifically, we cannot draw the inference that a difference between the two antiperspirants *caused* the difference in mean wetness.

Consider a different experimental design for this study, incorporating random assignment to groups. The 100 Chicago residents and 100 Atlanta residents could be combined to form a single pool of 200 experimental units. This combined sample could then be *randomly* divided into two groups, one group for each antiperspirant. The use of random assignment should result in each group having about the same mix of Chicago and Atlanta residents. Now, if one group has lower mean wetness, the result can be attributed to the antiperspirant—that is, we could conclude that the difference in antiperspirants caused the difference in mean wetness.

Even better, we could guarantee that each antiperspirant is used by exactly equal numbers of Chicago and Atlanta residents by performing the random assignment within each city. The 100 Chicago residents could be randomly divided into two groups of 50, and the same could be done in Atlanta. Each antiperspirant would be used by a total of 100 people, 50 from each city. We stress that within each city the division into two groups should still be done randomly. In the original experiment, we noticed an obvious confounding effect, namely the cities. But perspiration is probably also affected by an individual's weight, the kind of work performed, clothes worn, and a host of other factors. To validly compare the two anti-perspirants, we must start with two groups equal on *all* these variables. Random assignment allows all such factors of influence to balance out across the two groups, whether we noticed them in advance or not.

In short, random assignment is the key to drawing a causal conclusion. This basic principle of experimental design is relevant in all statistical tests, whether there are two groups or several groups. Random assignment also provides the ability to draw causal conclusions in regression (prediction) situations, to be discussed in later chapters. To be sure, in some settings it is impossible to use random assignment to groups. We cannot randomly assign people to be male or female, for example. But when possible, it is sound experimental practice to employ random assignment. Whatever flexibility you have in the design of your research, use wisely.

One of the best known examples of the random assignment principle involves the relationship between cigarette smoking and lung cancer. For many years, tobacco companies objected to the implication that smoking causes cancer. Smokers also tend to be drinkers, to eat less nutritious foods, to exercise less, and, in general, to live less healthy lifestyles than nonsmokers. The difference in cancer rates between smokers and nonsmokers could just as validly be attributed to any of these other factors as to smoking. It is clearly impossible to perform the definitive experiment necessary to establish a causal link between smoking and lung cancer. People cannot be randomly assigned to smoking and nonsmoking groups. But rats can. We now have proof that smoking causes cancer in rats. We also know that respiratory systems of rats and humans function in much the same fashion. Common sense suggests that cigarette smoking also causes lung cancer in humans. In some situations, we must be satisfied with evidence that only implies, rather than establishes, a causal link.

Before leaving this topic, we wish to emphasize the difference between random assignment and random sampling. Random assignment is necessary to determine causality. The ability to conclude that experimental manipulation *caused* the results is referred to as **internal validity**. An experiment is internally valid if we know *why* the results occurred. Random sampling, on the other hand, is needed to enable *generalization* from the sample to the population. This is known as **external validity**. An experiment is externally valid if sample results are likely to hold in the population. Both kinds of randomness are important for different reasons. The best researches are both internally and externally valid.

4.7 PAIRED DATA

We might have envisioned another way of designing the antiperspirant study of Example 4.2. Each person could use both of the antiperspirants, one under each arm. Random assignment could still be employed by having each person flip a coin to determine which antiperspirant is used under which arm. In this paradigm, each person serves as their own control. We no longer have two *independent* samples, since the same people are using each antiperspirant. The pooled t test and other

Example 4.3

A company executive was concerned with the tread life of tires on the fleet of cars used by its salespeople. Ten cars were randomly selected from the fleet and fitted with two tires of type 1 and two tires of type 2. The tires were rotated periodically to avoid bias in the comparison. The tread life of each type of tire was determined when the tread wear indicator for either tire of that type showed that replacement was advisable. The useful tread life (in thousands of miles) is shown below for each type of tire on each of the ten test cars. Do the two types of tires have the same tread life?

car	1	2	3	4	5	6	7	8	9	10
tire 1	17	21	34	19	21	36	30	23	20	26
tire 2	21	25	33	24	22	41	36	26	23	25

procedures discussed in the last few sections are not appropriate without independent samples, but there is still a way to validly compare the antiperspirants.

The method for analyzing paired data is simple and logical. For each person, a difference score, D, is computed by subtracting the measurements for the two treatments. The mean of the difference scores is the difference of the means—that is, $\mu_D = \mu_1 - \mu_2$. Our original hypothesis of H_0: $\mu_1 = \mu_2$ can be rewritten as H_0: $\mu_D = 0$. In short, we transform the paired data situation to a one-sample test on the mean of the difference scores. If the two separate measurements are each normally distributed, so is their difference, and a one-sample t test can be employed. This test is referred to as a **paired t test**. If the data do not appear normal, one sample nonparametric procedures, such as the signed rank and sign test, are available. Testing with paired data is demonstrated in connection with Example 4.3.

1. State the null and alternative hypotheses.
   We want to know if either tire has longer tread life than the other. Accordingly, a two-sided test is in order:

$$H_0: \mu_1 = \mu_2$$
$$H_1: \mu_1 \neq \mu_2.$$

TABLE 4.2 Difference Scores Between Two Tire Types

car	1	2	3	4	5	6	7	8	9	10
tire 1	17	21	34	19	21	36	30	23	20	26
tire 2	21	25	33	24	22	41	36	26	23	25
D	-4	-4	1	-5	-1	-5	-6	-3	-3	1

Equivalently, we can state the hypotheses in terms of the mean difference score as

$$H_0: \mu_D = 0$$
$$H_1: \mu_D \neq 0.$$

2. Determine the rejection region.
   Ten difference scores are computed from the ten test cars, resulting in $n - 1 = 9$ degrees of freedom. Splitting the 5% rejection region for a two-sided paired t test, we will reject H_0 if $|t|$ is greater than 2.262.
3. Obtain the data.
   For each car, we compute a difference score as D = tire 1 −tire 2. These values are shown in Table 4.2. The sample mean of these difference scores is $\overline{D} = -2.9$, and the sample standard deviation is $S_D = 2.47$.
4. Compute the test statistic.
   Using the one-sample t statistic from (3.8.1) (on the variable D instead of X), we obtain

$$t = \frac{\overline{D} - \mu_D}{S_D/\sqrt{n}} = \frac{-2.9 - 0}{2.47/\sqrt{10}} = -3.71.$$

5. Draw the conclusion.
   We reject the null hypothesis since the observed t value falls beyond ±2.262. As discussed in Section 3.7, a significant result in a two-sided test leads to a one-sided conclusion. Here, based on the negative t value, we conclude that $\mu_D < 0$, or, equivalently, $\mu_1 < \mu_2$. The second tire type has longer tread life than the first.

As alternatives to the paired t test, both the signed rank and sign test can be used for paired data. Each test is employed as in the one-sample case (see Section 3.13), except for using the difference scores as the raw data. Both these tests assume a continuous distribution, and the signed rank test requires symmetry. When applied with paired data, however, the symmetry requirement is usually satisfied. If the direction of skewness is the same in the two measurements, it will be essentially subtracted out in computation of the difference scores, which will be symmetric.

4.8 PAIRED DATA IN COMPUTER PROGRAMS

The tests performed on paired data are the same ones demonstrated in Section 3.15 for the one-sample case. Our discussions here will be relatively brief; refer to Section 3.15 for additional details. All three programs perform the paired t test, signed rank test, and sign test.

SAS Input

```
OPTIONS LS = 80;
TITLE 'EXAMPLE 4.3 - TREAD LIFE OF TIRES';
DATA;
INPUT TIRE1 TIRE2;
D = TIRE1 - TIRE2;
LINES;
17      21
21      25
34      33
19      24
21      22
36      41
30      36
23      26
20      23
26      25
PROC UNIVARIATE;
VAR D;
```

Discussion of SAS Input

For each car, tread life is entered for both types of tires (TIRE1 and TIRE2 on the INPUT statement). The difference score, D = TIRE1 - TIRE2, is computed by SAS for each observation, and is added to the dataset. The UNIVARIATE procedure and its associated VAR statement request descriptive information for the difference scores, including paired t, signed rank, and sign tests. These tests assume a hypothesized mean of 0, which is appropriate when testing on difference scores.

SAS Output

```
               EXAMPLE 4.3 - TREAD LIFE OF TIRES                    1
                      Univariate Procedure

Variable=DIFF

                            Moments

          N                10      Sum Wgts          10
          Mean           -2.9      Sum              -29
          Std Dev    2.469818      Variance         6.1
          Skewness   0.686982      Kurtosis    -0.76196
```

```
USS                 139      CSS                 54.9
CV             -85.1661      Std Mean        0.781025
T:Mean=0       -3.71307      Prob>|T|          0.0048
Num ~= 0             10      Num > 0                2
M (Sign)             -3      Prob>|M|          0.1094
Sgn Rank          -23.5      Prob>|S|          0.0137

                  Quantiles (Def=5)

100% Max              1      99%                    1
 75% Q3              -1      95%                    1
 50% Med           -3.5      90%                    1
 25% Q1              -5      10%                 -5.5
  0% Min             -6       5%                   -6
                              1%                   -6
Range                 7
Q3-Q1                 4
Mode                 -5

                      Extremes

Lowest        Obs            Highest          Obs
-6(           7)                -3(             8)
-5(           6)                -3(             9)
-5(           4)                -1(             5)
-4(           2)                 1(             3)
-4(           1)                 1(            10)
```

Discussion of SAS Output

Under the heading Moments, the seventh line shows the computed t (-3.71307), and its two-tailed p-value (0.0048), for testing that the mean of the D scores is 0. Since the two-tailed p-value is less than 5 percent, we reject the null hypothesis in our two-sided test. Based on the negative value of t, we conclude that the second type of tire has longer tread life than the first.

Three lines below this we see the signed rank statistic and its two-tailed p-value of 0.0137. Again, our test is significant because this is less than 5 percent. The signed rank statistic is negative, leading to the same one-sided conclusion as reached in the paired t test. The p-value obtained for the paired t test is smaller than that for the signed rank test, reflecting the greater power of the t test when the difference scores are close to normally distributed.

The sign test, shown directly above the signed rank test, has a two-tailed p-value of 0.1094, and results in acceptance of the null hypothesis. As noted in earlier examples, the sign test is generally less powerful than the paired t and signed rank tests.

SPSS Input

```
SET WIDTH = 80
TITLE 'EXAMPLE 4.3 - TREAD LIFE OF TIRES'
DATA LIST FREE / TIRE1 TIRE2
T-TEST PAIRS = TIRE1 TIRE2
BEGIN DATA
17 21
21 25
34 33
19 24
21 22
36 41
30 36
23 26
20 23
26 25
END DATA
NPAR TESTS WILCOXON = TIRE1 TIRE2 / SIGN = TIRE1 TIRE2
FINISH
```

Discussion of SPSS Input

For each car, tread life is entered in free format for both types of tires (TIRE1 and TIRE2 on the DATA LIST statement). The T-TEST procedure with the keyword PAIRS requests a paired t test on the mean difference for these two variables. SPSS always uses a hypothesized mean of 0 for the difference scores in this test. After the data are entered, the NPAR TESTS procedure asks for both the signed rank test (keyword WILCOXON) and sign test. Again, the program assumes a hypothesized mean difference of 0.

SPSS Output

```
              EXAMPLE 4.3 - TREAD LIFE OF TIRES                    Page 1
               ---t-tests for paired samples---

Variable      Number                  Standard     Standard
             of Cases       Mean      Deviation      Error
----------------------------------------------------------

TIRE1           10        24.7000      6.567        2.077
TIRE2           10        27.6000      6.703        2.120
```

```
(Difference) Standard Standard|       2-tail|t     Degrees of 2-tail
    Mean      Deviation   Error | Corr. Prob.|Value  Freedom    Prob.
- - - - - - - - - - - - - - - - |- - - - - - |- - - - - - - - - - - -
  -2.9000       2.470.     781  |  .931  .000|-3.71       9      .005
```

```
            EXAMPLE 4.3 - TREAD LIFE OF TIRES           Page 2
        Wilcoxon Matched-Pairs Signed-Ranks Test
     TIRE1
with TIRE2

     Mean Rank    Cases
        2.00        2     - Ranks (TIRE2 LT TIRE1)
        6.38        8     + Ranks (TIRE2 GT TIRE1)
                    0       Ties  (TIRE2 EQ TIRE1)
                   --
                   10       Total
         Z = -2.3953             2-Tailed P = .0166

                          Sign Test
     TIRE1
with TIRE2

          Cases
            2 - Diffs  (TIRE2 LT TIRE1)
            8 + Diffs  (TIRE2 GT TIRE1)            (Binomial)
            0   Ties                           2-Tailed P =.1094
           --
           10   Total
```

Discussion of SPSS Output

The first page of SPSS output contains the results of the t test, including descriptive information for each variable. At the bottom right of this page are printed the t value (-3.71), its degrees of freedom (9), and two-tailed p-value (0.005). The two-tailed p-value is less than 5 percent, allowing rejection of the null hypothesis in our two-sided test. Based on the negative value of t, we conclude that the second type of tire has longer tread life than the first.

Results of the signed rank and sign tests are reported on the second page of output. For the signed rank test, the two-tailed p-value of 0.0166 is less than 5 percent, so we reject the null hypothesis. Eight of the ten cars had greater tread life for the second tire than for the first, leading to the same one-sided conclusion as in the paired t test. The p-value obtained for the paired t test is smaller than that for the signed rank test, reflecting the greater power of the t test when the difference scores are close to normally distributed.

The sign test shows a two-tailed p-value of 0.1094, and results in acceptance of the null hypothesis. As noted in earlier examples, the sign test is generally less powerful than the paired t and signed rank tests.

MINITAB Transcript

```
MTB > NAME C1 = 'TIRE1'
MTB > NAME C2 = 'TIRE2'
MTB > READ 'TIRE1' 'TIRE2'
DATA > 17 21
DATA > 21 25
DATA > 34 33
DATA > 19 24
DATA > 21 22
DATA > 36 41
DATA > 30 36
DATA > 23 26
DATA > 20 23
DATA > 26 25
DATA > END
     10 ROWS READ

MTB > NAME C3 = 'D'
MTB > LET 'D' = 'TIRE1' - 'TIRE2'
MTB > TTEST 0 'D'

TEST OF MU = 0.000 VS MU N.E. 0.000

             N        MEAN      STDEV    SE MEAN        T     P VALUE
    D       10      -2.900      2.470      0.781    -3.71      0.0048
MTB > WTEST 0 'D'

TEST OF MEDIAN = 0.000000000 VERSUS MEDIAN N.E. 0.000000000

                  N FOR     WILCOXON                    ESTIMATED
             N     TEST     STATISTIC   P-VALUE           MEDIAN
    D       10       10          4.0      0.019           -3.000
```

```
MTB > STEST 0 'D'

SIGN TEST OF MEDIAN = 0.000000000 VERSUS N.E. 0.000000000
           N    BELOW   EQUAL  ABOVE  P-VALUE    MEDIAN
   D      10      8       0      2     0.1094    -3.500

MTB > STOP
```

Discussion of MINITAB Transcript

For each car, tread life is entered in column C1 for TIRE1 and in column C2 for TIRE2. As in Example 4.1, a READ command is used to input the data, since there are two values for each observation. The difference score, D = TIRE1 - TIRE2, is computed and stored in column C3 for each car.

The TTEST command specifies the hypothesized mean of 0 for the D scores. Along with some descriptive information, MINITAB prints the t value (-3.71) and its two-tailed p-value (0.0048). Since the two-tailed p-value is less than 5 percent, we reject the null hypothesis in our two-sided test. Based on the negative value of t, we conclude that the second type of tire has longer tread life than the first.

The WTEST command requests the (Wilcoxon) signed rank test on the difference scores. The test is reported as a test for a *median* of 0, but in a symmetric distribution, the mean and median are equal. The two-tailed p-value of 0.019 is less than 5 percent, so the signed rank test also rejects the null hypothesis. The estimated median of the difference scores is negative, leading to the same one-sided conclusion reached in the paired t test. The p-value obtained for the paired t test is smaller than that for the signed rank test, reflecting the greater power of the t test when the difference scores are close to normally distributed.

The sign test is obtained using the STEST command, and shows a two-tailed p-value of 0.1094. Since this is greater than 5 percent, we must accept the null hypothesis. As noted in earlier examples, the sign test is generally less powerful than the paired t and signed rank tests.

4.9 TO PAIR OR NOT TO PAIR

In Example 4.3, ten cars were used to compare two tire types by equipping each car with two tires of each type. Because each car provided a measurement for each type of tire, the data were paired, or dependent, and the paired t test was employed. However, using the same materials, ten cars and 40 tires, this study could have been designed with independent groups. The cars could have been randomly divided into two groups of five, with each group using one type of tire on all four wheels. Data from this experiment would have been analyzed with the pooled t test. Is there any advantage to pairing?

Examine the data from Example 4.3. For each type of tire, there is large variability in tread life across the ten cars. Using the sample range as a rough estimate of spread, we see a range of 36 - 17 = 19 for the first type of tire, and 41 - 21 = 20 for the second. On the other hand, the range of difference scores is much smaller, 1- (-6) = 7. The sample variances would show the same pattern—large for each tire individually, much smaller for the difference scores. The large individual variances are pooled in the denominator of the pooled t statistic, whereas the smaller variance of the difference scores is employed in the denominator of the paired t test. Because of its larger denominator, the pooled t would be much *smaller* than the paired t. With a smaller observed t value, the pooled t test is less likely to reject the null hypothesis than the paired t test—that is, the paired t test has greater power.

Many factors contribute to variability in tread life besides the type of tire. The weight and alignment of the car, the speed it is driven, the weather it is subjected to, the roads it is driven on, and the style of the driver all have obvious effects on tread life. These factors lead to large variability of each tire type across all ten cars. But in computing difference scores, we essentially subtract out all these sources of variability. Each tire is only compared with its counterpart on the same car. Common sense tells us that the variability of the difference scores should be much smaller. Recall the discussion of power in Section 3.5. When testing on means, one of the ways to increase power is to reduce variability. In many circumstances, pairing reduces variability and is a logical and practical method of increasing power.

In some situations, paired data can even be obtained from independent samples, by matching on some other relevant variable. For example, suppose we wish to compare two teaching techniques using a pool of 20 students. Clearly, we cannot use both techniques for each person, since some learning would be carried over from the first technique to the second. Instead, the students can be randomly divided into two groups of ten persons each. The same material would be taught to the two groups by different techniques, followed by a common test to assess learning. This design suggests use of the pooled t test to compare the results, but, with a bit of additional information, pairing can be employed.

Suppose, in addition to the learning test scores, we also have an intelligence measure available for each person. Pairs can be constructed by matching on intelligence. The person with the highest intelligence score in one group can be matched with the person with the highest score in the other group, to form one paired observation. The persons with the second highest intelligence scores in each group can be matched to form a second pair. This process is continued until all 20 people have been matched, yielding ten paired observations. The paired t test can now be used on these matched pairs.

Within each group, large variability on learning scores is likely, due in part to individual differences in intelligence. Without pairing, this variance would appear in the denominator of the pooled t ratio and adversely affect power. But when difference scores are constructed, the effects of intelligence are essentially removed. The variability among the difference scores should be much smaller than the variability within each group separately. The paired t test for this experiment should be substantially more powerful than the pooled t test.

In constructing matched pairs from independent samples, we should be careful to match on some logical basis. Suppose, instead of pairing on intelligence, we formed pairs on the basis of shoe size—that is, the people with the largest shoe size in each group are matched to form one paired observation. Those with the second biggest shoes in each group form the second pair, and so forth. Now when difference scores are constructed, we remove the effects of shoe size on learning. Since shoe size probably does not influence learning, we would not have reduced variability in the paired t test.

But has this ineffective pairing cost us anything? The answer is yes, it has cost us degrees of freedom. Without pairing, we would perform a pooled t test with $n_1 + n_2 - 2 = 10 + 10 - 2 = 18$ degrees of freedom. With pairing, we have ten paired observations and the paired t test has $n - 1 = 10 - 1 = 9$ degrees of freedom. Fewer degrees of freedom is equivalent to having fewer observations—it results in lower power. The paired t test has fewer degrees of freedom and, therefore, lower power than the pooled t test.

To pair or not to pair? The answer is obtained by applying common sense. If we can construct pairs on some basis logically related to the variable under test, we should do so. Pairing effectively halves the degrees of freedom, which reduces power. But rationally constructed pairs reduce variance on the relevant variable, which increases power. Usually, the lower variance more than offsets the loss in degrees of freedom, and the net effect of pairing is to improve power substantially. But we should not construct pairs arbitrarily or capriciously. We will only lose degrees of freedom, without reducing variance, and have a weaker test.

4.10 THE F DISTRIBUTION

Four distributions are used heavily in statistics: normal, t, chi-square, and F. We have learned about the first three of these, including how to use their tables in the Appendix. In order to compare variances for two populations, the topic of the next section, we need to use the **F distribution**. For this purpose, the F statistic is the ratio of the two sample variances. As we will see in later chapters, the F distribution is also used in many other testing procedures, and always appears as the ratio of two variance-type terms.

Like the normal, t, and chi-square, the F is a family of distributions. The various curves that make up this family are indexed by two parameters, the degrees of freedom for the numerator (df_{num}), and the degrees of freedom for the denominator

(df_{den}). These two parameters are always listed in order—numerator df before denominator df. For comparing two population variances, F is the ratio of the sample variances. Each sample variance has n - 1 degrees of freedom, and the numerator and denominator degrees of freedom for the F ratio are the same as those of the corresponding sample variances.

Since sample variances are never negative, neither is their ratio. The F distribution is bounded below by 0. On the other hand, there is no upper limit to the size of a sample variance, so the F distribution is unbounded at the upper end. As a result, the shape of the F curve is asymmetric, regardless of the numerator and denominator degrees of freedom. For moderate to large values of df_{num} and df_{den}, the shape of the F curve is much like that of the chi-square distribution. It begins at 0, has a lump near the lower end, and asymptotically approaches the axis at the upper end. (See the chi-square curve shown in Figure 2.3.) For very small values of df_{num} and df_{den}, the shape of the F curve can be rather different, but the tables automatically account for the shape of the curve.

Tabling the F distribution is more difficult than any of our earlier distributions. For the normal distribution, only a single curve, the standard normal, is tabled. Tail areas are shown for many different Z cutoff values, at intervals of 0.01 on the Z scale. Areas under any other normal distribution can be found by converting to the Z scale. For both the t and chi-square distributions, no such conversion is possible. Each value for degrees of freedom (i.e., each t and chi-square curve) must be tabled separately. The tables show many different values of df, but only a few cutoffs for each curve. With the F distribution, this problem is doubled since two separate df parameters must be considered. As a result, F tables are constructed differently from those of other distributions. First we note that only upper tail cutoffs are tabled. Each page of Table 4 in the Appendix lists many different values of df_{num} (top margin) and df_{den} (left margin), and shows cutoffs in the body of the table for a single upper tail area. Table 4a shows cutoffs for upper tail area of 0.05, Table 4b for upper tail area of 0.025, and Table 4c for 0.01.

The fact that only upper tail cutoffs are listed for an asymmetric distribution makes it difficult to perform two-sided tests. As we show in the next section, we can overcome this difficulty with careful construction of our F ratio. All applications of the F distribution in later chapters involve one-sided tests with upper tail rejection regions. For this reason, most statistics books include F tables listing only upper tail cutoffs.

4.11 COMPARISON OF VARIANCES

In Example 4.1, a building contractor compared the efficiencies of two types of heat pump. As we found in Section 4.2, the heat pumps do not differ significantly in *mean* efficiency. But it might also be useful to know if they differ in variability.

A lower variance would be desirable because it would indicate more *consistent* efficiency ratings. We wish to test

$$H_0: \sigma_1^2 = \sigma_2^2$$
$$H_1: \sigma_1^2 \neq \sigma_2^2 .$$

We will use an F test to compare these variances. Since our test is two-sided, we split the rejection region and put 0.025 in each tail of the F distribution. However, only the upper tail of the F distribution is tabled, so we must ensure that only the upper critical point is needed to reach a conclusion. The way to guarantee that our F value is closer to the upper cutoff than to the lower is to put the larger sample variance in the numerator. Thus, we form the F ratio

$$F = \frac{\text{larger } S^2}{\text{smaller } S^2} . \tag{4.11.1}$$

As shown in Section 4.2, the sample variances in this example are $S_1^2 = 9.55$ and $S_2^2 = 32.67$. Since S_2^2 is larger than S_1^2, our F ratio will have S_2^2 in the numerator and S_1^2 in the denominator, and, therefore, $df_{num} = n_2 - 1 = 5$ and $df_{den} = n_1 - 1 = 7$. From Table 4b, we find the upper 0.025 critical value of 5.29. We will reject H_0 if F is greater than this figure.

Substituting the sample variances into (4.11.1), we obtain

$$F = \frac{\text{larger } S^2}{\text{smaller } S^2} = \frac{32.67}{9.55} = 3.42.$$

This is less than the critical value of 5.29, so we accept the null hypothesis. The available evidence is insufficient to show that the two types of heat pump differ in variance.

In this example, one sample variance was more than three times as large as the other, but we still could not reject the null hypothesis of equal population variances. Larger sample sizes would have given our test more power and might have produced a significant result. In general, spread is more difficult to measure than center. We should use larger sample sizes when testing on variances than on means.

As we have just demonstrated, a two-sided comparison of variances is conducted by splitting the rejection region, and putting the larger sample variance in the numerator of the F ratio to ensure that we only need to compare our test statistic with the upper cutoff. To perform a one-sided test on two variances, we put in the numerator of the F ratio the sample variance that is *expected* to be larger, according to the alternative hypothesis. Then we locate all 5 percent of the rejection region in the upper tail of the F distribution.

Earlier in this chapter, we learned that two different t tests are available for comparing means, depending on whether or not we believe the variances are equal. Now we have a way of testing to see if the variances are equal. The F test on variances can be used to decide which t test to use for comparing means. If the F test suggests the variances are equal, we should use the pooled t test to compare means. On the other hand, if the F test is significant, we conclude that the variances are different, and we should use the unequal variance t test to compare means. This strategy is logical, and most computer programs routinely print the F test whenever a t test is requested.

We caution, however, that the F test is *not* the perfect tool for deciding between the two t tests. Unfortunately, the F test has its power under the wrong conditions. If the sample sizes are large, the F test is quite powerful and will probably be significant even when the population variances are nearly equal. A significant F test indicates that the unequal variance t test should be used to compare means. But with large sample sizes and nearly equal variances, the pooled t test is perfectly adequate and is more powerful than the unequal variance t test. The F test might direct us to employ the wrong t test. Conversely, if the sample sizes are small, the F test is quite weak and will probably accept the null hypothesis of equal variances even when they are rather different. Accepting equality of the variances suggests that we should use the pooled t test to compare means. But with small sample sizes and rather unequal variances, we might be better served using the unequal variance t test. Again, the F test may route us to the wrong t test. In short, the F test is quite powerful with large sample sizes, precisely when we need be least concerned with the equal variance assumption of the pooled t test.

These comments should not be construed as a condemnation of the F test. It *may* be helpful in deciding which t test to employ for comparing means, but it is only part of the picture. We should consider as well the sizes of the two samples and how similar they are. The pooled t test is valid if the sample variances are nearly equal, or if the sample sizes are large or nearly equal.

4.12 COMPARISON OF VARIANCES IN COMPUTER PROGRAMS

Both SAS and SPSS automatically perform the F test for equality of variances whenever a t test on independent samples is requested. Section 4.5 contains inputs and outputs from these two programs for Example 4.1, which was used to demonstrate the t test. The reader is referred to these outputs in conjunction with comments below about the F test.

MINITAB requires separate commands to produce the pooled and unequal variance t tests, and therefore does not automatically perform the F test. In order to test for equality of variances in MINITAB, it is necessary to construct the F ratio and then use built-in functions to determine its p-value.

SAS

At the bottom of the first page of output for Example 4.1, SAS prints the F test for equality of variances. The observed F value (3.42) and its degrees of freedom (5 in the numerator and 7 in the denominator) agree with our hand computations of the previous section. The probability value shown (0.1399) is a two-tailed p-value, since SAS assumes a two-sided test. Technically, this is the area above F = 3.42, plus an equivalent area in the lower tail. Since the F is asymmetric, this two-tailed p-value cannot be indicated with the absolute value notation, as is used for a t test (PROB >|T|). SAS uses a prime on the F statistic to signify that it is used for a two-sided test. Since the two-tailed p-value is greater than 5 percent, we accept the null hypothesis. The evidence is not strong enough to conclude that the population variances differ.

SPSS

At the bottom left of the first page of output for Example 4.1, SPSS prints the F test for equality of variances. The observed F value (3.42) agrees with our hand computations of the previous section. The two-tailed p-value is shown as 0.140. Technically, this is the area above F = 3.42, plus an equivalent area in the lower tail. Since the two-tailed p-value is greater than 5 percent, we accept the null hypothesis. The evidence is not strong enough to conclude that the population variances differ.

MINITAB Transcript

```
MTB > NAME C1 = 'TYPE1EFF'
MTB > SET 'TYPE1EFF'
DATA > 72 78 73 69 75 74 69 75
DATA > END
MTB > NAME C2 = 'TYPE2EFF'
MTB > SET 'TYPE2EFF'
DATA > 82 74 81 71 85 73
DATA > END

MTB > LET K1 = N('TYPE1EFF')-1
MTB > LET K2 = N('TYPE2EFF')-1
MTB > LET K3 = STDEV('TYPE1EFF')**2
MTB > LET K4 = STDEV('TYPE2EFF')**2
MTB > PRINT K3 K4

K3          9.55356
K4          32.6667
```

```
MTB > LET K5 = K4/K3
MTB > CDF K5 K6;
SUBC > F K2 K1.
MTB > LET K6 = (1-K6)*2
MTB > PRINT K5 K6
K5        3.41932
K6        0.139895

MTB > STOP
```

Discussion of MINITAB Transcript

In the first eight lines of this transcript, the data for our two samples are entered in columns C1 and C2 of the worksheet, named TYPE1EFF and TYPE2EFF, respectively. Constants K1 and K2 are set equal to the degrees of freedom for each sample variance, and constants K3 and K4 are used to store the sample variances. Printing K3 and K4 enables us to determine which variance is larger, for use in the numerator of our F ratio. The F ratio is computed and stored in K5.

The CDF command and its F subcommand compute the area *below* the observed F value and store this area in K6. Degrees of freedom for the F ratio are in K2 (numerator) and K1 (denominator), since the second sample variance is used in the numerator of the F statistic and the first in the denominator. The tail area *above* our F ratio is found by subtracting the lower tail area from one. For our two-sided test, we compute the two-tailed p-value by adding an equivalent area in the lower tail (i.e., doubling the upper tail area). Printing K5 and K6 allows us to see the obtained F value and its two-tailed p-value. The F value of 3.41932 agrees with our computations of the previous section. The two-tailed p-value (0.139895) is greater than 5 percent, so we accept the null hypothesis. The evidence is not strong enough to conclude that the population variances differ.

4.13 COMPARISON OF PROPORTIONS

So far in this chapter we have discussed techniques designed for numeric variables. Now we consider the comparison of two proportions, which is appropriate for categorical data. This test is the logical extension of the one-sample test for a proportion presented in Section 3.18.

Consider independent random samples taken from two populations and measured on a categorical variable. Suppose we are interested in a particular category and wish to compare the proportions in this category for the two populations. Letting π_1 and π_2 represent the proportions in the desired category for the two populations, we wish to test

$$H_0: \pi_1 = \pi_2.$$

The alternative hypothesis may be one-sided or two-sided, depending on the situation, and is written as

$$H_1: \pi_1 < \pi_2,$$
$$H_1: \pi_1 > \pi_2, \text{ or}$$
$$H_1: \pi_1 \neq \pi_2.$$

For the first population, the estimate of the *population* proportion π_1 is the *sample* proportion of observations in the desired category, denoted by P_1. Similarly, for the second population we estimate π_2 with P_2, the proportion in the desired category for the second sample. To compare the population proportions, we look at the difference in sample proportions, $(P_1 - P_2)$. The test statistic also involves an overall estimate of the proportion in the desired category, obtained by combining the two samples. For the combined samples, let P (without a subscript) represent the overall proportion in the category of interest. The test statistic, which has a standard normal distribution, is given by

$$Z = \frac{(P_1 - P_2)}{\sqrt{P(1-P)\left(\frac{1}{n_1} + \frac{1}{n_2}\right)}}. \qquad (4.13.1)$$

The Z value computed from this expression is compared with critical values from the standard normal distribution. This test for comparing two proportions is demonstrated in Example 4.4.

1. State the null and alternative hypotheses.
   Let us label the females as population one, and the males as population two. We are interested in the proportion of supervisors (π) in these two populations. The null hypothesis states that the proportion of supervisors is the same in the

---

Example 4.4

A large corporation was sued for not promoting female employees as frequently as males. Since most of their employees were males, the corporation protested that most of their promotions should naturally go to males. The claimants countered with the argument that the proportion of supervisors was smaller among females than among males. To bolster their claim in court, separate samples of females and males were obtained, and the number of supervisors in each sample was recorded. A total of 60 females included only 19 supervisors, while there were 56 supervisors among a sample of 120 males. Is the claim justified?

two populations, while the alternative claims that it is smaller for the females. Accordingly, we write

$$H_0: \pi_1 = \pi_2$$
$$H_1: \pi_1 < \pi_2 .$$

2. Determine the rejection region.
   Putting 5 percent in the left tail of the standard normal distribution, we will reject the null hypothesis if Z is less than -1.645.
3. Obtain the data.
   For $n_1 = 60$ females, the sample proportion of supervisors is $P_1 = 19/60 = 0.32$.
   For $n_2 = 120$ males, the sample proportion of supervisors is $P_2 = 56/120 = 0.47$.
   Combining the two samples, there are a total of 19 + 56 = 75 supervisors out of 60 + 120 = 180 observations, yielding a combined estimate of $P = 75/180 = 0.42$.
4. Compute the test statistic.
   Substituting n_1, n_2, P_1, P_2, and P into (4.13.1), we obtain

$$Z = \frac{(P_1 - P_2)}{\sqrt{P(1-P)\left(\frac{1}{n_1} + \frac{1}{n_2}\right)}} = \frac{.32 - .47}{\sqrt{.42(1-.42)\left(\frac{1}{60} + \frac{1}{120}\right)}} = -1.92.$$

5. Draw the conclusion.
   As the computed Z is less than our cutoff of -1.645, we reject the null hypothesis and conclude that sex bias has been demonstrated. The proportion of supervisors among females is less than among males. The p-value for this test is the tail area below Z = -1.92. From Table 1 in the Appendix, we find this area to be 0.0274.

Computer programs typically perform a test for the comparison of proportions using a chi-square test instead of this Z test. The chi-square statistic used for this purpose is the square of our Z statistic. In squaring, the sign is lost and, consequently, computer programs present this as a two-sided test. Since the comparison of two proportions is often performed with a directional alternative hypothesis, we have chosen to present the test here in terms of Z.

The chi-square test can be viewed in terms of examining the relationship between two variables, each with only two categories. For our example, these variables are sex (female, male) and status (supervisor, non-supervisor). These two variables determine a 2×2 table, and the observations are sorted into the cells of this table. The chi-square procedure can also be used for categorical variables with more than two categories. In Chapter 7, we discuss the logic for the chi-square test and demonstrate its use with a larger table.

4.14 COMPARISON OF PROPORTIONS IN COMPUTER PROGRAMS

All three of the programs (SAS, SPSS, and MINITAB) perform the chi-square test for the comparison of proportions. Each computer output displays the 2×2 table of observations, sorted according to sex and status. Both SAS and SPSS automatically print both the regular chi-square statistic and a version that includes a continuity correction. The correction improves the accuracy of the test and is recommended for practical applications. But the continuity corrected version of chi-square is no longer exactly the square of Z.

In each of the three programs, two different methods are available for data entry, depending on whether or not observed cell counts are known. If they are not known, data is entered one line per observation, and the program tallies observed counts as it sorts observations into cells. This approach to data entry is demonstrated here for each program. Alternatively, when observed cell counts are known, only these counts need be entered. This second approach to data entry is demonstrated in Section 7.3, in connection with a chi-square test for association between two categorical variables.

SAS Input

```
OPTIONS LS = 80;
TITLE 'EXAMPLE 4.4 - SEX BIAS IN PROMOTIONS';
DATA;
INPUT SEX $ STATUS $;
LINES;
FEMALE SUPER
   .
   .    (19 times)
   .
FEMALE NONSUPER
   .
   .    (41 times)
   .
MALE SUPER
   .
   .      (56 times)
   .
MALE NONSUPER
   .
   .      (64 times)
   .
PROC FREQ;
TABLES SEX * STATUS / CHISQ;
```

Discussion of SAS Input

On the INPUT statement, dollar signs following the variables SEX and STATUS inform SAS that the values of these variables are alphabetic rather than numeric. As usual, there is one line of input per observation. Here we have 19 lines with the values FEMALE and SUPER for our two variables, 41 lines with the values FEMALE and NONSUPER, 56 lines with MALE and SUPER, and 64 with MALE and NONSUPER. Although the data are organized by group in our input, these 180 lines may be in any order. The program simply counts the number of observations in each cell to construct its 2×2 table.

The SAS procedure FREQ (frequencies) and its TABLES statement request a table of the observations sorted by the two variables, SEX and STATUS. On SAS, keywords following a slash request additional information, not computed and printed unless specifically requested. The CHISQ option asks for a chi-square test to be performed on this table.

SAS Output

```
                EXAMPLE 4.4 - SEX BIAS IN PROMOTIONS              1
                     TABLE OF SEX BY STATUS

SEX          STATUS

Frequency|
Percent  |
Row Pct  |
Col Pct  |SUPER    |NONSUPER|  Total
-----------------------------------
FEMALE   |    19   |    41  |    60
         | 10.56   | 22.78  | 33.33
         | 31.67   | 68.33  |
         | 25.33   | 39.05  |
----------------------------
MALE     |    56   |    64  |   120
         | 31.11   | 35.56  | 66.67
         | 46.67   | 53.33  |
         | 74.67   | 60.95  |
----------------------------

Total         75       105      180
           41.67     58.33   100.00
```

```
              STATISTICS FOR TABLE OF SEX BY STATUS

Statistic                     DF     Value          Prob
--------------------------------------------------------
Chi-Square                     1     3.703          0.054
Likelihood Ratio Chi-Square 1        3.768          0.052
Continuity Adj. Chi-Square  1        3.111          0.078
Mantel-Haenszel Chi-Square  1        3.682          0.055
Fisher's Exact Test (Left)                          0.038
                    (Right)                         0.982
                    (2-Tail)                        0.057
Phi Coefficient                     -0.143
Contingency Coefficient              0.142
Cramer's V                          -0.143

Sample Size = 180
```

Discussion of SAS Output

Each cell of the 2×2 table contains four numbers, which are identified in the key in the upper left corner. These are Frequency (the number of observations in the cell), Percent (frequency as a percentage of total sample size), Row Pct (frequency as a percentage of the number in the row), and Col Pct (frequency as a percentage of the number in the column). The percentage of supervisors among females (P_1) is shown in the upper left cell to be 31.67 percent (a *proportion* of approximately 0.32, as computed by hand in the previous section). In the lower left cell, we see that the percentage of supervisors among males (P_2) is 46.67 percent (a proportion of about 0.47, as computed by hand). Since $P_1 < P_2$, the sample results are in the direction predicted by our alternative hypothesis (H_1: $\pi_1 < \pi_2$). We will be able to reject the null hypothesis if the one-tailed p-value is less than 5 percent.

Below the table is a list of several statistics produced by the CHISQ option. The value 3.703 shown for Chi-Square on the first line is the square of our Z value (-1.9243, with a bit more accuracy than presented above). Its two-tailed p-value of 0.054 (labeled Prob) is double the one-tailed p-value of 0.0274 we found for our Z statistic, since the chi-square test is automatically two-sided.

For better accuracy we should examine the continuity corrected form of chi-square, presented on the third line of this listing. Its two-tailed p-value of 0.078 converts to a one-tailed p-value of 0.078/2 = 0.039. Since this is less than 5 percent, we reject the null hypothesis and conclude that this corporation is guilty of sex bias in promotions.

SPSS Input

```
SET WIDTH = 80
TITLE 'EXAMPLE 4.4 - SEX BIAS IN PROMOTIONS'
```

```
DATA LIST FREE / SEX (A6) STATUS (A8)
CROSSTABS TABLES = SEX BY STATUS
     / CELLS = COUNT ROW COLUMN TOTAL
     / STATISTICS = CHISQ
BEGIN DATA
FEMALE SUPER
   .
   .      (19 times)
   .
FEMALE NONSUPER
   .
   .      (41 times)
   .
MALE SUPER
   .
   .      (56 times)
   .
MALE NONSUPER
   .
   .      (64 times)
   .
END DATA
FINISH
```

Discussion of SPSS Input

On the DATA LIST statement, the notations (A6) following SEX and (A8) following STATUS inform SPSS that the values of these variables are alphabetic rather than numeric (with a maximum of six characters for SEX and eight for STATUS). Following the BEGIN DATA statement, there is one line of input per observation, as usual. Here we have 19 lines with the values FEMALE and SUPER for our two variables, 41 lines with the values FEMALE and NON-SUPER, 56 lines with MALE and SUPER, and 64 with MALE and NONSUPER. Although the data are organized by group in our input, these 180 lines may be in any order. The program simply counts the number of observations in each cell to construct its 2×2 table.

The procedure CROSSTABS requests a table of the observations sorted by the two variables, SEX and STATUS. SPSS statements may be continued on additional lines by beginning subsequent lines after the first column. Here the CROSSTABS statement is continued on two separate lines, each of which begins with a slash indicating a request for additional information. The CELLS option requests specific information to be printed in each cell of the table, including cell frequencies (COUNT), row percentages (ROW), column percentages (COL-

UMN), and overall percentages (TOTAL). The STATISTICS option asks for the chi-square test.

SPSS Output

```
                  EXAMPLE 4.4 - SEX BIAS IN PROMOTIONS       Page 1
SEX by STATUS
                     STATUS
           Count    |
         Row Pct    |
         Col Pct    |                          Row
         Tot Pct    | SUPER   |NONSUPER| Total
SEX   ----------------------------------------
         FEMALE     |    19   |    41   |    60
                    |  31.7   |  68.3   |  33.3
                    |  25.3   |  39.0   |
                    |  10.6   |  22.8   |
                     ------------------

         MALE       |    56   |    64   |   120
                    |  46.7   |  53.3   |  66.7
                    |  74.7   |  61.0   |
                    |  31.1   |  35.6   |
                     ------------------

           Column        75       105       180
            Total      41.7      58.3     100.0

Chi-Square                      Value     DF            Significance
----------                     --------   ---           ------------
Pearson                        3.70286     1               .05432
Continuity Correction          3.11143     1               .07774
Likelihood Ratio               3.76815     1               .05224

Minimum Expected Frequency - 25.000
Number of Missing Observations: 0
```

Discussion of SPSS Output

Each cell of the 2×2 table contains four numbers, which are identified in the key in the upper left corner. These are Count (the number of observations in the cell), Row Pct (count as a percentage of the number in the row), Col Pct (count as a percentage of the number in the column), and Tot Pct (count as a percentage of total sample size). The percentage of supervisors among females (P_1) is shown in the upper left cell to be 31.7 percent (a *proportion* of approximately 0.32, as computed by hand in the previous section). In the lower left cell we see that the percentage of supervisors among males (P_2) is 46.7 percent (a proportion of about 0.47, as computed by hand). Since $P_1 < P_2$, the sample results are in the direction predicted by our alternative hypothesis (H_1: $\pi_1 < \pi_2$). We will be able to reject the null hypothesis if the one-tailed p-value is less than 5 percent.

The information below the table is produced by the request STATISTICS = CHISQ. The value 3.70286 shown for the Pearson Chi-Square on the first line is the square of our Z value (−1.9243, with a bit more accuracy than presented above). Its two-tailed p-value of 0.05432 (labeled Significance) is double the one-tailed p-value of 0.0274 we found for our Z statistic, since the chi-square test is automatically two-sided.

For better accuracy we should examine the continuity corrected form of chi-square, presented in the second line of this table. Its two-tailed p-value of 0.07774 converts to a one-tailed p-value of 0.07774/2 = 0.03887. Since this is less than 5 percent, we reject the null hypothesis and conclude that this corporation is guilty of sex bias in promotions.

MINITAB Transcript

```
MTB > NAME C1 = 'SEX'
MTB > NAME C2 = 'STATUS'
MTB > READ 'SEX' 'STATUS'
DATA > 1 1
  .
  .  (19 TIMES)
  .
DATA > 1 2
  .
  .  (41 TIMES)
  .
DATA > 2 1
  .
  .  (56 TIMES)
  .
DATA > 2 2
  .
  .  (64 TIMES)
  .
DATA > END
```

```
     180 ROWS READ

MTB > TABLE 'SEX' 'STATUS';
SUBC > COUNTS;
SUBC > ROWPERCENTS;
SUBC > COLPERCENTS;
SUBC > TOTPERCENTS;
SUBC > CHISQUARE.
ROWS:      SEX          COLUMNS:     STATUS
                 1             2          ALL


  1             19            41           60
             31.67         68.33       100.00
             25.33         39.05        33.33
             10.56         22.78        33.33
                19            41           60


  2             56            64          120
             46.67         53.33       100.00
             74.67         60.95        66.67
             31.11         35.56        66.67
                56            64          120

                 1             2          ALL


 ALL            75           105          180
             41.67         58.33       100.00
            100.00        100.00       100.00
             41.67         58.33       100.00
                75           105          180

CHI-SQUARE = 3.703    WITH D.F. = 1

    CELL CONTENTS—
                       COUNT
                       % OF ROW
                       % OF COL
                       % OF TBL
                       COUNT

MTB > CDF 3.703 K1;
```

```
SUBC > CHISQUARE 1.
MTB > LET K1 = (1-K1)/2
MTB > PRINT K1
K1       0.0271581

MTB > STOP
```

Discussion of MINITAB Transcript

The first two columns on the worksheet are defined to be variables named SEX and STATUS, respectively. There is one line of input per observation, resulting in 180 lines of data for this example. For SEX, the value 1 is used to represent female, and 2 for male. For STATUS, the value 1 represents supervisor and 2 represents non-supervisor. The first 19 lines of data have values of 1 for both variables (female supervisors), the next 41 lines have 1 for SEX and 2 for STATUS (female non-supervisors), the next 56 lines have values 2 and 1 (male supervisors), and the last 64 lines have values 2 and 2 (male non-supervisors). Although the data are organized by group in our transcript, these 180 lines may be in any order. The program simply counts the number of observations in each cell to construct its 2×2 table.

The TABLE command is used to request the 2×2 table of SEX by STATUS. The first four subcommands ask for specific items to be included in each cell, and the last subcommand requests the chi-square test.

Each cell of the 2×2 table contains four numbers, which are identified in the key below the table. These are COUNT (the number of observations in the cell), % OF ROW (count as a percentage of the number in the row), % OF COL (count as a percentage of the number in the column), and % OF TBL (count as a percentage of total sample size). The percentage of supervisors among females (P_1) is shown in the upper left cell to be 31.67 percent (a *proportion* of approximately 0.32, as computed by hand in the previous section). In the lower left cell we see that the percentage of supervisors among males (P_2) is 46.67 percent (a proportion of about 0.47, as computed by hand). Since $P_1 < P_2$, the sample results are in the direction predicted by our alternative hypothesis (H_1: $\pi_1 < \pi_2$). We will be able to reject the null hypothesis if the one-tailed p-value is less than 5 percent.

The CHI-SQUARE value of 3.703 shown below the table is the square of our Z value (−1.9243, with a bit more accuracy than presented above). Chi-square has one degree of freedom when used to compare two proportions. To compute the p-value, we use MINITAB's built-in CDF function, storing the area below 3.703 in K1. The area *above* 3.703 is the two-tailed p-value, and halving this we obtain the one-tailed p-value of 0.0271581. Since this is less than 5 percent, we reject the null hypothesis and conclude that this corporation is guilty of sex bias in promotions.

4.15 SUMMARY

This chapter has introduced two-sample tests for comparing population means, variances, and proportions. For testing on means, two different experimental designs were considered, the first employing independent samples and the second based on paired data. Several testing procedures were discussed for each type of design.

For independent samples, we began the topic of comparing means with a Z statistic, appropriate when the population variances are known. For the more realistic case of unknown variances, t tests were presented, although different t ratios are employed, depending on whether or not the variances can be considered equal. When the variances are equal, a pooled t test is used, based on estimating the variance with a weighted average of the two sample variances. This test is valid if the sample variances are nearly equal, or if the sample sizes are large or nearly equal. The unequal variance t test uses separate estimates of the two population variances and is generally not quite as powerful as the pooled t test.

For data from independent samples that do not appear to be normally distributed, two nonparametric procedures were discussed: the rank sum (or U) test and the median test. The former requires identically shaped distributions in the two populations, while the latter does not. When normality is a reasonable assumption, the t tests are more powerful than the rank sum test, which is more powerful than the median test.

Random assignment to groups is necessary to draw a causal conclusion—that is, when experimental units are randomly divided into two groups, a difference between means can be attributed to differential treatment of the two groups. This is the crux of internal validity, as opposed to external validity, which is the ability to generalize conclusions from the sample to the population.

Paired data is treated by constructing a difference score for each paired observation. In the case of normality, the two means can be compared using a paired t test, which is a one-sample t test on the difference scores. In the non-normal case, nonparametric tests for a single sample are applied, specifically the signed rank and sign tests. Statistical tests based on paired data may be more powerful than those using independent samples, since variability related to extraneous factors is eliminated in computation of difference scores. Research employing independent samples can be analyzed with paired data techniques, if data are first paired on the basis of some other relevant variable. When the matched pairs are formed on a logical basis, the paired test is likely to be more powerful than the corresponding two-sample test.

The F distribution was introduced in this chapter, and used to test for equality of two population variances. This test is not ideal for deciding which t test (pooled or unequal variance) to apply for comparing means. The F test is weak with small samples, when we should be concerned about the equal variance assumption of the pooled t test. But it is powerful with large samples, when we are less concerned about the equality of variances.

A Z test for equality of two proportions was introduced, although most computer programs print the square of this Z statistic as a chi-square test. SAS, SPSS, and MINITAB computer programs were used to demonstrate all testing procedures discussed in this chapter.

5

Analysis of Variance

The past two chapters presented tests appropriate for one and two samples. Now we consider experimental designs involving several samples, and discuss tests for comparing the means of the corresponding populations. Only two commonly used statistical techniques exist for this purpose—one parametric (assuming normality) and the other nonparametric. The parametric approach, to which we devote considerable attention, has the strange name **one-way analysis of variance**. Despite the fact that we are comparing *means*, it is the analysis or breaking down of *variance* that enables us to perform this comparison.

Analysis of variance, frequently abbreviated **anova,** is used in a large number of statistical procedures, including more complex experimental designs and prediction situations. In each application, the overall variability in the data is partitioned into components reflecting the various influences in the experiment that may lead to differences among observations. Tests are conducted to see if these influences have a substantial effect on the data. In a one-way design, the samples are interpreted as different levels of a single classification variable, and a single test is performed to assess the effect of this classification variable on the results of the experiment.

5.1 HYPOTHESES AND ASSUMPTIONS

One-way analysis of variance is the extension of the pooled t test to several samples. Both the hypotheses and assumptions of the pooled t test are generalized to the case of several groups.

The pooled t test compares two means and tests the null hypothesis

$$H_0: \mu_1 = \mu_2.$$

We use the symbol $\boldsymbol{a}$ to represent the number of groups in a one-way study. The logical extension of the null hypothesis is

$$H_0: \mu_1 = \mu_2 = \mu_3 = \ldots = \mu_a.$$

The alternative hypothesis is the opposite of this null hypothesis. The opposite of saying that several quantities are all equal is to say that they are *not all equal*, that at least one of them is different. (The opposite is *not* that *all* the means are different.) Accordingly, we write our alternative hypothesis simply as

$$H_1\text{: the means are not all equal.}$$

In the pooled t test, we assume that independent samples are drawn from two normal distributions with equal variances. In a one-way design, the observations represent independent samples from several normal distributions, all with the same variance. As in the case of the pooled t test, the normality assumption may be largely ignored in practical applications, unless plots of the separate samples reveal distinctly abnormal shapes. The equal variance requirement is obviated when sample sizes are large or nearly equal. Although procedures exist for testing the equality of several variances prior to performing the one-way analysis of variance, they are usually not employed in practice. As in the case of the F test for comparing two variances, these procedures are powerful only with large samples, precisely when they are needed least.

As with the pooled t test, one-way analysis of variance is most powerful when sample sizes are equal. Equal sample sizes thus achieve two goals: they minimize the need for equality of variances, and they maximize power. In general, it is good experimental practice to use equal sample sizes whenever possible. Throughout most of the discussion in this chapter, we assume equal sample sizes, with n (lower case) observations per group. When each of the $\boldsymbol{a}$ groups has n observations, the total sample size, denoted by N (upper case), is $N = an$. Of course, an experiment may be performed with unequal sample sizes, and all standard computer programs can handle either equal or unequal sample sizes.

Suppose all the data from a one-way study are plotted on a single axis, with a separate curve drawn over the observations from each sample. According to the assumptions stated above, the data should appear as shown in Figure 5.1. Each sample resembles the normal shape, and the curves all have the same spread, but the locations may be different. When the null hypothesis is true, however, the means are equal and the curves are all coincident. We may envision our test graphically as a decision procedure for determining if different locations are necessary, or whether a single curve can be used to describe all the data.

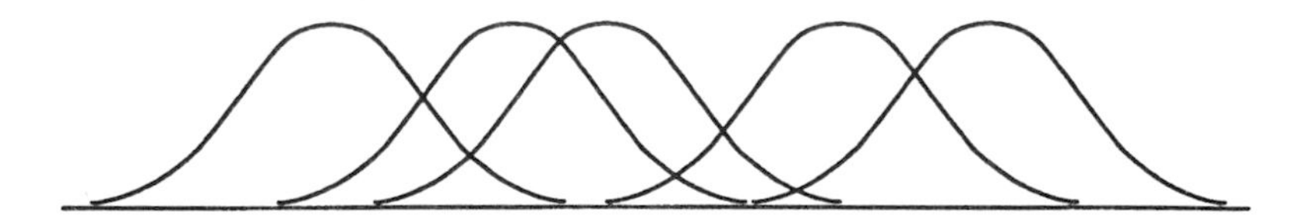

FIGURE 5.1. Data in one-way design.

5.2 PARTITIONS OF SUMS OF SQUARES AND DEGREES OF FREEDOM

In statistics, a **model** is a representation of the influences acting on individual observations. In the case of data from several groups, each observation reflects two influences. One factor is the population from which it is sampled. If the population means are not all equal, observations from different samples are expected to be different. Even within a single sample, however, we do not expect all observations to be identical due to variability in each population. In short, an observation has a particular value because of the influence of its group and because of random variability within that group.

The analysis of variance into components parallels the model. In a one-way design, **total variability** among all observations is partitioned into two components: variability **between groups** and variability **within groups**. In fact, the overall variance is not directly broken into two pieces, but instead separate partitions are performed of its numerator and denominator.

As we learned in Chapter 2, a sample variance is constructed by computing a sum of squared deviations form the average, and then dividing by the number of observations minus one:

$$S^2 = \frac{\Sigma(X-\overline{X})^2}{n-1}.$$

Clearly, the numerator of this ratio measures the actual spread of the data. But it also reflects sample size. We would expect a large sample to have a larger numerator than a small sample, because we would be summing over more terms. The denominator adjusts for the number of observations. As a result, the ratio S^2 is a good estimate of population variance, regardless of sample size.

The numerator of a sample variance is a **sum of squares** (SS), and the denominator is its degrees of freedom (df). The ratio of SS to df is frequently called a **mean square** (MS) and is written as MS = SS/df. That is, a mean square is the same as a sample variance.

Overall variability in a one-way design is measured by computing a sample variance among all observations, as if all the data were from a single sample. We refer to the numerator of this variance as SS_{TOTAL} and the denominator as df_{TOTAL}. We could write this overall sample variance as

$$MS_{TOTAL} = \frac{SS_{TOTAL}}{df_{TOTAL}} .$$

Since the total number of observations is N, $df_{TOTAL} = N - 1$. Instead of breaking this variance into two pieces, separate partitions are constructed for its numerator and denominator.

In the numerator, SS_{TOTAL} reflects the overall spread among all observations. This overall spread is partitioned into two kinds of spread: spread among the group averages, and spread within each group. The spread among group averages is called $SS_{BETWEEN}$, and the spread within groups is called SS_{WITHIN}. Thus,

$$SS_{TOTAL} = SS_{BETWEEN} + SS_{WITHIN}.$$

Figure 5.2 displays this partition graphically. The total spread, represented by the arrow underneath the axis, corresponds to overall variability among all the observations. At the top of each curve is a vertical line located at the average for that group. The spread among these group averages is the variability between groups. Arrows inside each curve refer to spread within each sample (i.e., within group variability). As shown in the equation above, total spread is the sum of between group spread and within group spread.

In the denominator of the overall variance ratio is $df_{TOTAL} = N - 1$. This term is also partitioned into pieces corresponding to between and within groups, called $df_{BETWEEN}$ and df_{WITHIN}. In the next few paragraphs we will discuss the logic for each of these df components, but for now we simply state that $df_{BETWEEN} = \boldsymbol{a} - 1$ and $df_{WITHIN} = N - \boldsymbol{a}$. Putting the pieces together, we can write

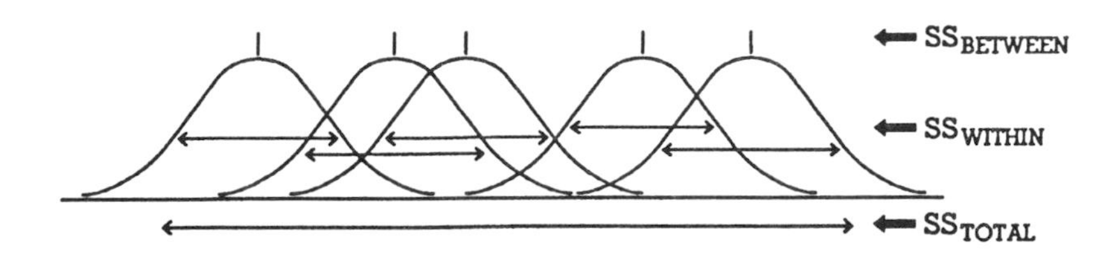

FIGURE 5.2. Partition of sum of squares.

$$df_{TOTAL} = df_{BETWEEN} + df_{WITHIN},$$

or, equivalently,

$$N - 1 = (\boldsymbol{a} - 1) + (N - \boldsymbol{a}).$$

Two new mean squares can now be written:

$$MS_{BETWEEN} = \frac{SS_{BETWEEN}}{df_{BETWEEN}},$$

and

$$MS_{WITHIN} = \frac{SS_{WITHIN}}{df_{WITHIN}}.$$

We note that separate partitions of the numerator and denominator of total variability do not result in a single partition for the whole ratio—that is, MS_{TOTAL} is not equal to the sum of $MS_{BETWEEN}$ and MS_{WITHIN}. This is because when we add fractions, we do not add their numerators and denominators separately.

The between group component reflects variability among the group means. Suppose the average is computed for each of the $\boldsymbol{a}$ groups, and these averages are used as the "observations" in calculating a sample variance. It can be shown that $MS_{BETWEEN}$ is simply n times this sample variance, where n is the number of observations per group. A sample variance computed using the group averages as "observations" has in its denominator the number of "observations" minus one, that is $\boldsymbol{a} - 1$. Therefore, the denominator of $MS_{BETWEEN}$ is $\boldsymbol{a} - 1$, which is $df_{BETWEEN}$.

In the case of unequal sample sizes, $MS_{BETWEEN}$ is a *weighted* variance of the group averages. Whether the sample sizes are equal or unequal, $MS_{BETWEEN}$ serves as a measure of how spread out the group averages are. In the pooled t test, we can measure spread between the group averages by subtracting $(\overline{X}_1 - \overline{X}_2)$. With more than two groups, however, we cannot measure spread by simple subtraction; we must compute a sample variance among the group averages. The more spread out the sample means are, the bigger their sample variance is.

The interpretation of MS_{WITHIN} is also straightforward. It is a generalization of the pooled variance estimate used in the pooled t test. The pooled t test assumes equal variances in two populations, and uses a weighted average of the two sample variances to estimate this common variance. In the one-way design, we assume equal variances in all $\boldsymbol{a}$ populations, and form a pooled variance estimate by taking the weighted average of all $\boldsymbol{a}$ sample variances. Again, we use degrees of freedom (i.e., sample sizes minus one) as weights. Generalizing the pooled variance estimate shown in (4.2.2) for the pooled t test, we can write MS_{WITHIN} as

$$MS_{WITHIN} = \frac{(n-1)\,S_1^2 + (n-1)\,S_2^2 + (n-1)\,S_3^2 + \ldots + (n-1)\,S_a^2}{(n-1) + (n-1) + (n-1) + \ldots + (n-1)} . \tag{5.2.1}$$

Since a mean square is a sum of squares divided by its degrees of freedom, the numerator above is equivalent to SS_{WITHIN}, and the denominator is df_{WITHIN}. In the denominator, we add the quantity (n - 1) exactly $\boldsymbol{a}$ times, yielding $\boldsymbol{a}(n-1) = \boldsymbol{a}n - \boldsymbol{a} = N - \boldsymbol{a}$. Thus, $df_{WITHIN} = N - \boldsymbol{a}$.

In the case of unequal sample sizes, the right side of (5.2.1) is replaced with the appropriate weighted average, using $(n_1 - 1)$ as the multiplier of S_1^2, $(n_2 - 1)$ as the multiplier of S_2^2, and so forth. The sum of weights, which appears in the denominator, is still $N - \boldsymbol{a} = df_{WITHIN}$. If the sample sizes are equal, as shown in (5.2.1), the weighted average reduces to a regular average—that is, MS_{WITHIN} is simply the average of the sample variances in the separate groups. In either case, MS_{WITHIN} is a pooled estimate of the variability within each group.

These two mean squares, $MS_{BETWEEN}$ and MS_{WITHIN}, are the crucial elements in the test of our null hypothesis. The first is a measure of variability among group averages, while the second is a pooled estimate of variability within groups. As we will see in the next section, their ratio is the test statistic used in one-way analysis of variance.

5.3 ANOVA TEST STATISTIC

In Section 4.10, we learned that the ratio of two variance-type terms has an F distribution. For testing the null hypothesis in one-way anova, the test statistic is

$$F = \frac{MS_{BETWEEN}}{MS_{WITHIN}} . \tag{5.3.1}$$

An F distribution has two separate df parameters, associated respectively with the numerator and denominator of the ratio. For this ratio, the numerator degrees of freedom are $df_{BETWEEN} = \boldsymbol{a} - 1$, and the denominator degrees of freedom are $df_{WITHIN} = N - \boldsymbol{a}$.

Our null hypothesis states that the means of all $\boldsymbol{a}$ populations are identical. When the null hypothesis is true, we expect the averages in our $\boldsymbol{a}$ samples to be similar, and their sample variance to be small. This leads to a small value for $MS_{BETWEEN}$ and, consequently, to a small F ratio. When the null hypothesis is true, the average value of the F ratio should be near one. On the other hand, if the alternative hypothesis is true, some of the sample means should be rather different from others. Their variance should be large, producing a large value of $MS_{BETWEEN}$ and, therefore, a large F ratio (substantially larger than one). Logically, if the computed F ratio is large, we are inclined to believe that the alternative hypothesis is true and should reject the null. Accordingly, we locate all 5 percent of our rejection region in the upper tail of the F distribution. Recall that the F table shows only upper tail cutoffs.

Example 5.1 has $\boldsymbol{a} = 5$ groups, with $n = 4$ observations per group, for a total of $N = 20$ observations. The degrees of freedom for our F test are $df_{BETWEEN} = \boldsymbol{a}-1 = 4$ in the numerator, and $df_{WITHIN} = N-\boldsymbol{a} = 15$ in the denominator. From Table 4a, we find the 5% cutoff of 3.06. We will reject the null hypothesis if the computed F value exceeds this critical value.

One-way analysis of variance involves a good deal of computation. In addition to sample means and sample variances in each group, we must also calculate the variance among the sample means, and a pooled estimate of within group variance. Fortunately, computers do all the work for us. They summarize results in the form of a table, with columns labeled for source of variation (between, within, and total), degrees of freedom, sums of squares, mean squares, and the final F ratio. The results for Example 5.1 are reported in the anova summary table shown in Table 5.1. MS_{TOTAL} is not usually included in the mean square column, as it is not used in construction of the F ratio. But all other terms discussed above are presented in this table.

Because the obtained F value of 15.758 exceeds the critical value of 3.06, we can reject the null hypothesis. We conclude that some differences exist among

Example 5.1

Five diffferent methods were compared for their abilities to retard spoilage in margarine. Two methods involved innovative processing techniques (proc. 1, proc. 2), while two used chemical additives (chem. 1, chem. 2). The fifth sample, without special treatment to retard spoilage, was used as a control. For each of these five methods, four specimens were prepared and stored in controlled conditions. Data below represent the degree of spoilage after a uniform time interval.

Group Method	1 proc. 1	2 proc. 2	3 chem. 1	4 chem. 2	5 control
	28	30	7	23	52
	37	19	16	23	42
	43	20	23	30	38
	31	18	11	20	54

TABLE 5.1 Anova Summary Table for Comparing Methods to Retard Spoilage in Margarine

Source	df	SS	MS	F
Between	4	2526.50	631.625	15.758
Within	15	601.25	40.083	
Total	19	3127.75		

these five groups in the degree of spoilage of margarine. Obviously, this conclusion is of limited utility. For practical application, we must be able to determine *which* groups differ. We will return to this issue in Section 5.5.

Before leaving the topic of the F test in anova, we should consider the issues of statistical significance and practical importance first raised in Section 3.10. Statistical significance addresses the question "How *sure* are we that H_0 is wrong?" The answer to this question is measured by how far out in the tail the test statistic falls. The smaller the p-value, the more convinced we are that H_0 is false. Computer programs generally print the obtained p-value. For Example 5.1, the tail area is approximately $p = 0.0001$. We are very sure these five groups do not all have the same degree of spoilage.

Practical importance, on the other hand, addresses the question "How *wrong* is H_0?" Our answer to this question must be based on some quantity that is independent of sample size, and preferably scale-free. In one-way anova, we assess practical importance with the **coefficient of determination,** denoted by R^2, and given by

$$R^2 = \frac{SS_{BETWEEN}}{SS_{TOTAL}}. \tag{5.3.2}$$

For Example 5.1, the value of R^2 is

$$R^2 = \frac{SS_{BETWEEN}}{SS_{TOTAL}} = \frac{2526.50}{3127.75} = 0.81.$$

Since R^2 is the ratio of one component of variability to total variance, it cannot be less than 0, nor greater than 1. R^2 is the proportion of total variability related to differences among groups. In our example, 81 percent of the variability in spoilage measurements is associated with differences among the five methods in their abilities to retard spoilage. Random fluctuation within groups accounts for the other 19 percent of total variability. From a practical standpoint, the method of treatment has a great influence on margarine spoilage.

The coefficient of determination satisfies all the desirable requirements for a measure of practical importance. It has a direct interpretation in terms of proportion of variance accounted for, it is insensitive to sample size, and it is scale-free. In addition, its values are restricted to the simple interval from 0 to 1. The higher the value of R^2, the greater the practical importance of our results.

Standards for values of R^2 depend on the field of study. In engineering and the physical sciences, where variability among items of the same type is relatively small, values of R^2 in the 80% to 90% range are not uncommon. Research in human behavior or the social sciences, on the other hand, rarely produces values of R^2 much above 30 percent. This is because people and social processes are subject to many influences that may cause individual differences. A relatively small proportion of this variability can be accounted for by any single factor.

5.4 ALTERNATIVE TESTING PROCEDURES

The nonparametric alternative to the F test is the **Kruskal-Wallis one-way analysis by ranks**. Just as the F ratio is an extension of the pooled t test, the Kruskal-Wallis statistic is an extension of the rank sum test (discussed in Section 4.4). Data for all samples are combined and ranked as a single group. Average ranks in each group are computed, and the variability among these average ranks is the key element in the Kruskal-Wallis statistic. For moderate to large samples, the distribution of the Kruskal-Wallis statistic is approximately the same as a chi-square distribution with $a - 1$ degrees of freedom, where a is the number of groups in the design. By the same logic used for the F test, the rejection region is located in the upper tail of the chi-square distribution.

Similar to the rank sum test, the Kruskal-Wallis test assumes that each population has the same shape, although it does not require that shape to be normal. It also assumes that the distributions are continuous, which technically prohibits ties. If ties occur in the actual data, mid-ranks are employed and a correction factor is incorporated in the Kruskal-Wallis statistic.

Omitting the actual computations, the data of Example 5.1 result in a value of 15.11 for the Kruskal-Wallis statistic, after correcting for ties. Our example involves $a = 5$ groups, and therefore df $= a - 1 = 4$. From Table 3b, we find the critical chi-square value of 9.488. Since the obtained test statistic exceeds this figure, we reject the null hypothesis. As with the F test, our conclusion is limited to stating that the five methods are not equal in retarding spoilage of margarine.

While the Kruskal-Wallis test is the only nonparametric procedure generally used for one-way designs, an alternative *parametric* technique exists if the experiment includes only two groups. The null hypothesis, H_0: $\mu_1 = \mu_2$, can be tested with either a pooled t test or the anova F test. Would different conclusions be reached? Fortunately, the answer is no. It can be shown that the F scale is the square of the t scale, and identical decisions must result. The test statistics and critical values are both related by the squaring rule. The obtained F ratio is the square of the pooled t statistic, and the critical F value is also the square of the critical t.

We can see the relationship between critical values by considering an example. Suppose we have n = 10 observations in each of two groups. For a pooled t test, the degrees of freedom are $n_1 + n_2 - 2 = 18$. A two-sided test at the 5% level results in rejecting the null hypothesis if |t| is greater than 2.101, as shown in Figure 5.3(a). If a one-way analysis of variance is performed, the degrees of freedom for the F test are $\boldsymbol{a} - 1 = 2 - 1 = 1$ in the numerator, and $N-\boldsymbol{a} = 20 - 2 = 18$ in the denominator. From Table 4a, we find the critical F value of 4.41, as shown in Figure 5.3(b). Note that $4.41 = (2.101)^2$. That is, the critical F value is the square of the critical t value. The same relationship holds for the observed test statistics. The computed F value is the square of the observed t. Therefore, the two tests must reach the same conclusion. In essence, the t distribution is "folded" at the origin to produce the F

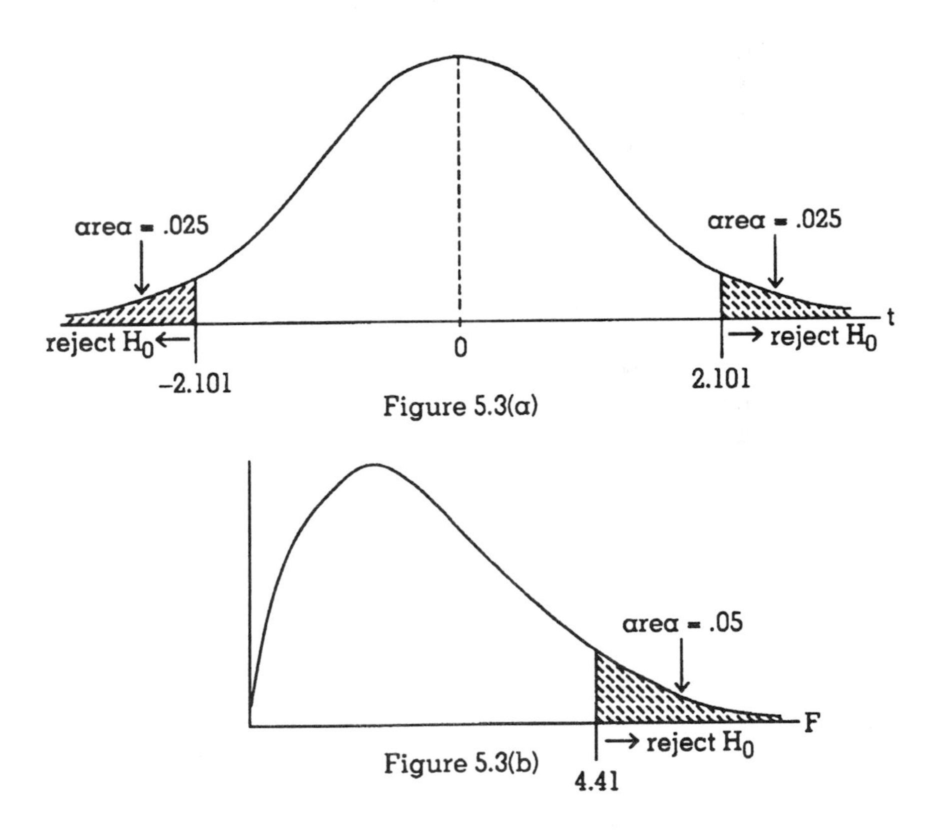

FIGURE 5.3. Rejection regions for t and F tests.

distribution. Both tails of the rejection region for the t wind up in the upper tail of the F curve.

In general, the square of a t statistic can be interpreted as an F ratio with $df_{num} = 1$, and df_{den} equal to the degrees of freedom of the t test. Equivalence of rejection regions, however, holds only for two-sided t tests. With a one-sided test, the rejection region for the t test is located in only one tail and does not satisfy the squaring rule. To perform a one-sided test with an F statistic, we must *halve* the obtained p-value to obtain the area in only one tail of the t distribution. Because of this equivalence, different computer programs sometimes present results of certain tests differently, one printing a t test and another employing an F test.

Logically, the t and F statistics are rather similar. The numerator of each ratio pertains directly to the null hypothesis, while the denominator is a measure of random error. For the t test, the numerator is $(\overline{X}_1-\overline{X}_2)$, which relates directly to the null hypothesis $(\mu_1 = \mu_2)$. The key term in the denominator is the pooled variance estimate S^2_p. For the F test, the numerator is $MS_{BETWEEN}$, a function of variability among sample averages. When the null hypothesis is true, this term should be small, and when H_0 is false, it should be large. The denominator of the F is MS_{WITHIN}, a pooled variance estimate.

This same theme—numerator directly related to the null hypothesis, denominator a measure of random error—is evident in most parametric tests. For example, a one-sample t statistic has $\overline{X}-\mu$ in the numerator, where μ is the hypothesized mean, and $S/\sqrt{n}$, the standard error of $\overline{X}$, in the denominator. Thus, the numerator pertains directly to the null hypothesis (it measures how close $\overline{X}$ is to the hypothesized mean), while the denominator is a measure of random variability. In later chapters, we will see other test statistics presented as ratios that can be interpreted in this same common sense manner.

5.5 FISHER'S LEAST SIGNIFICANT DIFFERENCE

In Example 5.1, we are able to conclude that the five methods studied do not all retard spoilage equally. But the F test does not indicate *which* groups differ. In this section, we discuss techniques for determining where the differences occur.

One approach to this problem is to perform pooled t tests among pairs of groups. However, we can improve on the usual pooled t test, and increase its power, by employing a variance estimate based on more information. In the regular pooled t test, the denominator contains the pooled variance estimate, S^2_p. This estimate, as shown in (4.2.2), is the weighted average of sample variances in the two groups whose means we are comparing. But in one-way anova, we require that all ***a*** groups have the same variance, and there is no reason to limit our estimate of this common variance to information from only two groups. We can replace S^2_p in the denominator of our t statistic with MS_{WITHIN}, the pooled variance estimate based on all ***a*** sample variances.

Recall that degrees of freedom for a t statistic are the same as the degrees of freedom of the variance estimate used in its denominator. The estimate S_p^2 has $n_1 + n_2 - 2$ degrees of freedom. In Example 5.1, each group has four observations, and S_p^2 has $4 + 4 - 2 = 6$ degrees of freedom, no matter which pair of means is compared. MS_{WITHIN}, on the other hand, has $N\text{-}a = 20 - 5 = 15$ degrees of freedom. Consequently, a t statistic employing MS_{WITHIN} to estimate variance has more degrees of freedom than one using S_p^2. More degrees of freedom is equivalent to more observations—it results in greater power.

Suppose we use this approach to compare two of the means in Example 5.1. Letting μ_i and μ_j represent any two of the means in our example, we wish to test H_0: $\mu_i = \mu_j$ against H_1: $\mu_i \neq \mu_j$. Replacing S_p^2 with MS_{WITHIN} in the pooled t statistic shown in (4.2.3), we can write

$$t = \frac{(\overline{X}_i - \overline{X}_j)}{\sqrt{MS_{WITHIN}\left(\frac{1}{n_i} + \frac{1}{n_j}\right)}} = \frac{(\overline{X}_i - \overline{X}_j)}{\sqrt{40.083\left(\frac{1}{4} + \frac{1}{4}\right)}} = \frac{(\overline{X}_i - \overline{X}_j)}{4.477}.$$

This t ratio has 15 degrees of freedom because it uses MS_{WITHIN} to estimate variance. For a two-sided t test with 15 degrees of freedom, we would reject H_0 if $|t|$ is greater than 2.131. Since the denominator of the t ratio is positive, we would reject H_0 if

$$\frac{|\overline{X}_i - \overline{X}_j|}{4.477} > 2.131.$$

Multiplying both sides of this inequality by 4.477, we can state the decision rule: reject H_0 if

$$|\overline{X}_i - \overline{X}_j| > 2.131(4.477) = 9.540.$$

The quantity 9.540 is referred to as **Fisher's Least Significant Difference** (LSD). For any one-way anova, we need to compute LSD only once. No matter which pair of means we compare, H_0 is rejected if the absolute difference between the two group averages exceeds LSD. This is a perfectly logical decision procedure. We conclude that two population means differ if their estimates are far enough apart. The quantity LSD simply tells us *how far apart* the sample averages must be. By the argument above, the LSD procedure is equivalent to performing a pooled t test, except that it uses MS_{WITHIN} to estimate variance.

5.6 MULTIPLE RANGE PROCEDURES

Suppose we wish to compare *all possible pairs of means* in a one-way design. Fisher's LSD approach is one way to accomplish this task, but a large number of tests are involved. A **multiple range procedure** is a way of presenting the results of these tests in a simplified manner. The results are displayed in terms of a rank ordering of the sample means in the study. For Example 5.1, the ordered means are shown below.

Group	3	2	4	1	5
Average	14.25	21.75	24	34.75	46.5

A line is drawn under any pair of sample means for which we accept the null hypothesis of equality for the corresponding population means. If the means are not adjacent, an extended line is used. Lines connect means that are equal. For example, the line drawn below would indicate that $\mu_2 = \mu_4 = \mu_1$.

Group	3	2	4	1	5
Average	14.25	21.75	24	34.75	46.5

Each line drawn connects a range of means that are considered equal. More than one line, or multiple ranges, may be required to summarize the results of all possible pairwise comparisons. For example, if two separate lines are drawn, such as shown below, we conclude that $\mu_2 = \mu_4$ and $\mu_4 = \mu_1$, but $\mu_2 \neq \mu_1$, since the means for groups two and one are not connected by a single line.

Group	3	2	4	1	5
Average	14.25	21.75	24	34.75	46.5

In a strict sense, overlapping lines, such as those above, imply a logical inconsistency. If μ_2 is equal to μ_4, and μ_4 is equal to μ_1, it should follow that μ_2 is equal to μ_1. But this inference is not appropriate when using a multiple range procedure. Two small differences, each insignificant, add to a bigger difference, which may be large enough for significance.

When Fisher's LSD is used for all pairwise comparisons in Example 5.1, the results are summarized in the display below. Each pair of sample means connected by a line differs by less than LSD (9.540). Each pair not connected differs by more than this quantity.

Group	3	2	4	1	5
Average	14.25	21.75	24	34.75	46.5

The large number of tests involved in using Fisher's LSD in a multiple range procedure is cause for some concern. Each pairwise comparison is equivalent to a t test with a 5% chance of a type one error (i.e., rejecting H_0 when it is true). Fisher's LSD is said to use **comparisonwise** control of type one error rate, since each comparison has a 5% chance of producing a type one error. If a great many tests are performed, the probability of making at least one type one error is rather large. This probability does not accumulate additively (if it did, 21 tests would yield a 105% chance of at least one type one error), but it does increase with the number of tests. Simply stated, small risks taken repeatedly become big risks.

Other multiple range procedures have been developed to minimize this overall risk by controlling type one error rate in a different manner. **Tukey's Honestly Significant Difference** (HSD) approach controls type one error rate on an **experimentwise** basis. All possible pairwise comparisons may be performed, but the chance of *ever* making a type one error is 5 percent. Naturally, if the probability of *ever* making a type one error is 5 percent, the chance on any particular comparison is much less. The reduced probability of a type one error for individual tests is obtained by using a larger critical value. While we will not discuss the theory involved in determining this critical value, we can see the results in terms of Example 5.1. The value of HSD is 13.824, which is a much more stringent criterion for significance than LSD (9.540). The display below summarizes results of Tukey's HSD for Example 5.1. Each pair of sample means connected by a line differs by less than HSD. Each unconnected pair differs by more than HSD.

Group	3	2	4	1	5
Average	14.25	21.75	24	34.75	46.5

Comparison with the display for Fisher's LSD reveals substantial differences between these two approaches. For Tukey's HSD, many more means are connected by lines than for Fisher's LSD. Connected means are deemed equal. In brief, Tukey's HSD finds far *fewer* significant differences among pairs of means than Fisher's LSD.

Many researchers feel that Fisher's LSD approach is too liberal, while Tukey's HSD is too conservative. The first allows too much chance of making at least one type one error, but the second allows too little. An alternative approach, **Duncan's multiple range test**, lies somewhere between these two in its conservatism. The Duncan procedure does not use the same critical value for each pairwise comparison. Instead, different critical values are used, depending on how far apart in the rank ordering are the means being compared. For comparing adjacent means, the criterion for significance is the same as LSD. For means further apart, the critical value increases. The further apart in the rank ordering that the means being compared are, the larger the critical value is. But the critical value is never as great as HSD. The net result is that Duncan's procedure is somewhere between Fisher's LSD and Tukey's HSD in conservatism. The display below summarizes results of Duncan's multiple range procedure applied to the data of Example 5.1.

Group	3	2	4	1	5
Average	14.25	21.75	24	34.75	46.5

Note that Duncan's procedure finds one more pair of means equal (μ_3 and μ_4) than the LSD approach. But it still finds many more significant differences than the HSD method, since far fewer means are connected than with HSD. In general, the fewer means connected by lines, the greater is the number of significant differences among pairs of means, and the more powerful the testing procedure is. Keep in mind, however, that this power is achieved simply by allowing greater chance of type one error.

Fisher's LSD, Tukey's HSD, and Duncan's procedure are three of the most commonly employed methods for performing multiple range tests. But several other techniques exist. The SPSS program includes 4 additional techniques; SAS has a total of 17 multiple range tests. With such a welter of procedures available, the reader may rightly feel confused in deciding which approach to use. One guideline is to be aware of methods used in various areas of research. Different fields of research tend to allow different levels of risk. In areas where theories can be tested fairly precisely, such as the physical sciences or engineering, higher levels of risk may be tolerated, and the Duncan approach is acceptable. In fields where knowledge is more tentative and temporal, such as the social sciences, more conservatism is in order, and the HSD method is more commonly used.

5.7 LINEAR CONTRASTS

Tests for equality of *pairs* of means are summarized through multiple range procedures. Occasionally, however, questions about individual groups in a one-way design are more complex than whether a pair of means is equal. For instance, in Example 5.1, the first two methods for retarding spoilage are based on processing techniques, while the next two involve chemical additives. We might want to compare these two general approaches—processing and additives. More precisely, we might ask whether the average spoilage for groups one and two is the same as the average spoilage for groups three and four. The null hypothesis (that there is no difference between these averages) is

$$H_0\text{: } \frac{\mu_1 + \mu_2}{2} = \frac{\mu_3 + \mu_4}{2}.$$

Expanding each fraction and subtracting the two terms on the right side of this equation, this null hypothesis can be written

$$H_0\text{: } \left(\frac{1}{2}\right)\mu_1 + \left(\frac{1}{2}\right)\mu_2 + \left(-\frac{1}{2}\right)\mu_3 + \left(-\frac{1}{2}\right)\mu_4 = 0.$$

On the left side of this equation, each mean is multiplied by a coefficient, and the results are summed. As long as the coefficients add to 0, this combination is called a **linear contrast** among the means. In our example, the coefficients are 1/2, 1/2,−1/2, and−1/2, which sum to 0, as required. Our null hypothesis states that this linear contrast is equal to 0. The test of this hypothesis is based on an estimate of the contrast, computed as the same combination of the *sample* means. In our example, the estimated contrast is

$$\left(\frac{1}{2}\right)\overline{X}_1 + \left(\frac{1}{2}\right)\overline{X}_2 + \left(-\frac{1}{2}\right)\overline{X}_3 + \left(-\frac{1}{2}\right)\overline{X}_4$$

$$= \left(\frac{1}{2}\right)34.75 + \left(\frac{1}{2}\right)21.75 + \left(-\frac{1}{2}\right)14.25 + \left(-\frac{1}{2}\right)24$$

$$= 9.125.$$

The test of a contrast may be constructed as either a t test or an F test. As discussed in Section 5.4, the F test is simply the square of the t test. Omitting the details of computation, the obtained t statistic in this example is 2.883, while the F ratio is $(2.883)^2 = 8.31$. For the t test, the degrees of freedom are $df_{WITHIN} = 15$, while the F has $df_{num} = 1$ and $df_{den} = df_{WITHIN} = 15$. For a two-sided test, as might be appropriate in our example, the two-tailed p-value for the t statistic is identical to the one-tailed p-value for the F ratio. As will be seen in computer outputs shown in the next section, this p-value is 0.0114. Since this is less than 5 percent, we reject the null hypothesis. We learned earlier (Section 3.7) that significance in a two-sided test should be accompanied by the appropriate *one-sided* conclusion. Because the estimated contrast is positive (9.125), the average spoilage for the processing techniques is *greater* than that for the chemical additives.

To demonstrate testing for a one-sided question, let us contrast the average spoilage for methods one through four with the control group. The null hypothesis is

$$H_0: \frac{\mu_1 + \mu_2 + \mu_3 + \mu_4}{4} = \mu_5,$$

or, equivalently,

$$H_0: \left(\frac{1}{4}\right)\mu_1 + \left(\frac{1}{4}\right)\mu_2 + \left(\frac{1}{4}\right)\mu_3 + \left(\frac{1}{4}\right)\mu_4 + (-1)\mu_5 = 0.$$

Note again that the sum of coefficients in this linear combination is 0, as required.

Clearly, we expect the first four methods to have less spoilage, on average, than the control group. Our alternative hypothesis is

$$H_1: \frac{\mu_1 + \mu_2 + \mu_3 + \mu_4}{4} < \mu_5,$$

which becomes

$$H_1: \left(\frac{1}{4}\right)\mu_1 + \left(\frac{1}{4}\right)\mu_2 + \left(\frac{1}{4}\right)\mu_3 + \left(\frac{1}{4}\right)\mu_4 + (-1)\mu_5 < 0.$$

The estimated contrast is

$$\left(\frac{1}{4}\right)\overline{X}_1 + \left(\frac{1}{4}\right)\overline{X}_2 + \left(\frac{1}{4}\right)\overline{X}_3 + \left(\frac{1}{4}\right)\overline{X}_4 + (-1)\overline{X}_5$$

$$= \left(\frac{1}{4}\right)34.75 + \left(\frac{1}{4}\right)21.75 + \left(\frac{1}{4}\right)14.25 + \left(\frac{1}{4}\right)24 + (-1)46.5$$

$$= -22.8125.$$

This quantity is negative, as predicted by our alternative hypothesis. Since the result is in the proper direction, we should halve the obtained two-tailed p-value to see if the test statistic is far enough out in the tail to reject the null hypothesis. For this contrast, the t statistic is -6.446 and the F ratio is $(-6.446)^2 = 41.55$. The two-tailed p-value for the t statistic is identical to the one-tailed p-value for the F ratio. SAS output, presented in the next section, shows this p-value to be 0.0001. (The true p-value may be smaller than this, due to a restriction on the number of digits printed.) Halving this, we find the one-tailed p-value of 0.00005, substantially less than 5 percent. Accordingly, we are *very sure* that the first four methods, on average, produce less spoilage than the control group.

5.8 ANOVA IN COMPUTER PROGRAMS

Each of the programs—SAS, SPSS, and MINITAB—performs the one-way analysis of variance with either equal or unequal sample sizes. The Kruskal-Wallis test is also available in each program. SAS and SPSS include multiple range procedures and tests of linear contrasts, while MINITAB does not include these features. The data of Example 5.1 are used to demonstrate available options.

SAS Input

```
OPTIONS LS = 80;
TITLE 'EXAMPLE 5.1 - METHODS TO RETARD SPOILAGE';
DATA;
INPUT METHOD SPOILAGE;
LINES;
1    28
1    37
1    43
1    31
2    30
2    19
2    20
```

```
2   18
3    7
3   16
3   23
3   11
4   23
4   23
4   30
4   20
5   52
5   42
5   38
5   54
PROC ANOVA;
CLASS METHOD;
MODEL SPOILAGE = METHOD;
MEANS METHOD / LSD TUKEY DUNCAN;
PROC GLM;
CLASS METHOD;
MODEL SPOILAGE = METHOD;
CONTRAST 'PROC 1&2 VS CHEM 1&2' METHOD .5 .5 -.5 -.5 0;
CONTRAST 'METH. 1-4 VS CONTROL' METHOD .25 .25 .25 .25 -1;
PROC NPAR1WAY WILCOXON;
CLASS METHOD;
VAR SPOILAGE;
```

Discussion of SAS Input

For each observation, two variables are entered—METHOD (group number) and SPOILAGE. The input lines shown here are ordered by group, but this is not required.

The SAS procedure ANOVA performs analysis of variance and requires accompanying CLASS and MODEL statements. The CLASS statement identifies METHOD as the classification variable, and the MODEL statement reflects the influences on individual observations. As discussed in Section 5.2, two influences affect the data in one-way analysis of variance, group differences, and random variability within groups. On the left side of the equation in the MODEL statement, we name the variable to be analyzed (SPOILAGE). On the right side we list the influences acting on this variable, except that SAS automatically assumes random variability as an influence, and this term is not listed. Therefore, only the group influence remains, and this is referenced by the name of the grouping variable

(METHOD). Below these lines is a MEANS command, which asks the ANOVA procedure to print the means for each group (METHOD). Optional requests, following the slash, are for LSD, HSD (TUKEY), and DUNCAN multiple range procedures.

The ANOVA procedure does not handle linear contrasts, but such tests are available as an option to another SAS procedure, GLM (general linear models). GLM performs a wide variety of statistical analyses, including analysis of variance, regression, analysis of covariance, and multivariate analysis. Like ANOVA, GLM requires both CLASS and MODEL statements to perform an analysis of variance. Each linear contrast is requested on a CONTRAST statement, which includes a title (up to 20 characters, enclosed in single quotes), the name of the grouping variable (METHOD), and coefficients for each of the groups. Since our first contrast does not involve the control group, a coefficient of 0 is listed for the fifth group.

The NPAR1WAY procedure, demonstrated in Section 4.5 for the rank sum test, also performs the Kruskal-Wallis test. Recall that the Kruskal-Wallis test is an extension of the rank sum test to experiments with more than two samples. It is therefore requested with the same keyword (WILCOXON). CLASS and VAR statements are required to identify the classification variable and the variable to be analyzed.

SAS Output

```
           EXAMPLE 5.1 - METHODS TO RETARD SPOILAGE            1
               Analysis of Variance Procedure

                   Class Level Information

              Class        Levels        Values
              METHOD          5          1 2 3 4 5

Number of observations in data set = 20
```

```
             EXAMPLE 5.1 - METHODS TO RETARD SPOILAGE                    2
                 Analysis of Variance Procedure
Dependent Variable: SPOILAGE
                               Sum of            Mean
Source               DF       Squares          Square     F Value    Pr > F
Model                 4  2526.5000000     631.6250000       15.76    0.0001
Error                15   601.2500000      40.0833333
Corrected Total      19  3127.7500000
```

R-Square	C.V.	Root MSE	SPOILAGE Mean
0.807769	22.41111	6.3311400	28.250000

Source	DF	Anova SS	Mean Square	F Value	Pr > F
METHOD	4	2526.5000000	631.6250000	15.76	0.0001

EXAMPLE 5.1 - METHODS TO RETARD SPOILAGE 3
Analysis of Variance Procedure

T tests (LSD) for variable: SPOILAGE
NOTE: This test controls the type I comparisonwise
error rate, not the experimentwise error rate.

Alpha = 0.05 df = 15 MSE = 40.08333
Critical Value of T = 2.13
Least Significant Difference = 9.5421

Means with the same letter are not significantly different.

T Grouping		Mean	N	METHOD
	A	46.500	4	5
	B	34.750	4	1
	C	24.000	4	4
	C			
D	C	21.750	4	2
D				
D		14.250	4	3

EXAMPLE 5.1 - METHODS TO RETARD SPOILAGE 4
Analysis of Variance Procedure

Tukey's Studentized Range (HSD) Test for variable: SPOILAGE
NOTE: This test controls the type I experimentwise
error rate, but generally has a higher type II error
rate than REGWQ.

Alpha = 0.05 df = 15 MSE = 40.08333
Critical Value of Studentized Range = 4.367
Minimum Significant Difference = 13.824

Means with the same letter are not significantly different.

```
Tukey Grouping          Mean        N      METHOD

             A        46.500        4         5
             A
      B      A        34.750        4         1
      B
      B      C        24.000        4         4
      B      C
      B      C        21.750        4         2
             C
             C        14.250        4         3
```

```
        EXAMPLE 5.1 - METHODS TO RETARD SPOILAGE              5
              Analysis of Variance Procedure

      Duncan's Multiple Range Test for variable: SPOILAGE
    NOTE: This test controls the type I comparisonwise error
           rate, not the experimentwise error rate.

           Alpha = 0.05   df = 15   MSE = 40.08333
       Number of Means      2       3        4       5
       Critical Range     9.53    9.99    10.31   10.49

  Means with the same letter are not significantly different.

  Duncan Grouping         Mean        N      METHOD

                A       46.500        4         5

                B       34.750        4         1

                C       24.000        4         4
                C
                C       21.750        4         2
                C
                C       14.250        4         3
```

```
        EXAMPLE 5.1 - METHODS TO RETARD SPOILAGE              6
             General Linear Models Procedure
                 Class Level Information

             Class       Levels      Values
             METHOD         5        1 2 3 4 5

          Number of observations in data set = 20
```

```
              EXAMPLE 5.1 - METHODS TO RETARD SPOILAGE                7
                  General Linear Models Procedure

Dependent Variable: SPOILAGE
                                   Sum of          Mean
Source                  DF        Squares        Square    F Value    Pr > F
Model                    4   2526.5000000   631.6250000      15.76    0.0001
Error                   15    601.2500000    40.0833333
Corrected Total         19   3127.7500000

R-Square           C.V.             Root MSE           SPOILAGE Mean
0.807769       22.41111            6.3311400               28.250000

Source      DF      Type I SS     Mean Square      F Value      Pr > F
METHOD       4   2526.5000000     631.6250000        15.76      0.0001

Source      DF    Type III SS     Mean Square      F Value      Pr > F
METHOD       4   2526.5000000     631.6250000        15.76      0.0001

Contrast                DF   Contrast SS    Mean Square  F Value   Pr > F
PROC 1&2 VS CHEM 1&2     1   333.0625000    333.0625000     8.31   0.0114
METH. 1-4 VS CONTROL     1  1665.3125000   1665.3125000    41.55   0.0001
```

```
           EXAMPLE 5.1 - METHODS TO RETARD SPOILAGE              8
               N P A R 1 W A Y   P R O C E D U R E

       Wilcoxon Scores (Rank Sums) for Variable SPOILAGE
                 Classified by Variable METHOD

                  Sum of      Expected       Std Dev          Mean
METHOD     N      Scores      Under H0       Under H0         Score
  1        4       58.0         42.0       10.5591068    14.5000000
  2        4       28.0         42.0       10.5591068     7.0000000
  3        4       15.0         42.0       10.5591068     3.7500000
  4        4       37.0         42.0       10.5591068     9.2500000
  5        4       72.0         42.0       10.5591068    18.0000000
            Average Scores were used for Ties

       Kruskal-Wallis Test (Chi-Square Approximation)
   CHISQ= 15.111        DF= 4        Prob > CHISQ= 0.0045
```

Discussion of SAS Output

In this output, we have renumbered the pages to correspond with the order of coverage of these topics in previous sections. The first page of output shows the Class Level Information relevant to analysis of variance. Our example involves a single classification variable (METHOD), which has five levels labeled by group numbers 1 through 5.

Page 2 displays the anova summary table and associated p-value. SAS uses the words Model and Error to label the sources of variation, instead of BETWEEN and WITHIN, respectively. Corrected Total is the same as TOTAL variability. Entries in this table agree with the anova summary table shown in Section 5.3. The p-value (Pr > F) of 0.0001 is less than 5 percent, leading to rejection of the null hypothesis. These five groups do not have the same degree of spoilage.

An F value of 15.76 is extremely far out in the tail of the F distribution. (Recall that under the null hypothesis, we expect F to be near one.) The actual p-value may be less than 0.0001, but SAS cannot display smaller values within its allotted number of digits.

Below the anova table are printed the values of R^2 and the coefficient of variation (C.V.). The latter was introduced in Section 2.2 as a measure of spread relative to the size of the numbers. In that section, we examined a single sample and computed the coefficient of variation as 100 times $(S/\overline{X})$. In anova, we use MS_{WITHIN} to estimate σ^2, and its square root (Root MSE) instead of S to estimate σ. The overall average (SPOILAGE Mean) is used in the denominator of the coefficient of variation. Thus, the value shown for C.V. is 100 times 6.3311400/28.25 = 22.41111.

At the bottom of page 2, SAS prints a line of information redundant with the anova table. In the next section, when we discuss blocking, this region of the printout will contain useful information not presented elsewhere.

Page 3 of the output reports results of the LSD multiple range procedure. The value of LSD shown by SAS (9.5421) differs slightly from the figure (9.540) we calculated in Section 5.5, since SAS can compute cutoffs for the t distribution with greater accuracy than shown in Table 2 of the Appendix. Instead of a horizontal display, SAS shows the results vertically. A string of letters is used in place of a line. The string of Ds connects the means 14.250 and 21.750, while the string of Cs connects the means 21.750 and 24.000. These results parallel those presented above in Section 5.6. In essence, SAS draws separate lines under the means 34.750 and 46.500, represented by the letters B and A, respectively. Since no other means are connected to these two, they are each significantly different from all other means.

Tukey's HSD is presented on the fourth page of output, and shows the same three lines drawn in our horizontal display in Section 5.6. Duncan's multiple range test is reported on page 5. The note that Duncan's procedure controls type one error rate comparisonwise is erroneous. Perhaps we might call the control rangewise, as different critical values are used for ranges of different lengths. SAS shows critical values that increase from 9.53, for adjacent means, to 10.49, for comparing the extremes in the rank ordering.

Pages 6 and 7 of output are produced by the GLM procedure. As with ANOVA, this routine prints class level information and the analysis of variance summary table. Type I and type III sums of squares are relevant only in experiments with more than one classification variable and unequal sample sizes, which will be discussed in Section 6.8. The bottom section on page 7 reports tests of our two contrasts. SAS uses an F ratio for each contrast, so the p-values shown (Pr > F) correspond to two-sided tests. The first contrast, for which we planned a two-sided test, is significant, since the two-tailed p-value is less than 5 percent. As discussed earlier, we would like to draw a one-sided conclusion, but SAS does not print the estimate of the contrast. To determine the direction of difference, we would have to compute the estimate by hand. For the second contrast, we planned a one-sided test. Halving the two-tailed p-value (0.0001), we obtain the one-tailed p-value of 0.00005. This figure is less than 5 percent, but we cannot be sure that we are in the predicted direction unless we compute the estimated contrast by hand.

Results of the Kruskal-Wallis test are reported on page 8. The Kruskal-Wallis statistic, which is approximately distributed as a chi-square variable, has the value 15.111. The p-value is shown as 0.0045, enabling us to conclude that these five groups do not have the same degree of spoilage of margarine.

SPSS Input

```
SET WIDTH = 80
TITLE 'EXAMPLE 5.1 - METHODS TO RETARD SPOILAGE'
DATA LIST FREE / METHOD SPOILAGE
ONEWAY SPOILAGE BY METHOD (1,5)
      / RANGES = LSD
      / RANGES = TUKEY
      / RANGES = DUNCAN
      / CONTRAST = .5 .5 -.5 -.5 0
      / CONTRAST = .25 .25 .25 .25 -1
BEGIN DATA
1    28
1    37
1    43
1    31
2    30
2    19
2    20
2    18
3     7
3    16
3    23
3    11
4    23
4    23
4    30
```

```
4    20
5    52
5    42
5    38
5    54
END DATA
NPAR TESTS K-W = SPOILAGE BY METHOD (1,5)
FINISH
```

Discussion of SPSS Input

For each observation, two variables are entered in free format, METHOD (group number), and SPOILAGE. The input lines shown here, between the BEGIN DATA and END DATA statements, are ordered by group, but this is not required.

The ONEWAY command requests analysis of variance for the variable SPOILAGE, using groups indexed by values one through five on METHOD. Optional requests, each following a slash, ask for LSD, HSD (TUKEY), and DUNCAN multiple range procedures, and tests of two separate contrasts. Each CONTRAST command specifies the coefficients for each group in the experiment. Since our first contrast does not involve the control group, a coefficient of 0 is listed for the fifth group.

After the data, NPAR TESTS requests the Kruskal-Wallis test (K-W), performed on the variable SPOILAGE, using groups indexed by values 1 through 5 on METHOD.

SPSS Output

```
         EXAMPLE 5.1 - METHODS TO RETARD SPOILAGE     Page 1
                           O N E W A Y
   Variable SPOILAGE
By Variable METHOD
                    ANALYSIS OF VARIANCE
                         SUM OF        MEAN         F        F
     SOURCE       D.F.   SQUARES      SQUARES     RATIO    PROB.
BETWEEN GROUPS      4   2526.5000    631.6250   15.7578   .0000
WITHIN GROUPS      15    601.2500     40.0833
TOTAL              19   3127.7500
--------------------------------------------------------------
         EXAMPLE 5.1 - METHODS TO RETARD SPOILAGE     Page 2
                           O N E W A Y
       Variable SPOILAGE
    By Variable METHOD
MULTIPLE RANGE TEST
```

```
LSD PROCEDURE
RANGES FOR THE 0.050 LEVEL -
          3.01  3.01  3.01  3.01
THE RANGES ABOVE ARE TABLE RANGES.
THE VALUE ACTUALLY COMPARED WITH MEAN(J) -MEAN(I) IS..
        4.4768 * RANGE * DSQRT (1/N(I) + 1/N(J))
    (*) DENOTES PAIRS OF GROUPS SIGNIFICANTLY DIFFERENT AT
THE 0.050 LEVEL

                        G G G G G
                        r r r r r
                        p p p p p

   Mean      Group      3 2 4 1 5
 14.2500     Grp 3
 21.7500     Grp 2
 24.0000     Grp 4      *
 34.7500     Grp 1      * * *
 46.5000     Grp 5      * * * *

HOMOGENEOUS SUBSETS     (SUBSETS OF GROUPS, WHOSE HIGHEST
                            AND LOWEST MEANS DO NOT DIFFER
                            BY MORE THAN THE SHORTEST
                            SIGNIFICANT RANGE FOR A SUBSET
                            OF THAT SIZE)

SUBSET 1
GROUP         Grp 3          Grp 2
MEAN        14.2500        21.7500
- - - - - - - - - - - - - - - - - -
SUBSET 2
GROUP         Grp 2          Grp 4
MEAN        21.7500        24.0000
- - - - - - - - - - - - - - - - - -
SUBSET 3
GROUP         Grp 1
MEAN        34.7500
- - - - - - - - - - - -
SUBSET 4
GROUP         Grp 5
MEAN        46.5000
- - - - - - - - - - - -
```

```
          EXAMPLE 5.1 - METHODS TO RETARD SPOILAGE     Page 3
                          O N E W A Y
       Variable SPOILAGE
    By Variable METHOD
MULTIPLE RANGE TEST

TUKEY-HSD PROCEDURE
RANGES FOR THE 0.050 LEVEL-
          4.37  4.37  4.37  4.37
THE RANGES ABOVE ARE TABLE RANGES.
THE VALUE ACTUALLY COMPARED WITH MEAN(J) -MEAN(I) IS..
        4.4768 * RANGE * DSQRT(1/N(I) + 1/N(J))
    (*) DENOTES PAIRS OF GROUPS SIGNIFICANTLY DIFFERENT AT
THE 0.050 LEVEL

                          G G G G G
                          r r r r r
                          p p p p p

   Mean      Group        3 2 4 1 5
  14.2500    Grp 3
  21.7500    Grp 2
  24.0000    Grp 4
  34.7500    Grp 1        *
  46.5000    Grp 5        * * *

HOMOGENEOUS SUBSETS  (SUBSETS OF GROUPS, WHOSE HIGHEST AND
                       LOWEST MEANS DO NOT DIFFER BY MORE
                       THAN THE SHORTEST SIGNIFICANT RANGE
                       FOR A SUBSET OF THAT SIZE)

SUBSET 1
GROUP       Grp 3        Grp 2        Grp 4
MEAN       14.2500      21.7500      24.0000
-----------------------------------------
SUBSET 2
GROUP       Grp 2        Grp 4        Grp 1
MEAN       21.7500      24.0000      34.7500
-----------------------------------------
SUBSET 3
GROUP       Grp 1        Grp 5
MEAN       34.7500      46.5000
------------------------------
```

```
          EXAMPLE 5.1 - METHODS TO RETARD SPOILAGE     Page 4
                             O N E W A Y
       Variable SPOILAGE
    By Variable METHOD
MULTIPLE RANGE TEST

DUNCAN PROCEDURE
RANGES FOR THE 0.050 LEVEL -
          3.01  3.16  3.26  3.31
THE RANGES ABOVE ARE TABLE RANGES.
THE VALUE ACTUALLY COMPARED WITH MEAN(J) -MEAN(I) IS..
        4.4768 * RANGE * DSQRT(1/N(I) + 1/N(J))
    (*) DENOTES PAIRS OF GROUPS SIGNIFICANTLY DIFFERENT AT
THE 0.050 LEVEL

                              G G G G G
                              r r r r r
                              p p p p p

   Mean      Group            3 2 4 1 5
  14.2500    Grp 3
  21.7500    Grp 2
  24.0000    Grp 4
  34.7500    Grp 1            * * *
  46.5000    Grp 5            * * * *

HOMOGENEOUS SUBSETS  (SUBSETS OF GROUPS, WHOSE HIGHEST AND
                        LOWEST MEANS DO NOT DIFFER BY MORE
                        THAN THE SHORTEST SIGNIFICANT RANGE
                        FOR A SUBSET OF THAT SIZE)

SUBSET 1
GROUP       Grp 3        Grp 2       Grp 4
MEAN       14.2500      21.7500     24.0000
- - - - - - - - - - - - - - - - - - - - - - - - - -
SUBSET 2
GROUP       Grp 1
MEAN       34.7500
- - - - - - - - - - -
SUBSET 3
GROUP       Grp 5
MEAN       46.5000
- - - - - - - - - - -
```

```
          EXAMPLE 5.1 - METHODS TO RETARD SPOILAGE       Page 5
                             O N E W A Y
        Variable SPOILAGE
     By Variable METHOD
CONTRAST COEFFICIENT MATRIX
                    Grp 1     Grp 2     Grp 3     Grp 4     Grp 5
CONTRAST 1           0.5       0.5      -0.5      -0.5       0.0
CONTRAST 2           0.3       0.3       0.3       0.3      -1.0

                              POOLED VARIANCE ESTIMATE
                 VALUE  S. ERROR     T VALUE  D.F.       T PROB.
CONTRAST 1     9.1250   3.1656        2.883  15.0        0.011
CONTRAST 2 -22.8125   3.5392       -6.446  15.0        0.000

                             SEPARATE VARIANCE ESTIMATE
                 VALUE  S. ERROR     T VALUE  D.F.       T PROB.
CONTRAST 1     9.1250   2.9660        3.077  10.8        0.011
CONTRAST 2 -22.8125   4.1371       -5.514   3.9        0.006
```

```
           EXAMPLE 5.1 - METHODS TO RETARD SPOILAGE     Page 6
                    Kruskal-Wallis 1-Way Anova
   SPOILAGE
by METHOD

    Mean Rank       Cases
      14.50           4  METHOD =  1
       7.00           4  METHOD =  2
       3.75           4  METHOD =  3
       9.25           4  METHOD =  4
      18.00           4  METHOD =  5
                     --
                     20  Total
                                             Corrected for ties
Cases  Chi-Square  Significance        Chi-Square  Significance
  20     15.0429       .0046             15.1110        .0045
```

Discussion of SPSS Output

In this output, we have renumbered the pages to correspond with the order of coverage of these topics in previous sections. The first page of output displays the anova summary table, which agrees with that shown in Section 5.3. An F value of 15.7578 is extremely far out in the tail of the F distribution. (Recall that under the null hypothesis, we expect F to be near one.) But clearly, there is still *some* tail area beyond F. The printed p-value (F PROB.) of 0.0000 indicates that the true p-value is closer to 0.0000 than to 0.0001. SPSS simply cannot display the exact p-value within its allotted number of digits. Since p is less than 5 percent, we reject the null hypothesis. We are very sure that these five groups do not have the same degree of spoilage.

Page 2 reports results of the LSD multiple range procedure in two different formats. First, a table of asterisks indicates all pairs of groups that differ according to the LSD criterion. The second method of presenting results, labeled HOMOGENEOUS SUBSETS, parallels the set of lines drawn in the rank ordering of means in Section 5.6. Subset 1 corresponds to the line drawn from 14.25 to 21.75, while subset 2 equates with the line drawn from 21.75 to 24. SPSS lists the means of 34.75 and 46.5 as separate subsets, since each of these is significantly different from all other means.

Tukey's HSD is presented on the third page of output, and Duncan's multiple range test is reported on page 4. Both pages show results equivalent to those presented in Section 5.6.

Page 5 reports tests of our two contrasts. For the second contrast, the CONTRAST COEFFICIENT MATRIX shows coefficients of 0.3 for the first four groups, because SPSS rounds the true coefficients of 0.25 to a single decimal place for printing. The correct coefficients, however, are used in performing the test. SPSS performs two types of t test for each contrast. The first, printed under the heading POOLED VARIANCE ESTIMATE, is a typical t test as discussed in Section 5.7. The second t test for each contrast (SEPARATE VARIANCE ESTIMATE) employs individual estimates of variance in each group, similar to the unequal variance t test discussed in Section 4.3 for comparing two means. This approach is used infrequently, since we have already assumed equal variances for all groups in performing one-way anova. Our discussion focuses on the first approach to testing each contrast.

For each contrast, SPSS prints the estimated contrast (VALUE), obtained by substituting the sample average for each population mean in the contrast. This estimate is the numerator of the t statistic, while the standard error (S. ERROR) is the denominator. The observed T VALUE is reported, along with its degrees of freedom (D.F., equal to df_{WITHIN}) and two-tailed p-value (T PROB.). The first contrast, for which we planned a two-sided test, is significant since the two-tailed p-value of 0.011 is less than 5 percent. Noting that the estimated value of the contrast is positive, we draw the one-sided conclusion that the two processing techniques, on average, result in more spoilage than the two chemical additives.

We planned a one-sided test of the second contrast, which compares the average of methods 1 through 4 with the control group. The estimated value is negative, in accordance with our alternative hypothesis. Ordinarily, we would halve the two-tailed p-value and compare this figure with 5 percent. However, SPSS prints a two-tailed p-value of 0.000, making this a moot point. We reject H_0 and conclude that the control group has more spoilage than the average of the first four methods.

Results of the Kruskal-Wallis test are reported on page 6. The Kruskal-Wallis statistic, which is approximately distributed as a chi-square variable, has the value 15.1110, when corrected for ties. The p-value is shown as 0.0045, enabling us to conclude that these five groups do not have the same degree of spoilage of margarine.

MINITAB Transcript

```
MTB > NAME C1 = 'METHOD'
MTB > NAME C2 = 'SPOILAGE'
MTB > READ 'METHOD' 'SPOILAGE'
DATA > 1  28
DATA > 1  37
DATA > 1  43
DATA > 1  31
DATA > 2  30
DATA > 2  19
DATA > 2  20
DATA > 2  18
DATA > 3   7
DATA > 3  16
DATA > 3  23
DATA > 3  11
DATA > 4  23
DATA > 4  23
DATA > 4  30
DATA > 4  20
DATA > 5  52
DATA > 5  42
DATA > 5  38
DATA > 5  54
DATA > END
     20 ROWS READ
MTB > ANOVA SPOILAGE = METHOD;
```

```
SUBC > MEANS METHOD.

Factor        Type        Levels        Values
METHOD        fixed          5            1  2  3  4  5

Analysis of Variance for SPOILAGE

Source      DF          SS           MS           F          P
METHOD       4     2526.50       631.62       15.76      0.000
Error       15      601.25        40.08
Total       19     3127.75

MEANS

METHOD      N      SPOILAGE
  1         4       34.750
  2         4       21.750
  3         4       14.250
  4         4       24.000
  5         4       46.500

MTB > KRUSKAL-WALLIS 'SPOILAGE' 'METHOD'

LEVEL          NOBS        MEDIAN         AVE. RANK        Z VALUE
  1              4          34.00            14.5             1.51
  2              4          19.50             7.0            -1.32
  3              4          13.50             3.8            -2.55
  4              4          23.00             9.2            -0.47
  5              4          47.00            18.0             2.83
OVERALL         20                           10.5

H = 15.04    d.f. = 4    p = 0.005
H = 15.11    d.f. = 4    p = 0.005 (adj. for ties)
* NOTE * One or more small samples

MTB > STOP
```

Discussion of MINITAB Transcript

The first two lines of this transcript define columns C1 and C2 on the worksheet to be variables named METHOD and SPOILAGE, respectively. Since more than one value is entered for each observation, a READ command is used to input the data. Although our transcript shows observations entered by group, the data may be read in any order.

The ANOVA command contains an equation, or model, reflecting the influences on individual observations. As discussed in Section 5.2, two influences affect the data in one-way analysis of variance, group differences, and random variability within groups. On the left side of the equation, we name the variable to be analyzed (SPOILAGE). On the right side, we list the influences acting on this variable, except that MINITAB automatically assumes random variability as an influence, and this term is not listed. Therefore, only the group influence remains, and this is referenced by the name of the grouping variable (METHOD). Single quotes around variable names in the model portion of the ANOVA command may be omitted. The MEANS subcommand asks for means computed separately for each METHOD. As noted at the beginning of this section, MINITAB does not perform multiple range procedures or tests for linear contrasts.

Output from this command includes a brief description of the experiment, which contains one factor (METHOD), having five levels numbered 1 through 5. In the anova summary table, MINITAB uses the words METHOD and Error to label the sources of variation, instead of BETWEEN and WITHIN, respectively. Entries in this table agree with the anova summary table shown in Section 5.3.

An F value of 15.76 is extremely far out in the tail of the F distribution. (Recall that, under the null hypothesis, we expect F to be near one.) But, clearly, there is still *some* tail area beyond F. The printed p-value of 0.000 indicates that the true p-value is closer to 0.000 than to 0.001. MINITAB simply cannot display the exact p-value within its allotted number of digits. Since p is less than 5 percent, we reject the null hypothesis. We are very sure that these five groups do not have the same degree of spoilage. Below the anova table, the sample mean is printed for each group.

The KRUSKAL-WALLIS command indicates that the test is to be performed on the variable SPOILAGE, for groups identified by the variable METHOD. Adjusted for ties, the Kruskal-Wallis statistic (labeled H by MINITAB), which is approximately distributed as a chi-square variable, has the value 15.11. The p-value is shown as 0.005, enabling us to conclude that these five groups do not have the same degree of spoilage of margarine. MINITAB's note concerning small samples sizes suggests that the p-value produced by the chi-square approximation may be slightly inaccurate. In fact, the sample sizes in this example are large enough to employ the chi-square approximation without difficulty, and the warning may be ignored.

5.9 BLOCKING IN ANOVA

Chapter 4 discussed two different experimental designs for comparing two means—independent samples and paired data. With independent samples, we usually employ the pooled t test, while paired data is analyzed using the paired t test. As discussed in Section 4.9, pairing can be an effective way of reducing variance and, consequently, increasing the power of our test. The paired t test is performed on difference scores, in which variability due to extraneous factors is removed by subtraction. We also noted that independent samples can sometimes be matched to form paired data. In this case, matched pairs are constructed on the basis of some variable influencing the results, whose effects we wish to remove by subtraction.

Blocking is a technique that can be employed in anova to reduce extraneous variability in comparing a means, just as pairing reduces variance in comparing two means. Instead of using pairs of observations, blocks are constructed with each block consisting of a observations. Each block includes observations alike in some manner, and these observations are randomly assigned to the a treatments to be compared. Research planned along these lines is said to employ a **randomized complete blocks** design.

In Example 5.2, we wish to compare mean power consumption for the five air conditioner brands. Each room orientation is a block containing one measurement for each mean to be compared, just as a matched pair represents both groups to be compared in the paired data design. In the paired t test, variability related to the

---

Example 5.2

Five different brands of air conditioners were to be compared for their power consumption. Four units of each brand were available for testing in a special laboratory building, with test rooms laid out as shown in Figure 5.4. Each test room has a single window facing the outside of the building.

While the 20 total units to be tested could have been assigned to these 20 test rooms completely at random, it was clear that the orientation of the test rooms (N, E, S, W) would have an effect on power consumption. Thus, each orientation was designated as a block, to contain one unit of each of the five brands. Within each block, the five units were assigned to the five test rooms randomly. Power consumption was monitored and recorded in the table below.

	Orientation			
AC Brand	N	E	S	W
1	685	792	875	838
2	722	806	953	893
2	733	802	941	880
4	811	888	1005	952
5	828	920	1023	978

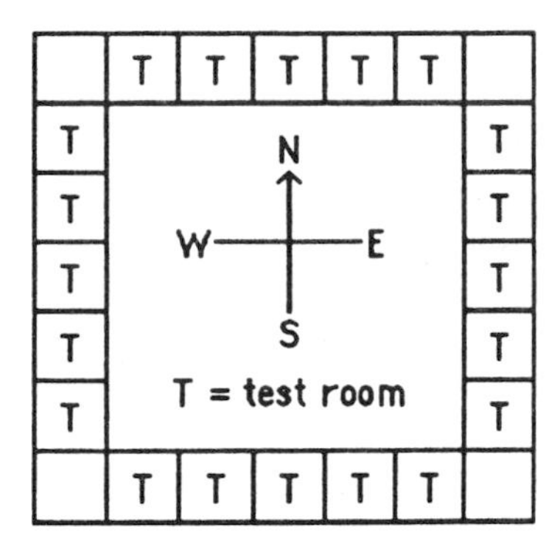

FIGURE 5.4. Test rooms for comparing air conditioners.

matching variable is removed by subtraction. But subtraction cannot be used with more than two groups, so variability related to room orientation cannot be removed entirely from the data to be analyzed in this example. Instead, we estimate how much influence room orientation is having on power consumption separately from that due to air conditioner brand or random variability. Total variability is analyzed into *three* components, related to the influences of air conditioner brand, room orientation, and random fluctuations.

As in the one-way design, total variability is expressed as MS_{TOTAL}, the ratio of SS_{TOTAL} to df_{TOTAL}. Also, as in one-way anova, we do not decompose MS_{TOTAL}, but perform separate partitions of SS_{TOTAL} and df_{TOTAL}. The total sum of squares is broken into sums of squares for **treatments, blocks,** and "**error.**" In Example 5.2, the treatments are the five air conditioner brands, the blocks are the four room orientations, and error is the remaining variability that cannot be accounted for by treatments or blocks.

Total variability among N observations has N - 1 degrees of freedom. $SS_{TREATMENTS}$, which measures spread among the ***a*** treatment means, has ***a*** - 1 degrees of freedom, similar to a sample variance computed using the ***a*** treatment means as "observations." In the same way, SS_{BLOCKS} reflects differences among the block means. Using the symbol ***b*** to represent the number of blocks, this term has ***b*** - 1 degrees of freedom. The error degrees of freedom can be found by subtracting degrees of freedom for treatments and blocks from total degrees of freedom. For Example 5.2, these values are

$$df_{TOTAL} = N - 1 = 20 - 1 = 19,$$
$$df_{TREATMENTS} (= df_{ACBRANDS}) = a - 1 = 5 - 1 = 4,$$
$$df_{BLOCKS} (= df_{ORIENTATION}) = b - 1 = 4 - 1 = 3, \text{ and}$$
$$df_{ERROR} = 19 - 4 - 3 = 12.$$

Mean squares are computed for treatments, blocks, and error by dividing each sum of squares by its degrees of freedom. The F statistic for testing equality of treatment means is constructed as the ratio of $MS_{TREATMENTS}$ to MS_{ERROR}. Numerator and denominator degrees of freedom for this F ratio are the degrees of freedom for treatments and error, respectively.

Technically, the equality of block means cannot be validly tested with an F test. This is because randomization is restricted by the requirement that each block include each treatment. In Example 5.2, for instance, the 20 air conditioners are not assigned to the 20 test rooms completely at random; each room orientation must contain one unit from each of the five air conditioner brands. Inability to test the block effect is generally not a concern. Our main interest lies in comparing treatment means, and the blocking factor is included in the design only to increase power in this test.

Nonetheless, most computer programs print a "test" for equivalence of blocks as the ratio of MS_{BLOCKS} to MS_{ERROR}. This F ratio has numerator degrees of freedom equal to df_{BLOCKS} and denominator degrees of freedom equal to df_{ERROR}. Although we should not interpret this statistic as testing for equality of blocks, it does measure the efficiency of blocking in reducing error variance. The smaller the p-value is for this statistic, the more useful the blocking factor is in reducing variability and increasing power in the test on treatment means.

The anova table summarizing results of this analysis for Example 5.2 is shown in Table 5.2. The F test for comparing the five air conditioner brands has $df_{num} = df_{ACBRAND} = 4$, and $df_{den} = df_{ERROR} = 12$. From Table 4a, we find the 5% critical value of 3.26. Our obtained F value of 95.567 is highly significant. If the null hypothesis were true, we would expect an F value near 1.0. As we will see from the computer programs presented in the next section, the tail area beyond 95.567 is less than 0.0001. We are very sure that these five air conditioner brands differ in mean power consumption.

Room orientation—the blocking factor in this example—is extremely useful in accounting for variability in power consumption among the 20 air conditioning units. Its F value of 278.199 has 3 degrees of freedom in the numerator (= $df_{ORIENTATION}$), and 12 degrees of freedom in the denominator (= df_{ERROR}). The tail

TABLE 5.2 Anova Summary Table for Comparing Air Conditioner Brands with Blocking

Source	df	SS	MS	F
ACBrand	4	53,231.00	13,307.75	95.567
Orientation	3	116,217.75	38,739.25	278.199
Error	12	1,671.00	139.25	
Total	19	171,119.75		

area beyond 278.199 is virtually 0, indicating that inclusion of blocks in this analysis greatly increases power in the test for treatment effects.

To see *how* important room orientation is in this example, suppose we choose to ignore this classification variable in performing our analysis. That is, suppose we view our experiment as a one-way design comparing the five air conditioner brands, with four observations per group. The summary table for this one-way analysis is shown in Table 5.3. With $df_{num} = 4$ and $df_{den} = 15$, the 5% critical value is 3.06. Our F value of 1.693 is not even significant! When we ignore room orientation, we must accept the null hypothesis of equal mean power consumption among the five air conditioner brands.

Comparison of Table 5.3 with Table 5.2 underscores some interesting and logical points. First, we note that total degrees of freedom and total sum of squares are the same in the two tables. Total variability is computed as if all 20 observations comprised a single sample, ignoring any grouping variables. Clearly, this overall variability is unaffected by whether our classification scheme involves two variables or only one. Second, we see that degrees of freedom and sum of squares for ACBrand are the same in the two tables. Variability among the means for the five air conditioner brands is also unaffected by whether or not we consider room orientation.

Finally, we note that two sources of variability *with blocking*, namely room orientation and error, are essentially pooled to form the within term *without blocking*—that is, $SS_{ORIENTATION}$ plus SS_{ERROR} in Table 5.2 add up to SS_{WITHIN} in Table 5.3. Also, $df_{ORIENTATION}$ plus df_{ERROR} in Table 5.2 add to df_{WITHIN} in Table 5.3. When we include room orientation in the analysis with blocking, it accounts for a substantial amount of variability. This variability is still present when we ignore room orientation and analyze the data without blocking. But since we do not consider room orientation as a separate source of variability, it is automatically pooled with the error term and contributes to within groups variability.

In the paired t test, when there are only two treatments, variability related to the matching variable can be removed from the data by subtraction. But when more than two treatments exist, we cannot subtract out the effects of a matching variable. Randomized complete blocks allows us to assess the influence of the blocking

TABLE 5.3 **Anova Summary Table for Comparing Air Conditioner Brands without Blocking**

Source	df	SS	MS	F
ACBrand	4	53,231.00	13,307.75	1.693
Within	15	117,888.75	7,859.25	
Total	19	171,119.75		

variable, so that we do not include its variability in the random error term. This results in a reduced denominator mean square, and therefore a more powerful test, when we test for treatment differences.

For non-normal data, ***Friedman's test*** is a nonparametric technique appropriate for randomized complete blocks. The treatments are ranked separately within each block, and the sum of ranks across blocks is computed for each treatment. When the null hypothesis is true, we expect roughly equal sums of ranks for the various treatments and, therefore, small variance among these rank sums. On the other hand, if the alternative hypothesis is true, some treatments should be consistently ranked above others, resulting in rather different rank sums and a large variance among them. Friedman's statistic,which is a function of the variance of the rank sums, has a distribution approximately the same as a chi-square distribution with $a - 1$ degrees of freedom.

For the data of Example 5.2, Friedman's statistic is 15.40, with $a - 1 = 5 - 1 = 4$ degrees of freedom. The upper 5% chi-square value is found from Table 3b to be 9.488. As with the parametric test, we reject the null hypothesis and conclude that the five air conditioner brands are not equivalent in mean power consumption.

5.10 BLOCKING IN COMPUTER PROGRAMS

Randomized complete blocks can be analyzed by SAS, SPSS, and MINITAB. For each program, the data of Example 5.2 are analyzed both with and without blocking, to reproduce the results of the previous section. Each program also performs the Friedman test, as illustrated here.

SAS Input

```
OPTIONS LS = 80;
TITLE 'EXAMPLE 5.2 - POWER CONSUMPTION OF AIR CONDITIONERS';
DATA;
INPUT ACBRAND ORIENT $ POWER;
LINES;
1   N    685
1   E    792
1   S    875
1   W    838
2   N    722
2   E    806
2   S    953
```

```
2   W    893
3   N    733
3   E    802
3   S    941
3   W    880
4   N    811
4   E    888
4   S   1005
4   W    952
5   N    828
5   E    920
5   S   1023
5   W    978
PROC ANOVA;
CLASS ACBRAND ORIENT;
MODEL POWER = ACBRAND ORIENT;
PROC ANOVA;
CLASS ACBRAND;
MODEL POWER = ACBRAND;
PROC FREQ;
TABLES ORIENT * ACBRAND * POWER / NOPRINT SCORES=RANK CMH2;
```

Discussion of SAS Input

Three variables are entered for each observation: ACBRAND, ORIENT (in alphabetic form, as indicated by the dollar sign), and POWER. In the ANOVA procedure, two classification variables, ACBRAND and ORIENT, are specified. As discussed in Section 5.8, the MODEL statement names the dependent variable, POWER, on the left side of the equation, and its influences on the right side. For a randomized complete blocks design, total variability is partitioned into three sources: treatments, blocks, and error. We list ACBRAND and ORIENT, but do not list error, since SAS includes this source automatically.

A second call to PROC ANOVA is used to perform the one-way analysis of variance without blocking. Note that ACBRAND is the only classification variable, and that the MODEL statement lists only this source of variability on the right side.

The Friedman test is produced by PROC FREQ, using the CMH2 (Cochran-Mantel-Haenszel) option, after first converting the scores to ranks. The NOPRINT option has been used to suppress printing of the meaningless three-way frequency table of ORIENT by ACBRAND by POWER that would otherwise be printed as a by-product of this analysis.

SAS Output

```
EXAMPLE 5.2 - POWER CONSUMPTION OF AIR CONDITIONERS    1
          Analysis of Variance Procedure
             Class Level Information

          Class       Levels    Values
          ACBRAND       5       1 2 3 4 5
          ORIENT        4       E N S W

     Number of observations in data set = 20
```

```
   EXAMPLE 5.2 - POWER CONSUMPTION OF AIR CONDITIONERS       2
               Analysis of Variance Procedure

Dependent Variable: POWER
                             Sum of          Mean
Source              DF      Squares        Square      F Value    Pr > F
Model                7   169448.75000   24206.96429     173.84    0.0001
Error               12     1671.00000     139.25000
Corrected Total     19   171119.75000

R-Square            C.V.          Root MSE        POWER Mean
0.990235         1.362242        11.800424         866.25000

Source     DF         Anova SS       Mean Square     F Value    Pr > F
ACBRAND     4      53231.00000       13307.75000       95.57    0.0001
ORIENT      3     116217.75000       38739.25000      278.20    0.0001
```

```
EXAMPLE 5.2 - POWER CONSUMPTION OF AIR CONDITIONERS    3
          Analysis of Variance Procedure
             Class Level Information

          Class       Levels    Values
          ACBRAND       5       1 2 3 4 5

     Number of observations in data set = 20
```

```
   EXAMPLE 5.2 - POWER CONSUMPTION OF AIR CONDITIONERS       4
               Analysis of Variance Procedure

Dependent Variable: POWER
                            Sum of            Mean
Source              DF      Squares          Square      F Value   Pr > F
Model                4   53231.000000   13307.750000       1.69   0.2038
Error               15  117888.750000    7859.250000
Corrected Total     19  171119.750000
```

```
R-Square          C.V.          Root MSE       POWER Mean
0.311075      10.23404        88.652411       866.25000

Source      DF      Anova SS        Mean Square      F Value     Pr > F
ACBRAND      4    53231.000000    13307.750000        1.69      0.2038
```

```
      EXAMPLE 5.2 - POWER CONSUMPTION OF AIR CONDITIONERS      5

           SUMMARY STATISTICS FOR ACBRAND BY POWER
                  CONTROLLING FOR ORIENT

  Cochran-Mantel-Haenszel Statistics (Based on Rank Scores)

Statistic     Alternative Hypothesis    DF      Value      Prob
---------------------------------------------------------------
    1         Nonzero Correlation        1     13.690     0.000
    2         Row Mean Scores Differ     4     15.400     0.004

Total Sample Size = 20
```

Discussion of SAS Output

Output pages 1 and 2 pertain to the first call to PROC ANOVA (analysis *with* blocking), while pages 3 and 4 are from the second (*without* blocking). The Class Level Information shown on page 1 lists our two classification variables—ACBRAND, with five levels numbered one through five, and ORIENT, with four levels labeled N, E, S, and W (but listed alphabetically by SAS). The anova table at the top of page 2 pools the sources of variation ACBRAND and ORIENT, and labels this combined term Model. Typically, we are not interested in performing a single test for all influences at once, so the F test reported in this table is of little use. However, the Error term shown is the correct denominator for testing individual effects in the model.

At the bottom of page 2, SAS prints tests for each model effect separately. As discussed in the previous section, the comparison of air conditioner brands is highly significant. The actual p-value for this test may be less than 0.0001, but SAS cannot display smaller values within its allotted number of digits. We are very certain that these five air conditioner brands differ in mean power consumption. The small p-value shown on the line for ORIENT indicates that blocking has greatly increased the power of our test comparing the air conditioner brands.

On page 3, only ACBRAND is listed as a classification variable. The associated one-way anova reported on page 4 displays an F statistic of 1.69, for which the p-value is 0.2038. Since this is greater than 5 percent, we would accept the null hypothesis of equal mean power consumption for the five air conditioner brands in an analysis without blocking.

The fifth page of output reports the results of the Friedman test as Statistic 2, including the approximate chi-square value of 15.400 and its associated p-value of 0.004. Again, we conclude that the five air conditioner brands differ in mean power consumption because p is less than 5 percent.

SPSS Input

```
SET WIDTH = 80
TITLE 'EXAMPLE 5.2 - POWER CONSUMPTION OF AIR CONDITIONERS'
DATA LIST FREE / ACBRAND ORIENT POWER
ANOVA VARIABLES = POWER BY ACBRAND (1,5) ORIENT (1,4)
     / MAXORDERS = NONE
     / VARIABLES = POWER BY ACBRAND (1,5)
BEGIN DATA
1   1    685
1   2    792
1   3    875
1   4    838
2   1    722
2   2    806
2   3    953
2   4    893
3   1    733
3   2    802
3   3    941
3   4    880
4   1    811
4   2    888
4   3   1005
4   4    952
5   1    828
5   2    920
5   3   1023
5   4    978
END DATA
DATA LIST FREE / ORIENT ACBRAND1 ACBRAND2 ACBRAND3 ACBRAND4
   ACBRAND5
NPAR TESTS FRIEDMAN = ACBRAND1 TO ACBRAND5
BEGIN DATA
```

```
1   685   722   733   811   828
2   792   806   802   888   920
3   875   953   941  1005  1023
4   838   893   880   952   978
END DATA
FINISH
```

Discussion of SPSS Input

The data are read into SPSS twice, since different organizations of the observations are required for analysis of variance and the Friedman test. For analysis of variance, data are input in the usual manner, with one observation per line. Three variables—ACBRAND, ORIENT, and POWER—are entered in free format for each observation.

The ANOVA command requests analysis of variance for the variable POWER, using groups determined by values 1 through 5 on ACBRAND and 1 through 4 on ORIENT. The first optional request (MAXORDERS = NONE) specifies a model without interaction. (The concept of interaction will be discussed in detail in the next chapter. For the time being, we simply state that absence of interaction implies that room orientation affects each air conditioner brand equally.) Since both factors are included, this analysis corresponds to the randomized complete blocks design of our study.

Because the second optional request begins with the keyword VARIABLES, it requests a new analysis of the data. In this case, only the single classification variable ACBRAND is specified, corresponding to one-way analysis without blocking.

After END DATA for the first set of data, a new DATA LIST command indicates that a second dataset follows. To perform the Friedman test, the data must be entered with one line per block. Each line includes all five treatment values for the block (ACBRAND1 through ACBRAND5). NPAR TESTS requests the FRIEDMAN test, comparing the means on variables ACBRAND1 through ACBRAND5.

SPSS Output

```
EXAMPLE 5.2 - POWER CONSUMPTION OF AIR CONDITIONERS   Page 1

        * * * A N A L Y S I S   O F   V A R I A N C E * * *

             POWER
        BY   ACBRAND
             ORIENT
```

Source of Variation	Sum of Squares	DF	Mean Square	F	Sig of F
Main Effects	169448.750	7	24206.964	173.838	.000
ACBRAND	53231.000	4	13307.750	95.567	.000
ORIENT	116217.750	3	38739.250	278.199	.000
Explained	169448.750	7	24206.964	173.838	.000
Residual	1671.000	12	139.250		
Total	171119.750	19	9006.303		

20 cases were processed.
0 cases (.0 pct) were missing.

EXAMPLE 5.2 - POWER CONSUMPTION OF AIR CONDITIONERS Page 2

* * * A N A L Y S I S O F V A R I A N C E * * *

POWER
by ACBRAND

Source of Variation	Sum of Squares	DF	Mean Square	F	Sig of F
Main Effects	53231.000	4	13307.750	1.693	.204
ACBRAND	53231.000	4	13307.750	1.693	.204
Explained	53231.000	4	13307.750	1.693	.204
Residual	117888.750	15	7859.250		
Total	171119.750	19	9006.303		

20 cases were processed.
0 cases (.0 pct) were missing.

EXAMPLE 5.2 - POWER CONSUMPTION OF AIR CONDITIONERS Page 3

Friedman Two-Way Anova

Mean Rank	Variable
1.00	ACBRAND1
2.75	ACBRAND2
2.25	ACBRAND3
4.00	ACBRAND4
5.00	ACBRAND5

Cases	Chi-Square	D.F.	Significance
4	15.4000	4	.0039

Discussion of SPSS Output

Page 1 of the output contains the results of the analysis *with* blocking, while page 2 relates to analysis *without* blocking. On the first page, the line labeled Main

Effects pools the sources of variation ACBRAND and ORIENT. Typically, we are not interested in performing a single test for all influences at once, so the F test reported on this line is of little use. (This test is repeated in the line labeled Explained.) SPSS uses the word Residual to identify the error term.

Also included in this table are tests for each model effect separately. As discussed in the previous section, the comparison of air conditioner brands is highly significant. The printed p-value of 0.000 indicates that the true p-value is closer to 0.000 than to 0.001. SPSS cannot display the exact p-value within its allotted number of digits. Since p is less than 5 percent, we reject the null hypothesis. We are very certain these five air conditioner brands differ in mean power consumption. The small p-value shown on the line for ORIENT indicates that blocking has greatly increased the power of our test comparing the air conditioner brands.

On page 2, corresponding to one-way analysis of variance without blocking, the test for ACBRAND produces an F statistic of 1.693, for which the p-value is 0.204. Since this is greater than 5 percent, we would accept the null hypothesis of equal mean power consumption for the five air conditioner brands in an analysis without blocking.

The third page of output reports results of the Friedman test, including the approximate chi-square value of 15.40 and its associated p-value of 0.0039. Again we conclude that the five air conditioner brands differ in mean power consumption because p is less than 5 percent.

MINITAB Transcript

```
MTB > NAME C1 = 'ACBRAND'
MTB > NAME C2 = 'ORIENT'
MTB > NAME C3 = 'POWER'
MTB > READ C1-C3
DATA > 1  1   685
DATA > 1  2   792
DATA > 1  3   875
DATA > 1  4   838
DATA > 2  1   722
DATA > 2  2   806
DATA > 2  3   953
DATA > 2  4   893
DATA > 3  1   733
DATA > 3  2   802
DATA > 3  3   941
DATA > 3  4   880
DATA > 4  1   811
```

```
DATA > 4  2   888
DATA > 4  3  1005
DATA > 4  4   952
DATA > 5  1   828
DATA > 5  2   920
DATA > 5  3  1023
DATA > 5  4   978
DATA > END
    20 ROWS READ

MTB > ANOVA POWER = ACBRAND ORIENT

Factor          Type       Levels    Values
ACBRAND         fixed         5        1    2    3    4    5
ORIENT          fixed         4        1    2    3    4

Analysis of Variance for POWER

Source       DF          SS          MS          F           P
ACBRAND       4       53231       13308      95.57       0.000
ORIENT        3      116218       38739     278.20       0.000
Error        12        1671         139
Total        19      171120

MTB > ANOVA POWER = ACBRAND

Factor          Type       Levels    Values
ACBRAND         fixed         5        1    2    3    4    5

Analysis of Variance for POWER

Source       DF          SS          MS          F           P
ACBRAND       4       53231       13308       1.69       0.204
Error        15      117889        7859
Total        19      171120

MTB > FRIEDMAN 'POWER' 'ACBRAND' 'ORIENT'

Friedman test of POWER by ACBRAND blocked by ORIENT

S = 15.40 d.f. = 4 p = 0.004
```

```
                        Est.          Sum of
ACBRAND        N       Median          RANKS
   1           4       802.60            4.0
   2           4       850.70           11.0
   3           4       846.20            9.0
   4           4       921.80           16.0
   5           4       944.20           20.0

Grand median = 873.10

MTB > STOP
```

Discussion of MINITAB Transcript

In the first three lines of this transcript, columns C1, C2, and C3 are defined as variables named ACBRAND, ORIENT, and POWER, respectively. A READ command is used to enter these three values for each observation on the data lines.

The first ANOVA command requests analysis *with* blocking. As discussed in Section 5.8, the ANOVA command names the dependent variable, POWER, on the left side of the equation, and its influences on the right side. For a randomized complete blocks design, total variability is partitioned into three sources: treatments, blocks, and error. We list ACBRAND and ORIENT, but do not list error, since MINITAB includes this source automatically. Output from this command includes a brief description of the experiment, which contains two factors (ACBRAND with five levels, and ORIENT with four). The anova summary table displays an F statistic of 95.57 for the test of ACBRAND. The printed p-value of 0.000 indicates that the true p-value is closer to 0.000 than to 0.001. MINITAB cannot display the exact p-value within its allotted number of digits. Since p is less than 5 percent, we reject the null hypothesis. We are very certain that these five air conditioner brands differ in mean power consumption. The small p-value shown on the line for ORIENT indicates that blocking has greatly increased the power of our test comparing the air conditioner brands.

The second ANOVA command includes only ACBRAND as an influence on POWER, and corresponds to one-way analysis *without* blocking. The test for ACBRAND produces an F statistic of 1.69, and a p-value of 0.204. Since this is greater than 5 percent, we would accept the null hypothesis of equal mean power consumption for the five air conditioner brands in an analysis without blocking.

In the FRIEDMAN command, the first variable named is the dependent variable, the second is the treatment variable, and the third is the blocking variable. For Example 5.2 we request a comparison of mean POWER values for levels of ACBRAND, blocked by ORIENT. Printed results include the approximate chi-square value (S) of 15.40, and its associated p-value of 0.004. Again we conclude that the five air conditioner brands differ in mean power consumption because p is less than 5 percent.

5.11 SUMMARY

In this chapter, analysis of variance was introduced for testing equality of several means. The phrase "analysis of variance" refers to analyzing or partitioning variability into components that correspond to different influences on the data. In a one-way design, two such influences exist: the effect of the grouping variable, and random fluctuations within groups.

In fact, overall variance is not itself decomposed, but separate partitions are constructed for the sum of squares (numerator of variance) and degrees of freedom (denominator). For the between group component, the ratio of sum of squares to degrees of freedom is called the mean square between groups. This term measures variability among the group averages, and is expected to be small when the population means are equal. For the within group component, the ratio of sum of squares to degrees of freedom is the mean square within groups, and is a pooled estimate of the common variance in each group. The test statistic, an F ratio, is constructed by dividing mean square between by mean square within.

Results are summarized in an anova table, which lists columns for source of variation, degrees of freedom, sum of squares, mean square, and F value. The p-value associated with the observed F statistic measures statistical significance, while practical importance is assessed by the coefficient of determination (R^2).

For non-normal data, the Kruskal-Wallis test analyzes data from a one-way design. This test is an extension of the rank sum test to several groups.

In the case of normal data, but in only two groups, equality of means can be tested with either the pooled t test or the one-way anova F test. These two tests are equivalent; the F ratio is the square of the t statistic, and the critical F value is the square of the critical t value.

Rejecting the null hypothesis in one-way anova only tells us that the means are not *all* equal. To determine *which* means differ, several different procedures are available. Fisher's LSD approach uses a modified t test, including a pooled variance estimate based on sample variances in all the groups. This approach allows a 5 percent risk of type one error for each pairwise comparison. Tukey's HSD method, on the other hand, controls the chance of making a type one error at 5 percent over all possible pairwise comparisons. Duncan's test is somewhat between these two in its control of type one error probability. All three tests are generally presented in multiple range format, a method for summarizing results when performing all possible pairwise comparisons.

A linear contrast is a combination of means in which the coefficients sum to zero. Tests of contrasts allow us to answer more complicated questions, such as whether the average of one subset of groups is equal to the average of another subset. Such a test can be performed using either a t statistic or an F ratio; the F is again the square of the t value.

Blocking in anova is the extension of pairing in the comparison of two means. Each block, or matched set of observations, includes one measurement under each treatment. In this analysis, overall variability is broken into three components: for treatments, blocks, and random error. Similarly to the paired t test, blocking in anova reduces unexplained variability and makes the test for treatment effects more powerful. For non-normal data, the Friedman test is appropriate for randomized blocks experiments.

All statistical tests introduced in this chapter were demonstrated on SAS, SPSS, and MINITAB computer programs.

6

Factorial Designs

Chapter 5 discussed one-way analysis of variance, in which we compare means for several levels of a single classification variable. This chapter considers experiments employing two or more classification variables, or **factors**. We examine the influence of these factors on the mean of a variable measured for each observation. Our emphasis is on two-way analysis of variance, in which two classification variables are involved, although Sections 6.10 and 6.12 illustrate three-way and four-way designs, respectively. Many of the concepts of one-way anova are relevant to designs discussed in this chapter—for instance, the partitioning of sums of squares and degrees of freedom. But there are no accepted nonparametric techniques for analysis of data from designs involving more than one factor.

6.1 TWO-WAY DESIGN

In one-way analysis of variance, observations are sorted into cells that correspond to various levels of a single classification variable. For a two-way design, each cell represents the combination of two levels, one from each of the factors. The cells are usually displayed in a rectangular array, with rows corresponding to levels of one factor and columns corresponding to levels of the other. Each cell contains observations assumed to be drawn randomly from a population associated with its combination of levels. As in the one-way layout, these populations are technically assumed to be normal with equal variances.

If sample sizes are large enough, the data in each cell, plotted separately, should appear normal, with all plots having the same variance. In most practical applications, however, sample sizes are too small to make reasonable determinations about normality or equality of variances. Tests for normality and equality of variances exist, but are infrequently used. As in the one-way case, these tests are powerful only with large sample sizes, precisely when we need be *least* concerned with the validity of assumptions.

In Example 6.1, a two-way design, we wish to answer three questions of interest. First, we ask if there are row differences. Do these three coatings provide equal protection from corrosion? Second, we compare the columns. Are these four soil types different in terms of corroding pipe? Third, we look for interaction. Does the effect of a particular coating depend on the type of soil in which the pipe is buried?

The first two issues involve only single factors and are questions about **main effects**. The third concern addresses **interaction effects** between two factors. Each question is stated as a null hypothesis for testing purposes:

H_0: no row effects,
H_0: no column effects, and
H_0: no interaction effects.

Each null hypothesis requires its own statistical test. Thus, within the context of a single experiment, we perform three separate tests.

In this experiment, each cell has two observations. Designs with the same sample size in every cell are called **balanced** or **orthogonal**. Most of this chapter is limited to discussion of balanced designs. The difficulties inherent with unequal sample sizes, and one possible approach to analysis of such data, are examined in Section 6.8.

Example 6.1

Corrosion is a major problem for underground metal pipe. An investigation comparing three different coatings for corrosion protection also utilized four different soil mixtures, to simulate various building sites. For each coating, two pieces of pipe were buried in each soil type for a fixed time interval. Each piece of pipe was then examined to determine the amount of corrosion, operationally defined as the depth of the maximum pit in ten-thousandths of an inch. These data are displayed below.

	Soil			
Coating	S1	S2	S3	S4
C1	10	22	11	23
	16	20	11	14
C2	8	8	13	12
	7	16	5	18
C3	19	18	12	28
	15	27	18	18

Example 6.1 is a two-way design, with each coating used in each soil type. A three-way design is the logical extension of this scheme to include a third classification variable. For example, suppose we also considered two kinds of metal pipe. The study would be a three-way design if each combination of coating, soil type, and kind of pipe were examined. In general, a **factorial design** includes each possible combination of levels, where one level is chosen from each factor. Two-way and three-way are forms of factorial designs. In theory, there is no limit to the number of factors that can be included. Practical considerations and the need to interpret results, however, encourage us to keep experiments within reasonable bounds. We will discuss the relative advantages and disadvantages of complex designs later in this chapter. The next several sections are devoted to analysis in two-way settings.

6.2 PARTITIONS OF SUMS OF SQUARES AND DEGREES OF FREEDOM

In one-way anova, the overall variance, or total mean square, is defined as the ratio of total sum of squares to total degrees of freedom. Separate partitions of sums of squares and degrees of freedom are performed, each partition producing between and within groups components. The same strategy is employed in two-way anova, but a second-level breakdown divides the between groups component into portions corresponding to row effects, column effects, and interaction effects.

To demonstrate, suppose the 12 cells in Example 6.1 are stretched out in a line, instead of structured in a rectangle. The 4 cells in the first row are followed by the 4 cells in the second row, and then the 4 cells in the third row. As shown below, the experiment can be conceptualized as a one-way design with 12 treatments.

Treat.	1	2	3	4	5	6	7	8	9	10	11	12
Cell	C1S1	C1S2	C1S3	C1S4	C2S1	C2S2	C2S3	C2S4	C3S1	C3S2	C3S3	C3S4

In one-way analysis, the total sum of squares is broken into between and within components. The between groups sum of squares reflects all differences among the means of these 12 groups. Some of these differences correspond to row effects in the two-way structure, others to column effects, and still others to interaction effects. For example, if the first four treatments, on average, differ from the next four, then row one differs from row two. Similarly, if treatments 1, 5, and 9, on average, differ from treatments 2, 6, and 10, then column one differs from column two. Logically, the between groups sum of squares in one-way analysis includes row effects, column effects, and interaction effects in the two-way structure. To obtain these three separate terms, a second-level partition is performed, dividing the sum of squares between groups into portions for row effects, column effects, and interaction effects.

Although the between groups sum of squares from one-way analysis must be further partitioned in two-way anova, the within groups component is not subdivided. Recall that mean square within in one-way is the average of sample variances computed separately in each cell. Because each sample variance is computed *within* a cell, it does not matter whether the cells are structured in a rectangular array or strung out in a line. Within cells variability in two-way anova is identical to that in one-way.

The two-level partition of sums of squares is shown graphically in Figure 6.1. More than a conceptual tool, however, this two-step approach is actually employed in computer programs. Some programs even report results corresponding to the first stage of this partition, in addition to printing the second-level analysis needed to answer the three questions asked in a two-way design.

Degrees of freedom are also partitioned in a two-step process. We use the symbol $\boldsymbol{a}$ to represent the number of rows in a two-way design, $\boldsymbol{b}$ as the number of columns, and n as the number of observations per cell. Using these symbols, a two-way design is of order $\boldsymbol{a}\times\boldsymbol{b}$. In Example 6.1, the design is 3×4. Since there are $\boldsymbol{ab}$ cells, and each cell has n observations, the total number of observations is $N = \boldsymbol{ab}n$.

In two-way anova, as in one-way, total degrees of freedom are the total number of observations minus one. For two-way, this is $N - 1 = \boldsymbol{ab}n - 1$. The first step of the partition corresponds to one-way, with total degrees of freedom broken into degrees of freedom between groups and degrees of freedom within groups. As in one-way, degrees of freedom between is the number of groups minus one, in this case $\boldsymbol{ab} - 1$. Within each group, we have $n - 1$ degrees of freedom to estimate

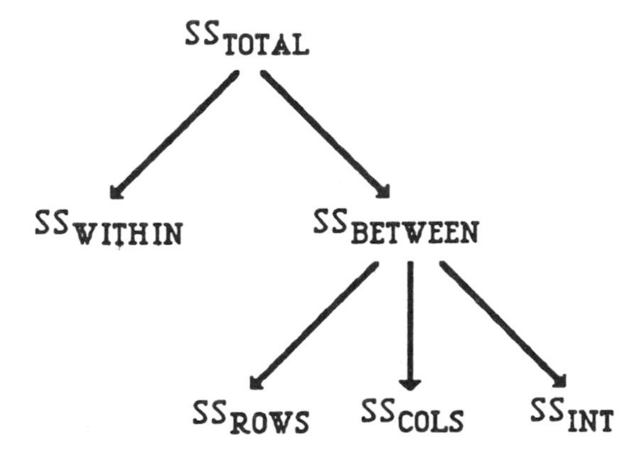

FIGURE 6.1. Two-step partition of sums of squares in two-way anova.

variance. Pooling variance estimates for all ab groups yields $ab(n - 1)$ degrees of freedom within groups. Putting the pieces together, we write the first level partition of degrees of freedom as

$$df_{TOTAL} = df_{BETWEEN} + df_{WITHIN},$$

or, equivalently,

$$abn - 1 = (ab - 1) + ab(n - 1).$$

In the second-level partition, degrees of freedom between is divided into degrees of freedom for rows, columns, and interaction. Since there are a rows, the sample variance among row means has $a - 1$ degrees of freedom. Similarly, the degrees of freedom for columns is $b - 1$. Interaction degrees of freedom can be found by multiplying to be $(a - 1)(b - 1)$. The second-level partition of degrees of freedom is

$$df_{BETWEEN} = df_{ROWS} + df_{COLS} + df_{INT},$$

or, equivalently,

$$(ab - 1) = (a - 1) + (b - 1) + (a - 1)(b - 1).$$

For Example 6.1, total degrees of freedom are $abn - 1 = (3)(4)(2) - 1 = 23$, degrees of freedom are within are $ab(n - 1) = (3)(4)(2 - 1) = 12$, degrees of freedom among coatings are $a - 1 = 3 - 1 = 2$, degrees of freedom among soils are $b - 1 = 4 - 1 = 3$, and interaction degrees of freedom are $(a - 1)(b - 1) = (3 - 1)(4 - 1) = 6$.

6.3 TESTING IN TWO-WAY ANOVA

As in one-way analysis of variance, each sum of squares is converted to a mean square by dividing by its degrees of freedom. In two-way, unlike one-way, three separate questions are asked, and thus three separate tests are made. To test each contribution to total variance, we form a ratio of mean squares, with the mean square being tested in the numerator, and mean square within in the denominator. Equality of row means is tested by the ratio of MS_{ROWS} to MS_{WITHIN}, the comparison of column means uses MS_{COLS} over MS_{WITHIN}, and the test for interaction uses MS_{INT} over MS_{WITHIN}.

As we already know, a ratio of mean squares has an F distribution. Each of our three F ratios has MS_{WITHIN} in the denominator and, therefore, each F test has denominator degrees of freedom $ab(n - 1)$. But each uses a different mean square in the numerator, and thus a different numerator degrees of freedom. As a result, the computed F ratios for testing row effects, column effects, and interaction effects must be compared against different critical values. In practice, this causes no difficulties. Computer programs automatically figure the proper degrees of freedom, and the p-value for each test is the upper tail area beyond the computed F ratio in the corresponding F distribution. For each test, we need merely examine the printed p-value. If it is less than 5 percent, we reject the null hypothesis.

For the data of Example 6.1, the anova summary table, including p-values, is shown in Table 6.1. We conclude that the three coatings do not provide equal protection from corrosion ($p = 0.011$), and the four soil types differ in terms of corroding metal pipe ($p = 0.034$). Interaction, however, is not present ($p = 0.959$).

Having found significant differences among the coatings and among the soil types, multiple range tests may be used for each main effect to determine which pairs of means differ. Applying Duncan's multiple range test to the three coatings produces the result:

Coating	2	1	3
Average	10.875	15.875	19.375

The second coating results in smaller mean pits than the third, and is therefore more effective in preventing corrosion. No other pair of coatings is significantly different.

Duncan's multiple range test on the four soil types yields:

Soil	3	1	2	4
Average	11.667	12.500	18.500	18.833

The third and first soil types produce less corrosion than the second and fourth. But the third and first are equally corrosive, as are the second and fourth.

TABLE 6.1 Anova Summary Table for Corrosion of Metal Pipe

Source	df	SS	MS	F	P
Coating	2	292.00	146.00	6.75	0.011
Soil	3	262.46	87.49	4.05	0.034
Interaction	6	29.67	4.94	0.23	0.959
Within	12	259.50	21.63		
Total	23	843.63			

6.4 TWO-WAY ANOVA IN COMPUTER PROGRAMS

Two-way analysis of variance can be performed with each of the programs, SAS, SPSS, and MINITAB. Multiple range tests, however, are available in factorial designs only on SAS. The data of Example 6.1 are used to demonstrate two-way anova in each package. Some of the input lines shown for SAS and SPSS are used to fit a two-way model without interaction, a topic discussed in Section 6.6. Output from these commands is deferred until Section 6.7.

SAS Input

```
OPTIONS LS = 80;
TITLE 'EXAMPLE 6.1 - METAL PIPE CORROSION';
DATA;
INPUT COATING SOIL CORRSN;
LINES;
1  1  10
1  1  16
1  2  22
1  2  20
1  3  11
1  3  11
1  4  23
1  4  14
2  1   8
2  1   7
2  2   8
2  2  16
2  3  13
2  3   5
2  4  12
2  4  18
3  1  19
3  1  15
3  2  18
3  2  27
3  3  12
3  3  18
3  4  28
3  4  18
```

```
PROC ANOVA;
CLASS COATING SOIL;
MODEL CORRSN = COATING SOIL COATING*SOIL;
MEANS COATING SOIL / DUNCAN;
PROC ANOVA;
CLASS COATING SOIL;
MODEL CORRSN = COATING SOIL;
```

Discussion of SAS Input

Three variables are entered for each observation, COATING (row number), SOIL (column number), and CORRSN (depth of maximum pit, in ten-thousandths of an inch). The input lines shown here are ordered by group, but this is not required. The first call to PROC ANOVA lists the classification variables COATING and SOIL in the CLASS statement. In the model statement, CORRSN is identified as the dependent variable, with components of variance for each main effect and the interaction. Interactions are indicated in SAS with an asterisk; COATING*SOIL is the interaction of our two factors. The MEANS statement accompanying this procedure requests Duncan's multiple range test for each main effect.

Input and output for the second PROC ANOVA command will be discussed in Section 6.7.

SAS Output

```
                EXAMPLE 6.1 - METAL PIPE CORROSION                 1
                  Analysis of Variance Procedure
                     Class Level Information

                    Class      Levels    Values
                    COATING      3       1 2 3
                    SOIL         4       1 2 3 4

              Number of observations in data set = 24
```

```
                  EXAMPLE 6.1 - METAL PIPE CORROSION                   2
                    Analysis of Variance Procedure

Dependent Variable:   CORRSN
                           Sum of           Mean
Source             DF      Squares          Square      F Value    PR > F
Model              11   584.12500000   53.10227273     2.46      0.0690
Error              12   259.50000000   21.62500000
Corrected Total    23   843.62500000
```

R-Square	C.V.	Root MSE	CORRSN Mean
0.692399	30.24565	4.6502688	15.375000

Source	DF	Anova SS	Mean Square	F Value	Pr > F
COATING	2	292.00000000	146.00000000	6.75	0.0109
SOIL	3	262.45833333	87.48611111	4.05	0.0335
COATING*SOIL	6	29.66666667	4.94444444	0.23	0.9593

EXAMPLE 6.1 - METAL PIPE CORROSION 3
Analysis of Variance Procedure

Duncan's Multiple Range Test for variable: CORRSN

NOTE: This test controls the type I comparisonwise
error rate, not the experimentwise error rate

Alpha= 0.05 df= 12 MSE= 21.625

Number of Means	2	3
Critical Range	5.056	5.297

Means with the same letter are not significantly different.

Duncan Grouping		Mean	N	COATING
	A	19.375	8	3
	A			
B	A	15.875	8	1
B				
B		10.875	8	2

EXAMPLE 6.1 - METAL PIPE CORROSION 4
Analysis of Variance Procedure

Duncan's Multiple Range Test for variable: CORRSN

NOTE: This test controls the type I comparisonwise
error rate, not the experimentwise error rate

Alpha = 0.05 df= 12 MSE = 21.625

Number of Means	2	3	4
Critical Range	5.839	6.116	6.302

```
Means with the same letter are not significantly different.

     Duncan Grouping          Mean        N        SOIL

                   A        18.833        6          4
                   A
                   A        18.500        6          2

                   B        12.500        6          1
                   B
                   B        11.667        6          3
```

Discussion of SAS Output

The Class Level Information shown on the first page of output lists our two classification variables: COATING with three levels and SOIL with four levels. At the top of page 2 is the first-level breakdown of variance, equivalent to a one-way analysis on all 12 cells. Recall that SAS uses the word Model instead of BETWEEN, and Error instead of WITHIN. Corrected Total is the same as TOTAL variability. The F value of 2.46 would be appropriate if we wanted to test the one-way hypothesis that all 12 cells in our design have the same mean. Typically, this is not a meaningful test in a two-way design. Values of R^2 and the coefficient of variation are printed for this one-way analysis.

At the bottom of this page is the second-level partition of $SS_{BETWEEN}$ into sums of squares for COATING, SOIL, and COATING*SOIL (interaction). Note that these sums of squares add to 584.125, which is $SS_{BETWEEN}$. F ratios for these terms are constructed by dividing each mean square by MS_{WITHIN}. Based on the p-values, we conclude that the three coatings do not provide equal protection from corrosion ($p = 0.0109$), and the four soil types differ in terms of corroding metal pipe ($p = 0.0335$). Interaction is not significant ($p = 0.9593$).

Duncan's multiple range test on the three coatings is shown on page 3. The string of As connects the means of coatings three and one, while the Bs connect the means of coatings one and two. Coatings three and two are the only pair that differ in corrosion protection. On page 4 are the results of Duncan's multiple range test among the four soil types. As discussed in Section 6.3, the fourth and second soil types are more corrosive than the first and third.

SPSS Input

```
SET WIDTH = 80
TITLE 'EXAMPLE 6.1 - METAL PIPE CORROSION'
DATA LIST FREE / COATING SOIL CORRSN
```

```
ANOVA VARIABLES = CORRSN BY COATING (1,3) SOIL (1,4)
BEGIN DATA
1  1  10
1  1  16
1  2  22
1  2  20
1  3  11
1  3  11
1  4  23
1  4  14
2  1   8
2  1   7
2  2   8
2  2  16
2  3  13
2  3   5
2  4  12
2  4  18
3  1  19
3  1  15
3  2  18
3  2  27
3  3  12
3  3  18
3  4  28
3  4  18
END DATA
ANOVA VARIABLES = CORRSN BY COATING (1,3) SOIL (1,4)
      / MAXORDERS = NONE
FINISH
```

Discussion of SPSS Input

Three variables are entered in free format for each observation—COATING (row number), SOIL (column number), and CORRSN (depth of maximum pit, in ten-thousandths of an inch). The input lines shown here are ordered by group, but this is not required.

The ANOVA command requests analysis of variance for the variable CORRSN using factors COATING, having levels one through three, and SOIL, with levels one through four. Interactions are automatically included by SPSS for anova designs up to five-way.

Multiple range tests are not available in SPSS, except in the ONEWAY procedure. At first thought, one might be inclined to use this approach to obtain multiple range tests for each of the main effects, run as separate one-way designs. However, this is definitely *not* advisable. Performing a one-way analysis on the three coatings, for example, implies pooling variability associated with soil types and interaction into the error term. As we will see in Section 6.6, inappropriate pooling inflates the error term and decreases the chance of finding significant results.

The second ANOVA command, after the data, and its associated output, will be discussed in Section 6.7.

SPSS Output

```
          EXAMPLE 6.1 - METAL PIPE CORROSION          Page 1

      * * * A N A L Y S I S   O F   V A R I A N C E * * *

          CORRSN
     by   COATING
          SOIL

                          Sum of              Mean               Sig
Source of Variation      Squares    DF      Square         F    of F
Main Effects             554.458     5     110.892     5.128    .010
   COATING               292.000     2     146.000     6.751    .011
   SOIL                  262.458     3      87.486     4.046    .034
2-Way Interactions        29.667     6       4.944      .229    .959
   COATING SOIL           29.667     6       4.944      .229    .959
Explained                584.125    11      53.102     2.456    .069
Residual                 259.500    12      21.625
Total                    843.625    23      36.679
24 cases were processed.
0 cases (.0 pct) were missing.
```

Discussion of SPSS Output

Page 1 of the output reports results from the first ANOVA command. At the bottom of the table is the first-level breakdown of variance, equivalent to a one-way analysis on all 12 cells. SPSS uses the word Explained instead of BETWEEN, and Residual instead of WITHIN. The F value of 2.456 would be appropriate if we wanted to test the one-way hypothesis that all 12 cells in our design have the same mean. Typically, this is not a meaningful test in a two-way design.

Terms higher in the table correspond to the second level partition of $SS_{BETWEEN}$ into sums of squares for COATING, SOIL, and interaction. Note that these sums of squares add to 584.125, which is $SS_{BETWEEN}$. F ratios are computed for these mean squares by dividing each by MS_{WITHIN}. Based on the p-values, we conclude that the three coatings do not provide equal protection from corrosion ($p = 0.011$), and the four soil types differ in terms of corroding metal pipe ($p = 0.034$). Interaction is not significant ($p = 0.959$). On the line labeled Main Effects, SPSS prints a pooled test for COATING and SOIL, although it is unlikely that this test is meaningful.

MINITAB Transcript

```
MTB > NAME C1 = 'COATING'
MTB > NAME C2 = 'SOIL'
MTB > NAME C3 = 'CORRSN'
MTB > READ 'COATING' 'SOIL' 'CORRSN'
DATA > 1   1  10
DATA > 1   1  16
DATA > 1   2  22
DATA > 1   2  20
DATA > 1   3  11
DATA > 1   3  11
DATA > 1   4  23
DATA > 1   4  14
DATA > 2   1   8
DATA > 2   1   7
DATA > 2   2   8
DATA > 2   2  16
DATA > 2   3  13
DATA > 2   3   5
DATA > 2   4  12
DATA > 2   4  18
DATA > 3   1  19
DATA > 3   1  15
DATA > 3   2  18
DATA > 3   2  27
DATA > 3   3  12
DATA > 3   3  18
DATA > 3   4  28
DATA > 3   4  18
DATA > END
     24 ROWS READ
```

```
MTB > ANOVA CORRSN = COATING SOIL COATING*SOIL

Factor       Type      Levels     Values
COATING     fixed        3          1   2   3
SOIL        fixed        4          1   2   3   4

Analysis of Variance for CORRSN

Source            DF        SS         MS         F        P
COATING            2    292.00     146.00      6.75    0.011
SOIL               3    262.46      87.49      4.05    0.034
COATING*SOIL       6     29.67       4.94      0.23    0.959
Error             12    259.50      21.62
Total             23    843.63
```

Discussion of MINITAB Transcript

In the first three lines of this transcript, columns C1, C2, and C3 on the worksheet are defined to be variables named COATING (row number), SOIL (column number), and CORRSN (depth of maximum pit, in ten-thousandths of an inch). A READ command is used to input the data, since more than one value is entered for each observation. Although our transcript shows observations entered by group, data may be read in any order.

The ANOVA command contains a model identifying CORRSN as the dependent variable, with components of variance for each main effect and the interaction. Interactions are indicated in MINITAB with an asterisk; COATING*SOIL is the interaction of our two factors. Recall that single quotes around variable names may be omitted in the model portion of the ANOVA command.

Output from this command includes a brief description of the experiment, which contains two factors: COATING (having three levels), and SOIL (with four levels). In the anova summary table, MINITAB uses the label "Error" to refer to the within groups source of variability. Based on the p-values, we conclude that the three coatings do not provide equal protection from corrosion ($p = 0.011$), and the four soil types differ in terms of corroding metal pipe ($p = 0.034$). Interaction is not significant ($p = 0.959$).

The STOP statement is not shown in this MINITAB transcript, since an additional command, used to fit a two-way model without interaction, was appended to the end of the transcript shown here. Analysis without interaction is discussed in Section 6.6, and the additional MINITAB command and its output are presented in Section 6.7.

6.5 INTERACTION

Interaction in a two-way design indicates that row and column effects do not operate independently. The desirability of a particular row treatment may depend on which column treatment is considered. Even more importantly, the presence of interaction shapes the way we interpret the tests of row and column effects. To illustrate this phenomenon, we will examine plots of cell means. Let us begin by plotting means for Example 6.1, in which interaction was absent. A second example, with significant interaction, will be used to show the pattern produced by interaction. Cell means for Example 6.1 are shown in Table 6.2.

In Figure 6.2(a), one line is drawn for each coating. Although these (broken) lines are not perfectly parallel, they show essentially the same pattern across the four soil types. Similarly, the broken lines drawn for the four soil types in Figure 6.2(b) each show a decrease in corrosion from the first coating to the second, and an increase from the second to the third.

Absence of interaction, as seen in this example, is reflected in nearly parallel lines. In fact, the test for interaction can be interpreted as a test for parallelism. The

TABLE 6.2 **Cell Means for Corrosion of Metal Pipe**

	Soil			
Coating	S1	S2	S3	S4
C1	13.0	21.0	11.0	18.5
C2	7.5	12.0	9.0	15.0
C3	17.0	22.5	15.0	23.0

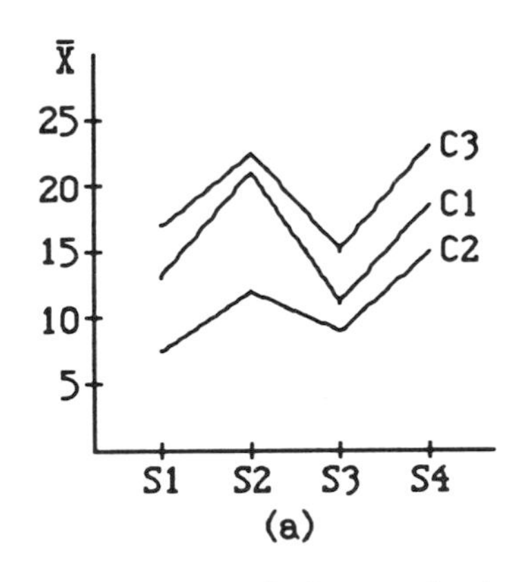

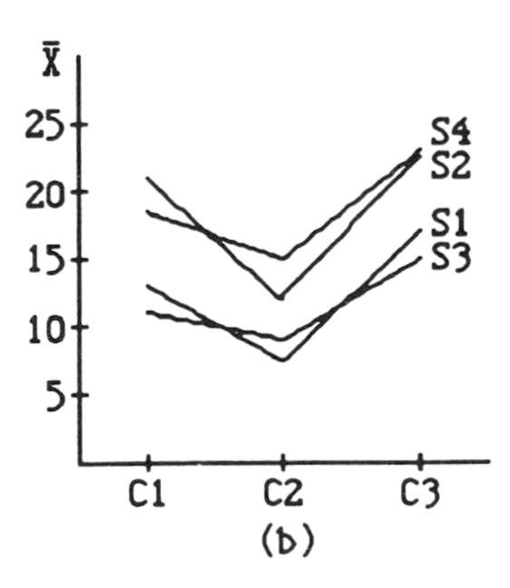

FIGURE 6.2. Plots of cell means when interaction is absent.

null hypothesis of no interaction implies parallel lines in the population. If lines drawn with sample data are close to parallel, we accept the null hypothesis of no interaction in the population. In Example 6.1, the sample lines are nearly parallel, and the test for interaction is nonsignificant.

Example 6.2, on the other hand, illustrates the effects of significant interaction. For these data, the anova summary table, including p-values, is shown in Table 6.3. Inspection of main effect tests suggests that the two construction designs differ (p = 0.0001), but plastic type has no effect on bursting strength (p = 0.4258).

The significant interaction (p = 0.0001) poses a problem. If there is interaction, then both main effects must exist, or else what is interacting? Plots of cell means help shed light on this enigma. For these data, the cell means are shown in Table 6.4, and plotted in Figure 6.3.

When lines are drawn for each construction design, as in Figure 6.3(a), we see distinct nonparallelism. Whether there is a difference between the two construction designs depends on which plastic type is employed. The two designs differ slightly in bursting strength for the first type of plastic, greatly for the second type, but not at all for the third. The test for construction design is significant, reflecting that line D1, *on average* across the three plastic types, is higher than line D2.

Figure 6.3(b) paints an even more complex picture. Here the *direction* of difference between any pair of plastic types changes from one construction design

---

Example 6.2

A study of the strength of plastic grocery bags involved two different construction designs and three types of plastic. For each combination of construction design and plastic type, five bags were tested for bursting strength, resulting in the following data.

Design	P1	Plastic P2	P3
	87	101	86
	92	93	91
D1	85	102	84
	88	95	87
	90	92	88
	80	77	84
	85	73	83
D2	87	72	87
	82	73	92
	85	75	90

TABLE 6.3 **Anova Summary Table for Plastic Grocery Bags**

Source	df	SS	MS	F	P
Design	1	616.53	616.53	59.95	0.0001
Plastic	2	18.20	9.10	0.88	0.4258
Interaction	2	713.27	356.63	34.68	0.0001
Within	24	246.80	10.28		
Total	29	1594.80			

TABLE 6.4 **Cell Means for Plastic Grocery Bags**

Design	P1	Plastic P2	P3
D1	88.4	96.6	87.2
D2	83.8	74.0	87.2

to the other. For example, P2 is higher than P1 with design D1, but lower with design D2. However, *on average* across the two construction designs, lines P1, P2, and P3 are approximately equal. As a result, the test of the plastic main effect is nonsignificant. This does not imply that the three plastic types are equivalent in bursting strength. They are greatly different. But the strongest plastic type cannot be chosen without knowing the construction design.

As this discussion implies, the test of each main effect compares means *averaged* across levels of the other main effect. When interaction exists, the averages

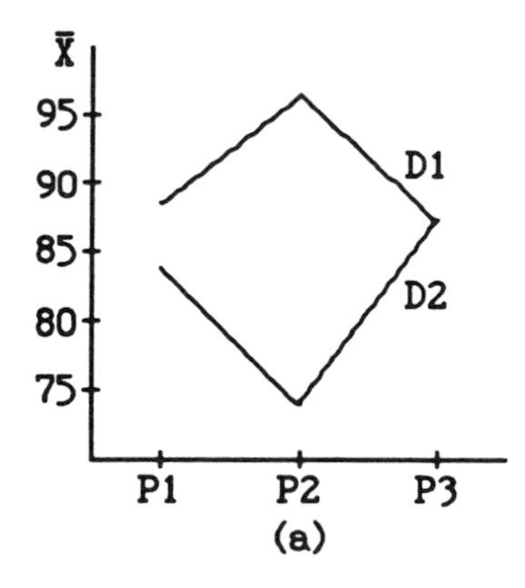

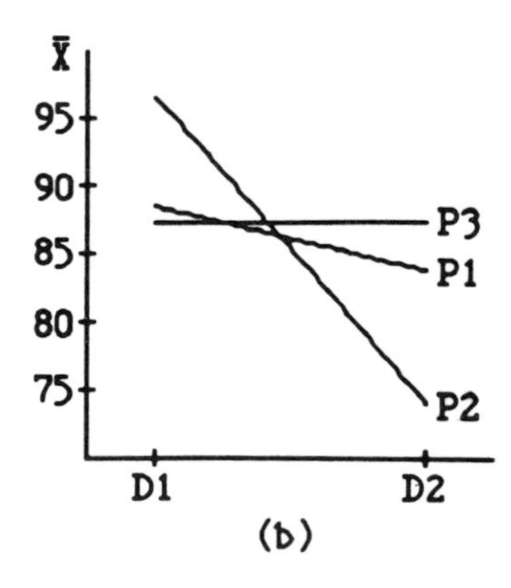

FIGURE 6.3. Plots of cell means when interaction is present.

may be nearly equal even though the lines are greatly different. Interaction may mask main effects.

In practical applications, we should examine the test for interaction before looking at main effects. If there is no interaction, the lines are at least roughly parallel. The difference between parallel lines is the same at any point and, therefore, the same as the *average* difference between the lines. Main effect tests, which look at averages, describe the differences between lines adequately. But if interaction is present, average differences may be misleading. The average difference between levels of one main effect may not be the same as the difference at *any single level* of the other main effect. In the presence of interaction, main effect tests are relevant only if we are interested in such averages. Depending on our objective, we may or may not wish to examine tests of main effects.

Consider the situation of Example 6.2, in which interaction is significant. Suppose differences in customer preferences require that grocery bags be available in both types of construction designs. (Some customers prefer bags with handles, others like drawstrings.) If retooling equipment is difficult enough to necessitate that a single type of plastic be used for both construction designs, we wish to choose the strongest plastic on average. In this case, we would be interested in the test of the plastic main effect, which compares the three plastic types *on average* across the two construction designs.

But if different types of plastic can be used for each construction design, we do not care which plastic is best on average. The test of the plastic main effect is irrelevant. We will use the strongest plastic for each construction design separately. To determine which plastic to use for each construction design, we could do two separate one-way anovas, one comparing the three plastic types for the first construction design, and one comparing them for the second design. These one-way anovas are called **simple effects tests.**

In summary, interaction signifies that something complex is present in our experiment. It is a red flag that says "tread with caution." We should always observe the test of interaction before looking at main effects, since the presence of interaction may alter the way we examine the main effects.

6.6 POOLING

Each mean square is similar to a sample variance and estimates some sort of variability. Mean square within estimates random variability, the mean square for rows estimates variability among row averages, the mean square for columns estimates variability among column averages, and the mean square for interaction estimates variability associated with interaction between row and column factors.

Suppose the test for interaction is *clearly* nonsignificant, as in Example 6.1. (The p-value of 0.959 is not even close to 0.05.) We are quite convinced that interaction does not exist in the population. But even when there is no interaction in the population, we do not expect the interaction mean square in our sample to be exactly 0 due to random sampling. In essence, the mean square for interaction is another estimate of random variability.

When two estimates of the same quantity are available, we can take a weighted average (i.e., **pool** them), to obtain a single estimate. If there is no interaction, both mean square within and the interaction mean square are estimates of random variability. Using degrees of freedom as weights, the weighted average of these terms is

$$\frac{(df_{WITHIN})MS_{WITHIN} + (df_{INT})MS_{INT}}{df_{WITHIN} + df_{INT}}.$$

Each mean square is a sum of squares divided by degrees of freedom. Multiplying a mean square by its degrees of freedom produces the sum of squares. Therefore, we can write our pooled estimate of random variability as

$$\frac{SS_{WITHIN} + SS_{INT}}{df_{WITHIN} + df_{INT}}. \tag{6.6.1}$$

A pooled estimate of random variability is generally called an **error term**. As discussed in earlier chapters, degrees of freedom are additive, so the error term in (6.6.1) has degrees of freedom equal to $df_{WITHIN} + df_{INT}$. This error term can be used as the denominator for row and column mean squares when testing main effects. Applying this technique to Example 6.1 yields the anova summary table shown in Table 6.5. Note that sum of squares for error is $SS_{WITHIN} + SS_{INT}$, and its degrees of freedom are $df_{WITHIN} + df_{INT}$. The ratio of sum of squares to degrees of freedom is the mean square for error.

Pooling can be advantageous. The error term in Table 6.5 has more degrees of freedom (18) than mean square within in Table 6.1 (which has only 12). More

TABLE 6.5 Anova Summary Table for Corrosion of Metal Pipe with Interaction Pooled Into Error

Source	df	SS	MS	F	P
Coating	2	292.00	146.00	9.09	0.002
Soil	3	262.46	87.49	5.45	0.008
Error	18	289.17	16.06		
Total	23	843.63			

degrees of freedom is equivalent to a larger sample size—it gives us greater power. The tests for row and column effects may be more powerful when we use a pooled error term. Logically, we are using more information to estimate random variability, and should have a greater chance of reaching correct decisions. In Table 6.5, p-values for tests of coating and soil are smaller than when we used mean square within as the denominator in Table 6.1. In some cases, a smaller p-value can mean the difference between rejecting and accepting the null hypothesis.

But pooling can also be detrimental. If interaction actually exists, the interaction mean square should be larger than mean square within. Pooling produces an error term that overestimates random variability. Using this inflated denominator in main effect tests yields smaller F ratios and less chance for significant results. The main effect tests may be less powerful.

To pool or not to pool? The solution is to apply common sense. When we are sure there is no interaction, it is advantageous to pool; otherwise, it may be hazardous. If the F ratio for testing interaction is less than 1, it is surely safe to pool. The pooled error term will be smaller than mean square within, and tests for main effects will definitely be more powerful. On the other hand, if the F ratio for testing interaction is greater than 1, the pooled error term will be larger than mean square within. Even in this case, however, it may be beneficial to pool, since pooling also increases degrees of freedom. When the error mean square increases only slightly, but degrees of freedom increase substantially, the resulting tests of main effects may be more powerful.

It is difficult to state a definitive rule for pooling when the F ratio for interaction is greater than 1. We suggest that two conditions be met. First, the test for interaction should be *clearly* nonsignificant, with a p-value *much* greater than 5 percent. Second, the *proportional* increase in degrees of freedom for the error term should be substantial. An increase in degrees of freedom from 10 to 20 produces a larger change in power than an increase from 110 to 120.

Pooling is a basic statistical tool that can be used in a variety of analytic procedures. It is applicable to two-way and higher order factorial designs, as well as regression techniques. Pooling is easily accomplished in computer programs. We simply omit the source to be pooled from our list of influences. Any term left out of the model is automatically pooled into the error term.

6.7 POOLING IN COMPUTER PROGRAMS

Virtually every statistical program can perform pooling. Data from Example 6.1 are used to demonstrate pooling on SAS, SPSS, and MINITAB. The SAS and SPSS commands for pooling are included in inputs shown in Section 6.4, but are repeated here for easy reference. For MINITAB, the transcript shown here was actually appended to the transcript presented in Section 6.4.

SAS Input

```
•
• (data input as in Section 6.4)
•
PROC ANOVA;
CLASS COATING SOIL;
MODEL CORRSN = COATING SOIL;
```

Discussion of SAS Input

This call to PROC ANOVA lists the classification variables COATING and SOIL in the CLASS statement, but omits interaction in the model statement.

SAS Output

```
            EXAMPLE 6.1 - METAL PIPE CORROSION                 5
              Analysis of Variance Procedure
                 Class Level Information

               Class      Levels     Values
               COATING      3        1 2 3
               SOIL         4        1 2 3 4

          Number of observations in data set = 24
```

```
             EXAMPLE 6.1 - METAL PIPE CORROSION                 6
               Analysis of Variance Procedure

Dependent Variable: CORRSN
                             Sum of            Mean
Source             DF       Squares          Square    F Value    Pr > F
Model               5   554.45833333   110.89166667     6.90     0.0009
Error              18   289.16666667    16.06481481
Corrected Total    23   843.62500000

R-Square             C.V.          Root MSE          CORRSN Mean
0.657233         26.06890         4.0080937            15.375000

Source        DF         Anova SS      Mean Square     F Value      Pr > F
COATING        2     292.00000000     146.00000000        9.09      0.0019
SOIL           3     262.45833333      87.48611111        5.45      0.0076
```

Discussion of SAS Output

Pages 1 through 4 of the output were presented and discussed in Section 6.4. Pages 5 and 6, produced by this PROC ANOVA command, are shown here. Class Level Information still includes our two classification variables: COATING with three levels and SOIL with four.

In the anova table at the top of page 6, the Error term is obtained by pooling interaction and within. The Model term pools row and column effects, and its test is generally not meaningful. At the bottom of this page are separate tests of COATING and SOIL, using the pooled error term as denominator.

SPSS Input

```
•
• (data input as in Section 6.4)
•
ANOVA VARIABLES = CORRSN BY COATING (1,3) SOIL (1,4)
     / MAXORDERS = NONE
FINISH
```

Discussion of SPSS Input

This ANOVA command requests analysis of the dependent variable CORRSN, using factors COATING and SOIL. The MAXORDERS = NONE subcommand deletes all interactions from the model.

SPSS Output

```
             EXAMPLE 6.1 - METAL PIPE CORROSION       Page 2
      * * * A N A L Y S I S   O F   V A R I A N C E * * *

             CORRSN
        by   COATING
             SOIL
                         Sum of             Mean              Sig
Source of Variation     Squares    DF     Square       F     of F
Main Effects            554.458     5    110.892   6.903     .001
   COATING              292.000     2    146.000   9.088     .002
   SOIL                 262.458     3     87.486   5.446     .008
Explained               554.458     5    110.892   6.903     .001
Residual                289.167    18     16.065
Total                   843.625    23     36.679
24 cases were processed.
0 cases (.0 pct) were missing.
```

Discussion of SPSS Output

Page 1 of output was presented and discussed in Section 6.4. Page 2, produced by this ANOVA command, is shown here. The Residual term in the anova table is obtained by pooling interaction and within. Tests of COATING and SOIL use the pooled error term as denominator. The line labeled Explained pools row and column effects, and its test is generally not meaningful.

MINITAB Transcript

```
•
• (data input as in Section 6.4)
•
MTB > ANOVA CORRSN = COATING SOIL

Factor       Type     Levels     Values
COATING      fixed       3         1   2   3
SOIL         fixed       4         1   2   3   4

Analysis of Variance for CORRSN

Source       DF        SS          MS          F          P
COATING       2      292.00      146.00      9.09      0.002
SOIL          3      262.46       87.49      5.45      0.008
Error        18      289.17       16.06
Total        23      843.63

MTB > STOP
```

Discussion of MINITAB Transcript

This ANOVA command identifies CORRSN as the dependent variable, and COATING and SOIL as factors, but does not list interaction in the model. Output again includes a brief description of factor levels and the anova summary table. Interaction and within are pooled to obtain the Error term in this table. Tests of COATING and SOIL use the pooled error term as denominator.

6.8 UNBALANCED DESIGNS

As mentioned in Section 6.1, designs with equal sample sizes are called balanced. For balanced designs, the two-level breakdown of variance described in Section

6.2 poses no logical problems. **Unbalanced designs,** however, have inherent problems with confounding. It is impossible to completely separate the effects of rows, columns, and interaction when sample sizes are unequal.

We can see the effects of unequal sample sizes most easily with a small design, so consider a 2×2 experiment. Suppose the sample sizes (n) and cell means ($\overline{X}$) are as shown in Table 6.6. The overall average for the first row is computed as the weighted average of the two cell means in that row:

$$\overline{X}_{row1} = \frac{(2)(10) + (200)(20)}{2 + 200} = 19.90.$$

Calculating the overall average for the second row and for each column in similar fashion produces the values shown in Table 6.7.

There is a difference of approximately ten units between the row means, so row effects exist. Also, the difference between the column means is approximately ten, so column effects exist. Although we will not go into the details of computation, there is an interaction effect of approximately the same magnitude. But all these effects are due to the difference between the two cells with large sample sizes. These two cells have means of 20 and 30, and this spread of ten units appears in row, column, and interaction calculations. As a result, the row, column, and interaction effects are confounded. In short, there is only *one* effect. Whether we choose to call it a row effect, a column effect, or an interaction effect is arbitrary, but we should not claim there is a row effect of ten units *and* a column effect of ten units *and* an interaction effect of ten units.

TABLE 6.6 Cell Means in 2×2 Design

n	Col. 1	Col. 2	$\overline{X}$	Col. 1	Col. 2
Row 1	2	200	row 1	10	20
Row 2	200	2	row 2	30	40

TABLE 6.7 Row and Column Means for 2×2 Design

$\overline{X}$	Col. 1	Col. 2	Row Ave.
Row 1	10	20	19.90
Row 2	30	40	30.10
Col. Ave.	29.80	20.20	

Due to extremely unequal sample sizes in this example, a single difference between means accounts almost entirely for row, column, and interaction effects. With more nearly equal sample sizes, a single difference is unlikely to have large influence on more than one effect. But the principle applies as long as sample sizes are at all unequal: differences in cell means contribute to more than one effect. It is impossible to determine how much of total variability should be attributed to the three separate sources.

This phenomenon is displayed graphically in Figure 6.4. Total sum of squares is represented by the overall box in each of Figures 6.4(a), (b), and (c). The first-level breakdown of variability—total into within and between components—is shown in Figure 6.4(a). No problems occur with this partition due to unequal

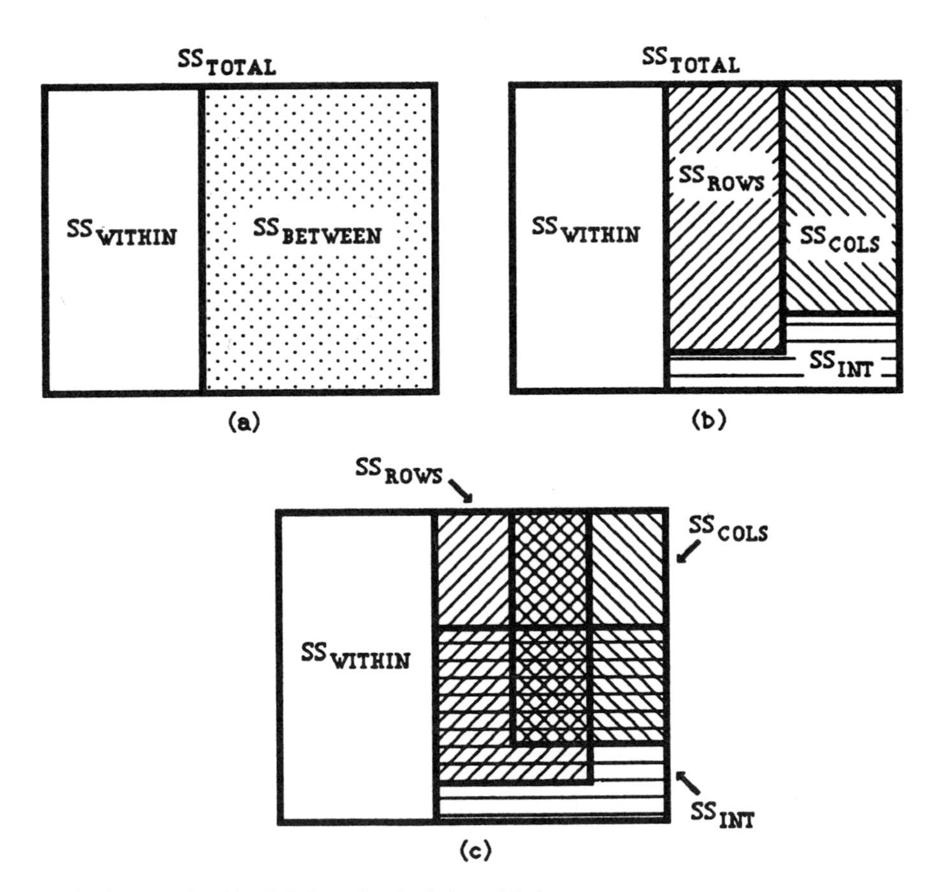

FIGURE 6.4. Partition in balanced and unbalanced designs.

sample sizes. (Recall that one-way anova can be performed with unequal sample sizes.) Since within group variability is properly separated from other sources, the *denominator* for F ratios is not a problem in unbalanced designs.

In the second level partition, between group variability is decomposed into row, column, and interaction sources. In balanced designs, these components are distinct, as depicted in Figure 6.4(b). Sums of squares for row, column, and interaction effects do not overlap. But with unequal sample sizes, regions for row, column, and interaction sums of squares overlap, as shown in Figure 6.4(c). The overlap portions of total variability cannot be uniquely assigned to specific effects.

We hasten to add that Figure 6.4(c) is a crude representation of the complexities posed by unequal sample sizes. In theory, proper pictorial representations can be drawn, but they require more dimensions. For a design of order a xb a figure of ab dimensions is needed. Even the smallest two-way design, 2×2, requires four dimensions and cannot be drawn or easily visualized. With balanced data, the axes of this ab-dimensional space are perpendicular. With unbalanced data, they are not.

Strange things happen in plots of high dimensionality with non-perpendicular axes! Some of these cannot be adequately represented in a two-dimensional approximation. For example, in Figure 6.4(c), due to overlaps, the between groups sum of squares is smaller than the sum of its parts (rows, columns, and interaction). But in non-perpendicular higher-dimensional space, it is possible for the between groups sum of squares to be *larger* than the sum of its parts! This utterly baffling phenomenon helps explain what we mean when we say that unbalanced designs are *inherently* complex.

Analysis of unbalanced data poses difficulties because there is no "correct" way to assign overlap (or underlap) variability to individual sources. Various strategies for attacking this problem revolve around different ways of dealing with overlap portions. The simplest approach is not to count overlaps at all. Each source is judged according to its unique contribution to total variability. Testing unique contributions is demonstrated in connection with Example 6.3.

The anova summary table in Table 6.8 shows unique sums of squares and associated tests. Adding sums of squares for gender, entry group, interaction, and within produces the value 371.12, which is not the same as total sum of squares. We have lost the difference, or 409.69 - 371.12 = 38.57, by ignoring overlap regions. Since the test for interaction is significant, we may or may not be interested in main effect results.

Cell and marginal means, shown in Table 6.9, reveal that the weight loss program was most effective for moderately overweight females and greatly overweight males. (This is the nature of interaction in these data.) Near normal participants tended to lose less weight, on a percentage basis, than those moderately or greatly overweight. On average, weight losses for females and males were about equal.

Example 6.3

The effectiveness of a weight loss program may depend on gender and the degree to which participants are overweight. According to standards for height and body type, persons can be categorized as near normal in weight, moderately overweight, or greatly overweight. A balanced 2×3 experiment was planned, using five females and five males in each of these three weight categories, for a total of 30 observations. But participants dropped out of the experiment unequally, and only 20 of the original 30 entrants completed the program. The dependent variable, shown below for each person who completed the program, is percentage weight loss (weight loss divided by initial weight, expressed as a percentage).

	Entry Group		
Gender	Near Normal	Moderately Overweight	Greatly Overweight
	8.3	13.5	12.2
	5.7	21.3	9.1
Female	11.6	18.1	
	8.6	12.9	
		14.4	
	4.3	10.8	15.2
		6.0	11.0
Male		12.3	19.4
			18.1
			14.5

TABLE 6.8 Anova Summary Table for Percentage Weight Loss

Source	df	Unique SS	MS	F	P
Gender	1	12.63	12.63	1.28	0.2768
Entry Group	2	111.24	55.62	5.64	0.0160
Interaction	2	109.17	54.58	5.53	0.0169
Within	14	138.08	9.86		
Total	19	409.69			

TABLE 6.9 Cell and Marginal Means for Percentage Weight Loss

	Entry Group			
Cell Means	Near Normal	Moderately Overweight	Greatly Overweight	Row Ave.
Female	8.55	16.04	10.65	12.34
Male	4.30	9.70	15.64	12.40
Col. Ave.	7.70	13.66	14.21	

In Example 6.3, the experiment began with five observations per cell, but unequal dropout rates resulted in an unbalanced design. When unequal sample sizes do not reflect unequal population proportions, as in this case, it is appropriate to ignore overlap regions and base our analysis on unique sums of squares. But suppose the weight loss program had been studied using 20 volunteers who were classified on entry according to gender and weight, with no entrants dropping out. Now the unequal sample sizes would probably reflect unequal proportions in the population, and a different method of analysis should be used. Alternative approaches are also in order if any of the cells are completely empty. Techniques for handling unbalanced data, other than ignoring overlaps, are discussed in more advanced books on analysis of variance. If you regularly analyze unbalanced data, we strongly encourage you to learn more about these other approaches.

6.9 UNBALANCED DESIGNS IN COMPUTER PROGRAMS

Each of the programs, SAS, SPSS, and MINITAB, computes unique sums of squares for unbalanced data. The data of Example 6.3 are used to demonstrate these tests. (Each program also includes other options for analysis, based on partial assignment of overlap regions, which are not demonstrated here.)

SAS Input

```
OPTIONS LS = 80;
TITLE 'EXAMPLE 6.3 - PERCENTAGE WEIGHT LOSS';
DATA;
INPUT GENDER ENTRYGRP PCTLOSS;
LINES;
1 1  8.3
1 1  5.7
```

```
1  1  11.6
1  1   8.6
1  2  13.5
1  2  21.3
1  2  18.1
1  2  12.9
1  2  14.4
1  3  12.2
1  3   9.1
2  1   4.3
2  2  10.8
2  2   6.0
2  2  12.3
2  3  15.2
2  3  11.0
2  3  19.4
2  3  18.1
2  3  14.5
PROC GLM;
CLASS GENDER ENTRYGRP;
MODEL PCTLOSS = GENDER ENTRYGRP GENDER*ENTRYGRP;
```

Discussion of SAS Input

GENDER, ENTRYGRP, and PCTLOSS are entered for each observation. PROC ANOVA is restricted to balanced data, but GLM (general linear models) can handle unbalanced designs. Like ANOVA, GLM requires CLASS and MODEL statements. In this example, GENDER and ENTRYGRP are identified as classification variables. The MODEL statement lists PCTLOSS as the variable to be analyzed, and includes the two factors and their interaction as influences.

SAS Output

```
            EXAMPLE 6.3 - PERCENTAGE WEIGHT LOSS              1
               General Linear Models Procedure
                   Class Level Information

                  Class        Levels    Values
                  GENDER          2      1 2
                  ENTRYGRP        3      1 2 3

           Number of observations in data set = 20
```

```
              EXAMPLE 6.3 - PERCENTAGE WEIGHT LOSS                     2
                General Linear Models Procedure

Dependent Variable: PCTLOSS
                              Sum of            Mean
Source               DF      Squares          Square      F Value    Pr > F
Model                 5   271.60650000     54.32130000       5.51    0.0052
Error                14   138.07900000      9.86278571
Corrected Total      19   409.68550000

R-Square            C.V.          Root MSE         PCTLOSS Mean
0.662963        25.39836         3.1405072            12.365000

Source               DF       Type I SS     Mean Square   F Value    Pr > F
GENDER                1      0.02004545      0.02004545      0.00    0.9647
ENTRYGRP              2    162.42012817     81.21006409      8.23    0.0043
GENDER*ENTRYGRP       2    109.16632637     54.58316319      5.53    0.0169

Source               DF     Type III SS     Mean Square   F Value    Pr > F
GENDER                1     12.62818792     12.62818792      1.28    0.2768
ENTRYGRP              2    111.23891123     55.61945561      5.64    0.0160
GENDER*ENTRYGRP       2    109.16632637     54.58316319      5.53    0.0169
```

Discussion of SAS Output

The Class Level Information shown on the first page of output lists our two classification variables: GENDER with two levels and ENTRYGRP with three levels.

At the top of page 2 is the first-level breakdown of variance, equivalent to a one-way analysis on all six cells. The Error term is within groups variability, and is used as the denominator in all tests. Two approaches to second-level partitioning are automatically printed by GLM. Type III SS correspond to unique sums of squares, as discussed above. Significant interaction warns us to be careful about interpreting main effect tests.

SPSS Input

```
SET WIDTH = 80
TITLE 'EXAMPLE 6.3 - PERCENTAGE WEIGHT LOSS'
DATA LIST FREE / GENDER ENTRYGRP PCTLOSS
ANOVA VARIABLES = PCTLOSS BY GENDER (1,2) ENTRYGRP (1,3)
     / METHOD = UNIQUE
BEGIN DATA
```

```
1 1  8.3
1 1  5.7
1 1 11.6
1 1  8.6
1 2 13.5
1 2 21.3
1 2 18.1
1 2 12.9
1 2 14.4
1 3 12.2
1 3  9.1
2 1  4.3
2 2 10.8
2 2  6.0
2 2 12.3
2 3 15.2
2 3 11.0
2 3 19.4
2 3 18.1
2 3 14.5
END DATA
FINISH
```

Discussion of SPSS Input

GENDER, ENTRYGRP, and PCTLOSS are entered in free format for each observation. The ANOVA command lists PCTLOSS as the variable to be analyzed, and specifies GENDER and ENTRYGRP as factors. The METHOD subcommand requests UNIQUE sums of squares.

SPSS Output

```
          EXAMPLE 6.3 - PERCENTAGE WEIGHT LOSS       Page 1
     * * * A N A L Y S I S  O F  V A R I A N C E * * *

          PCTLOSS
      by  GENDER
          ENTRYGRP
                           Sum of          Mean              Sig
Source of Variation        Squares   DF    Square      F     of F
Main Effects               112.250    3    37.417    3.794   .035
   GENDER                   12.628    1    12.628    1.280   .277
```

```
    ENTRYGRP               111.239    2    55.619    5.639    .016
2-Way Interactions         109.166    2    54.583    5.534    .017
    GENDER   ENTRYGRP      109.166    2    54.583    5.534    .017
Explained                  271.606    5    54.321    5.508    .005
Residual                   138.079   14     9.863
Total                      409.685   19    21.562
20 cases were processed.
0 cases (.0 pct) were missing.
```

Discussion of SPSS Output

At the bottom of the anova table, Explained (equivalent to BETWEEN) and Residual (equivalent to WITHIN), reflect the first level breakdown of variance. The Residual term is used as the denominator in all tests. Tests of GENDER, ENTRYGRP, and interaction are as reported in the previous section. Significant interaction warns us to be careful about interpreting main effect tests.

MINITAB Transcript

```
MTB > NAME C1 = 'GENDER'
MTB > NAME C2 = 'ENTRYGRP'
MTB > NAME C3 = 'PCTLOSS'
MTB > READ C1-C3
DATA > 1  1   8.3
DATA > 1  1   5.7
DATA > 1  1  11.6
DATA > 1  1   8.6
DATA > 1  2  13.5
DATA > 1  2  21.3
DATA > 1  2  18.1
DATA > 1  2  12.9
DATA > 1  2  14.4
DATA > 1  3  12.2
DATA > 1  3   9.1
DATA > 2  1   4.3
DATA > 2  2  10.8
DATA > 2  2   6.0
DATA > 2  2  12.3
DATA > 2  3  15.2
DATA > 2  3  11.0
DATA > 2  3  19.4
DATA > 2  3  18.1
```

```
DATA > 2  3  14.5
DATA > END
     20 ROWS READ

MTB > GLM PCTLOSS = GENDER ENTRYGRP GENDER*ENTRYGRP

Factor          Levels          Values
GENDER            2               1   2
ENTRYGRP          3               1   2   3

Analysis of Variance for PCTLOSS

Source              DF    Seq SS    Adj SS    Adj MS     F      P
GENDER               1     0.020    12.628    12.628   1.28  0.277
ENTRYGRP             2   162.420   111.239    55.619   5.64  0.016
GENDER*ENTRYGRP      2   109.166   109.166    54.583   5.53  0.017
Error               14   138.079   138.079     9.863
Total               19   409.685

MTB > STOP
```

Discussion of MINITAB Transcript

Values for GENDER, ENTRYGRP, and PCTLOSS are entered in columns C1-C3 on the worksheet. The ANOVA command is restricted to balanced data, but GLM (general linear models) can handle unbalanced designs. Like ANOVA, GLM contains a model of the experiment. In this example, PCTLOSS is the variable to be analyzed, and the two factors and their interaction are listed as influences.

Output from this command includes a brief description of the experiment, which contains two factors: GENDER (having two levels), and ENTRYGRP (with three levels). In the anova summary table, "Error" refers to within groups variability, and this mean square is used as the denominator in all tests. Two approaches to second-level partitioning are automatically printed by MINITAB. "Adj SS" (adjusted sums of squares) correspond to unique sums of squares, as discussed above. The F tests printed by MINITAB are based on these unique sums of squares. Significant interaction warns us to be careful about interpreting main effect tests.

6.10 THREE-WAY ANOVA

A three-way factorial design involves three classification variables, and includes observations for all possible combinations of levels. Tests are performed for each of the three main effects, all pairwise interactions and the **three-way interaction**.

Example 6.4 illustrates a three-way design. In this experiment, we test for differences in milk quality among the four breeds, among the three diets, and between the two machines. We also examine the data for interactions of breed by diet, breed by machine, and diet by machine. Each of these two-way interactions has the same interpretation as in two-way anova. If a two-way interaction is significant, it may mask the two main effects interacting, but will not obscure the third main effect. Three-way interaction occurs if the pattern of any two-way interaction changes over levels of the third factor. This definition of three-way interaction will be explained graphically in a few pages, in connection with Figure 6.5. If three-way interaction exists, it may mask some or all of the other effects in the experiment.

Example 6.4 is a 4 (breed) × 3 (diet) × 2 (machine) design with a total of (4)(3)(2) = 24 cells. In theory, the cells are structured in a three-dimensional rectangular box. But for display purposes, the data are presented as a series of layers of the box. As in two-way anova, partitions of sums of squares and degrees of freedom are performed in a two-step process. In the first step, we act as though all 24 cells were stretched out in a line, performing a one-way breakdown of total sum of squares into sums of squares between and within. The second step partitions

Example 6.4

An experiment was conducted to investigate the quality of milk produced from dairy cattle as a function of three factors: B = breed of cow (4 levels), D = dietary feed (3 levels), and M = type of milking machine (2 levels). For each breed × diet × machine combination, two independent observations of milk quality are presented below.

	Diet							
	D1			D2			D3	
	Machine			Machine			Machine	
Breed	M1	M2	Breed	M1	M2	Breed	M1	M2
B1	7	29	B1	23	40	B1	29	34
	16	25		25	41		13	23
B2	15	4	B2	16	5	B2	12	9
	7	9		22	7		6	9
B3	3	7	B3	11	15	B3	9	2
	11	1		7	6		2	12
B4	18	8	B4	11	11	B4	19	9
	23	7		25	15		19	15

the between groups sum of squares into components for each main effect, the three two-way interactions, and the three-way interaction.

Degrees of freedom are similarly partitioned. As a one-way design, this experiment would have 23 degrees of freedom between groups for the 24 cells. In the second level breakdown, each main effect has one fewer degrees of freedom than its number of levels, and degrees of freedom for interactions are found by multiplying. For the main effects in this design, breed has four levels and thus three degrees of freedom, diet has two degrees of freedom, and machine has one. The breed by diet interaction has (3)(2) = 6 degrees of freedom, breed by machine has (3)(1) = 3 degrees of freedom, and diet by machine has (2)(1) = 2 degrees of freedom. The three-way interaction has (3)(2)(1) = 6 degrees of freedom. Thus, the 23 degrees of freedom between cells is divided into seven terms, with degrees of freedom 3, 2, 1 (main effects), 6, 3, 2 (two-way interactions), and 6 (three-way interaction).

The within groups component measures variability within each cell and is unaffected by whether the cells are structured in a rectangular box or stretched out in a line. Consequently, the sum of squares within and degrees of freedom within from the one-way partition are used to form the error term for all tests in three-way analysis of variance.

Each sum of squares is divided by its degrees of freedom to form a mean square. And the mean square for each effect is tested using mean square within as the denominator in an F ratio. Since each F ratio has a different numerator mean square, critical values must be determined separately for each test. Computer programs automatically print the p-value for each F statistic adjusted for the proper degrees of freedom. Results are summarized in an analysis of variance table similar to that used in two-way anova.

For the data of Example 6.4, the anova summary table is shown in Table 6.10. Since the presence of three-way interaction may mask some or all of the other effects in the experiment, we should look at this test first. In our example we see that three-way interaction is absent ($p = 0.3786$), and it is appropriate to examine

TABLE 6.10 Anova Summary Table for Milk Quality

Source	df	SS	MS	F	P
Breed	3	2312.17	770.72	29.27	0.0001
Diet	2	260.17	130.08	4.94	0.0160
Machine	1	0.75	0.75	0.03	0.8674
Breed × Diet	6	162.83	27.14	1.03	0.4300
Breed × Machine	3	829.75	276.58	10.50	0.0001
Diet × Machine	2	6.50	3.25	0.12	0.8845
Breed × Diet × Machine	6	177.50	29.58	1.12	0.3786
Within	24	632.00	26.33		
Total	47	4381.67			

TABLE 6.11 Cell Means for Milk Quality

	Diet							
	D1			D2			D3	
	Machine			Machine			Machine	
Breed	M1	M2	Breed	M1	M2	Breed	M1	M2
B1	11.5	27.0	B1	24.0	40.5	B1	21.0	28.5
B2	11.0	6.5	B2	19.0	6.0	B2	9.0	9.0
B3	7.0	4.0	B3	9.0	10.5	B3	5.5	7.0
B4	20.5	7.5	B4	18.0	13.0	B4	19.0	12.0

tests of other effects. Two-way interactions should be checked next, as their presence may alter our interpretation of main effect tests. Interactions of breed × diet ($p = 0.4300$) and diet × machine ($p = 0.8845$) are nonsignificant, but the breed × machine interaction ($p = 0.0001$) is quite strong.

Cell means, shown in Table 6.11, are used to plot the interaction of breed × machine separately for each diet in Figure 6.5. In each plot of Figure 6.5, B2, B3, and B4 show downward trend or stay nearly horizontal from M1 to M2, while B1 increases. Three-way interaction does not exist because the pattern of two-way interaction between breed and machine is the same for each diet.

In the absence of three-way interaction, it is sufficient to examine the two-way interaction between breed and machine by averaging over the three diets. Means for each combination of breed and machine, averaged over diets, are shown in

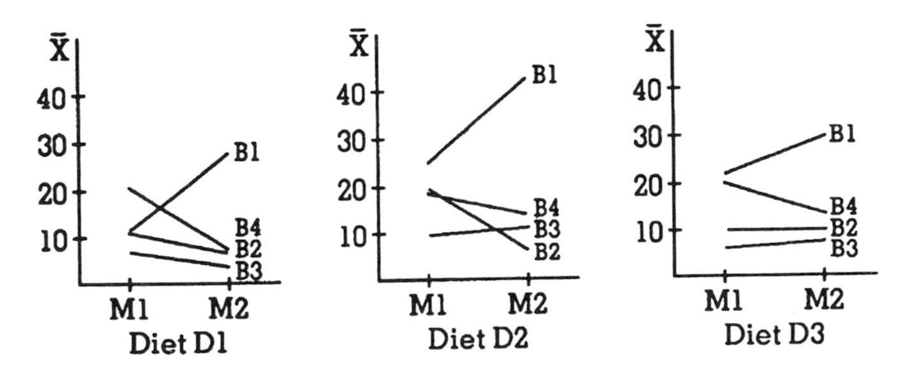

FIGURE 6.5. Breed × machine interaction for each diet.

TABLE 6.12 Mean Milk Quality for Each Breed × Machine Combination

	M1	M2	Row Ave.
B1	18.83	32.00	25.42
B2	13.00	7.17	10.08
B3	7.17	7.17	7.17
B4	19.17	10.83	15.00
Col. Ave.	14.54	14.29	

Table 6.12. The plot of these means, as seen in Figure 6.6, displays the same pattern as each of the plots in Figure 6.5. For breed 1, the second type of milking machine produces higher milk quality; for breeds 2 and 4, the first machine is preferable; and for breed 3, the two machines produce the same milk quality.

The main effect test for breed is significant ($p = 0.0001$), despite the interaction between breed and machine. That is, averaged over diets and machines, the four breeds do not produce the same quality of milk. Applying Duncan's multiple range test to the four breeds, we find the result:

Breed	B1	B4	B2	B3
Average	25.42	15.00	10.08	7.17

On average, the first breed produces superior milk quality to the fourth, which is better than the second and third breeds. There is no difference in milk quality between the second and third breeds. But the interaction between breed and

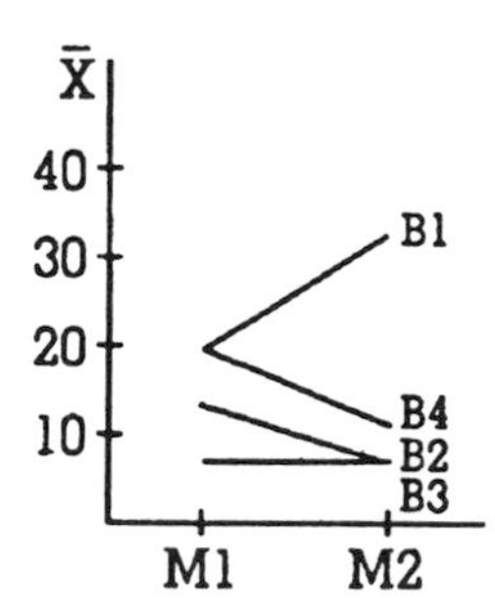

FIGURE 6.6. Breed×machine interaction averaged over diets.

machine cautions us to be wary of hasty interpretations. As shown in Figure 6.6, for the first type of milking machine, breed one is nearly equal to breed four, and breed two is higher than breed three.

The main effect test comparing the two machines is nonsignificant ($p = 0.8674$), and Duncan's multiple range test yields:

Machine	M1	M2
Average	14.54	14.29

On average, these two milking machines produce about the same quality of milk. But, again, interaction implies that there are indeed differences between the machines. For the first breed, machine two produces higher quality milk than machine one, as seen in Figure 6.6, but for the second and fourth breeds, machine one is preferred to machine two.

The test of the diet main effect is uncontaminated by interaction, since the only significant interaction (breed × machine) does not involve diet. The three diets are significantly different ($p = 0.0160$), and Duncan's multiple range test produces the result:

Diet	D2	D3	D1
Average	17.500	13.875	11.875

The second diet is preferable to the first, but neither of these is significantly different from the third.

As seen in this example, interpretation of results for a three-way design involves considerable effort, especially in the presence of interaction. We proceed by steps, beginning with the three-way interaction term, and working downward toward tests of main effects. Despite the complexity of the analysis above, the task could have been considerably worse—in our example, the three-way interaction was nonsignificant.

When three-way interaction is significant, we may or may not be interested in tests of other effects. In the presence of three-way interaction, the test of a two-way interaction pertains to the *average* pattern of the two-way interaction across levels of the third factor. The test of a main effect compares averages across levels of both other factors. If we are not interested in such averages, we may conduct simple effects tests by performing a series of two-way anovas. For example, if three-way interaction had been significant in Example 6.4, we might have conducted two separate two-way anovas of breed × diet—one for the first milking machine and one for the second. Alternatively, we could perform three separate two-way anovas of breed × machine—one for each diet—or four separate two-way anovas of diet × machine—one for each breed.

6.11 THREE-WAY ANOVA IN COMPUTER PROGRAMS

Each of the programs, SAS, SPSS, and MINITAB, can perform three-way analysis of variance, but only SAS does multiple range tests in factorial designs. Example 6.4 is demonstrated on each program.

SAS Input

```
OPTIONS LS = 80;
TITLE 'EXAMPLE 6.4 - MILK QUALITY';
DATA;
INPUT BREED DIET MACHINE MILKQUAL;
LINES;
1  1  1   7
1  1  1  16
1  1  2  29
1  1  2  25
1  2  1  23
1  2  1  25
1  2  2  40
1  2  2  41
1  3  1  29
1  3  1  13
1  3  2  34
1  3  2  23
2  1  1  15
2  1  1   7
2  1  2   4
2  1  2   9
2  2  1  16
2  2  1  22
2  2  2   5
2  2  2   7
2  3  1  12
```

```
2  3  1   6
2  3  2   9
2  3  2   9
3  1  1   3
3  1  1  11
3  1  2   7
3  1  2   1
3  2  1  11
3  2  1   7
3  2  2  15
3  2  2   6
3  3  1   9
3  3  1   2
3  3  2   2
3  3  2  12
4  1  1  18
4  1  1  23
4  1  2   8
4  1  2   7
4  2  1  11
4  2  1  25
4  2  2  11
4  2  2  15
4  3  1  19
4  3  1  19
4  3  2   9
4  3  2  15
PROC ANOVA;
CLASS BREED DIET MACHINE;
MODEL MILKQUAL = BREED DIET MACHINE BREED*DIET BREED*MACHINE
     DIET*MACHINE BREED*DIET*MACHINE;
MEANS BREED DIET MACHINE BREED*DIET BREED*MACHINE
     DIET*MACHINE BREED*DIET*MACHINE / DUNCAN;
```

Discussion of SAS Input

For each observation, BREED, DIET, MACHINE, and MILKQUAL are entered. Our three classification variables are identified in the CLASS statement of PROC ANOVA, and the MODEL statement lists all sources of variability in the experiment. Each interaction is indicated with asterisks between the factors involved. As part of PROC ANOVA, means are requested for each effect in the model. Duncan's multiple range test is specified on the MEANS statement, but SAS only performs multiple range tests for main effects.

SAS Output

```
                  EXAMPLE 6.4 - MILK QUALITY                  1
                Analysis of Variance Procedure
                   Class Level Information

                   Class     Levels     Values
                   BREED       4        1  2  3  4
                   DIET        3        1  2  3
                   MACHINE     2        1  2

           Number of observations in data set = 48
```

```
                   EXAMPLE 6.4 - MILK QUALITY                       2
                 Analysis of Variance Procedure
Dependent Variable:    MILKQUAL
                         Sum of            Mean
Source              DF      Squares          Square    F Value     Pr > F
Model               23   3749.6666667   163.0289855       6.19     0.0001
Error               24    632.0000000    26.3333333
Corrected Total     47   4381.6666667

R-Square               C.V.            Root MSE            MILKQUAL Mean
0.855763           35.59492           5.1316014                14.416667

Source                 DF        Anova SS   Mean Square    F Value     Pr > F
BREED                   3    2312.1666667   770.7222222      29.27     0.0001
DIET                    2     260.1666667   130.0833333       4.94     0.0160
MACHINE                 1       0.7500000     0.7500000       0.03     0.8674
BREED*DIET              6     162.8333333    27.1388889       1.03     0.4300
BREED*MACHINE           3     829.7500000   276.5833333      10.50     0.0001
DIET*MACHINE            2       6.5000000     3.2500000       0.12     0.8845
BREED*DIET*MACHINE      6     177.5000000    29.5833333       1.12     0.3786
```

```
                  EXAMPLE 6.4 - MILK QUALITY                  3
                Analysis of Variance Procedure

     Duncan's Multiple Range Test for variable: MILKQUAL

      NOTE: This test controls the type I comparisonwise
         error rate, not the experimentwise error rate
```

```
     Alpha= 0.05    df= 24    MSE= 26.33333

    Number of Means         2          3          4
    Critical Range      4.320      4.538      4.689

Means with the same letter are not significantly different.

      Duncan Grouping         Mean        N        BREED
                    A       25.417       12        1

                    B       15.000       12        4

                    C       10.083       12        2
                    C
                    C        7.167       12        3
```

```
                EXAMPLE 6.4 - MILK QUALITY                      4
               Analysis of Variance Procedure

    Duncan's Multiple Range Test for variable: MILKQUAL

     NOTE: This test controls the type I comparisonwise
       error rate, not the experimentwise error rate

      Alpha= 0.05    df= 24    MSE= 26.33333

            Number of Means         2          3
            Critical Range      3.741      3.930

Means with the same letter are not significantly different.

      Duncan Grouping         Mean        N       DIET

                    A       17.500       16       2
                    A
            B       A       13.875       16       3
            B
            B               11.875       16       1
```

```
                EXAMPLE 6.4 - MILK QUALITY                      5
               Analysis of Variance Procedure

   Duncan's Multiple Range Test for variable: MILKQUAL

     NOTE: This test controls the type I comparisonwise
       error rate, not the experimentwise error rate
```

```
          Alpha= 0.05    df= 24    MSE= 26.33333

                 Number of Means         2
                 Critical Range      3.054

Means with the same letter are not significantly different.

       Duncan Grouping      Mean       N      MACHINE

                     A    14.542      24      1
                     A
                     A    14.292      24      2
```

```
                 EXAMPLE 6.4 - MILK QUALITY                   6
               Analysis of Variance Procedure
```

Level of BREED	Level of DIET	N	------MILKQUAL------ Mean	SD
1	1	4	19.2500000	9.81070844
1	2	4	32.2500000	9.56991815
1	3	4	24.7500000	9.03234927
2	1	4	8.7500000	4.64578662
2	2	4	12.5000000	7.93725393
2	3	4	9.0000000	2.44948974
3	1	4	5.5000000	4.43471157
3	2	4	9.7500000	4.11298756
3	3	4	6.2500000	5.05799697
4	1	4	14.0000000	7.78888096
4	2	4	15.5000000	6.60807587
4	3	4	15.5000000	4.72581563

Level of BREED	Level of MACHINE	N	------MILKQUAL------ Mean	SD
1	1	6	18.8333333	8.25631072
1	2	6	32.0000000	7.58946638
2	1	6	13.0000000	6.00000000
2	2	6	7.1666667	2.22860195
3	1	6	7.1666667	3.92003401
3	2	6	7.1666667	5.49241902
4	1	6	19.1666667	4.83390801
4	2	6	10.8333333	3.48807492

EXAMPLE 6.4 - MILK QUALITY 7
Analysis of Variance Procedure

Level of DIET	Level of MACHINE	N	------MILKQUAL------ Mean	SD
1	1	8	12.5000000	6.6761837
1	2	8	11.2500000	10.0959681
2	1	8	17.5000000	7.1713717
2	2	8	17.5000000	14.6969385
3	1	8	13.6250000	8.5513157
3	2	8	14.1250000	10.0347610

Level of BREED	Level of DIET	Level of MACHINE	N	------MILKQUAL------ Mean	SD
1	1	1	2	11.5000000	6.3639610
1	1	2	2	27.0000000	2.8284271
1	2	1	2	24.0000000	1.4142136
1	2	2	2	40.5000000	0.7071068
1	3	1	2	21.0000000	11.3137085
1	3	2	2	28.5000000	7.7781746
2	1	1	2	11.0000000	5.6568542
2	1	2	2	6.5000000	3.5355339
2	2	1	2	19.0000000	4.2426407
2	2	2	2	6.0000000	1.4142136
2	3	1	2	9.0000000	4.2426407
2	3	2	2	9.0000000	0.0000000
3	1	1	2	7.0000000	5.6568542
3	1	2	2	4.0000000	4.2426407
3	2	1	2	9.0000000	2.8284271
3	2	2	2	10.5000000	6.3639610
3	3	1	2	5.5000000	4.9497475
3	3	2	2	7.0000000	7.0710678
4	1	1	2	20.5000000	3.5355339
4	1	2	2	7.5000000	0.7071068
4	2	1	2	18.0000000	9.8994949
4	2	2	2	13.0000000	2.8284271
4	3	1	2	19.0000000	0.0000000
4	3	2	2	12.0000000	4.2426407

Discussion of SAS Output

Page 1 of the output describes the experiment, and page 2 presents the anova summary table. Duncan's multiple range test is reported for the four breeds on page 3, for the three diets on page 4, and for the two machines on page 5. Pages 6 and 7 display means for each pair of factors averaged over the third factor, and for individual cells.

SPSS Input

```
SET WIDTH = 80
TITLE 'EXAMPLE 6.4 - MILK QUALITY'
DATA LIST FREE / BREED DIET MACHINE MILKQUAL
ANOVA VARIABLES = MILKQUAL BY BREED (1,4) DIET (1,3)
        MACHINE (1,2)
      / STATISTICS = MEAN
BEGIN DATA
1  1  1   7
1  1  1  16
1  1  2  29
1  1  2  25
1  2  1  23
1  2  1  25
1  2  2  40
1  2  2  41
1  3  1  29
1  3  1  13
1  3  2  34
1  3  2  23
2  1  1  15
2  1  1   7
2  1  2   4
2  1  2   9
2  2  1  16
2  2  1  22
2  2  2   5
2  2  2   7
2  3  1  12
2  3  1   6
2  3  2   9
2  3  2   9
3  1  1   3
3  1  1  11
3  1  2   7
3  1  2   1
```

```
3  2  1  11
3  2  1   7
3  2  2  15
3  2  2   6
3  3  1   9
3  3  1   2
3  3  2   2
3  3  2  12
4  1  1  18
4  1  1  23
4  1  2   8
4  1  2   7
4  2  1  11
4  2  1  25
4  2  2  11
4  2  2  15
4  3  1  19
4  3  1  19
4  3  2   9
4  3  2  15
END DATA
FINISH
```

Discussion of SPSS Input

For each observation, BREED, DIET, MACHINE, and MILKQUAL are entered in free format. Analysis of variance is requested for MILKQUAL, using the first three variables as factors. The optional STATISTICS command asks for means for each effect in the model.

SPSS Output

```
                    EXAMPLE 6.4 - MILK QUALITY           Page 1
                       * * * CELL MEANS * * *
            MILKQUAL
       BY   BREED
            DIET
            MACHINE

TOTAL POPULATION
    14.42
 (     48)
```

BREED

1	2	3	4
25.42	10.08	7.17	15.00
(12)	(12)	(12)	(12)

DIET

1	2	3
11.88	17.50	13.88
(16)	(16)	(16)

MACHINE

1	2
14.54	14.29
(24)	(24)

BREED	DIET 1	2	3
1	19.25	32.25	24.75
	(4)	(4)	(4)
2	8.75	12.50	9.00
	(4)	(4)	(4)
3	5.50	9.75	6.25
	(4)	(4)	(4)
4	14.00	15.50	15.50
	(4)	(4)	(4)

EXAMPLE 6.4 - MILK QUALITY Page 2

BREED	MACHINE 1	2
1	18.83	32.00
	(6)	(6)
2	13.00	7.17
	(6)	(6)
3	7.17	7.17
	(6)	(6)
4	19.17	10.83
	(6)	(6)

DIET	MACHINE 1	MACHINE 2
1	12.50	11.25
	(8)	(8)
2	17.50	17.50
	(8)	(8)
3	13.63	14.13
	(8)	(8)

EXAMPLE 6.4 - MILK QUALITY Page 3

MACHINE = 1

BREED	DIET 1	DIET 2	DIET 3
1	11.50	24.00	21.00
	(2)	(2)	(2)
2	11.00	19.00	9.00
	(2)	(2)	(2)
3	7.00	9.00	5.50
	(2)	(2)	(2)
4	20.50	18.00	19.00
	(2)	(2)	(2)

MACHINE = 2

BREED	DIET 1	DIET 2	DIET 3
1	27.00	40.50	28.50
	(2)	(2)	(2)
2	6.50	6.00	9.00
	(2)	(2)	(2)
3	4.00	10.50	7.00
	(2)	(2)	(2)
4	7.50	13.00	12.00
	(2)	(2)	(2)

```
                    EXAMPLE 6.4 - MILK QUALITY                    Page 4
               * * * ANALYSIS OF VARIANCE * * *

          MILKQUAL
     by   BREED
          DIET
          MACHINE
```

Source of Variation	Sum of Squares	DF	Mean Square	F	Sig of F
Main Effects	2573.083	6	428.847	16.285	.000
BREED	2312.167	3	770.722	29.268	.000
DIET	260.167	2	130.083	4.940	.016
MACHINE	.750	1	.750	.028	.867
2-Way Interactions	999.083	11	90.826	3.449	.005
BREED DIET	162.833	6	27.139	1.031	.430
BREED MACHINE	829.750	3	276.583	10.503	.000
DIET MACHINE	6.500	2	3.250	.123	.884
3-Way Interactions	177.500	6	29.583	1.123	.379
BREED DIET MACHINE	177.500	6	29.583	1.123	.379
Explained	3749.667	23	163.029	6.191	.000
Residual	632.000	24	26.333		
Total	4381.667	47	93.227		

```
48 cases were processed.
0 cases (.0 pct) were missing.
```

Discussion of SPSS Output

Output pages 1 through 3 report means on MILKQUAL for the whole population, by levels of each factor, for each pair of factors averaged over levels of the third factor, and for individual cells in our design. Page 4 contains the analysis of variance summary table.

MINITAB Transcript

```
MTB > NAME C1 = 'BREED'
MTB > NAME C2 = 'DIET'
MTB > NAME C3 = 'MACHINE'
MTB > NAME C4 = 'MILKQUAL'
MTB > READ C1-C4
DATA > 1  1  1   7
DATA > 1  1  1  16
DATA > 1  1  2  29
DATA > 1  1  2  25
```

```
DATA > 1  2  1  23
DATA > 1  2  1  25
DATA > 1  2  2  40
DATA > 1  2  2  41
DATA > 1  3  1  29
DATA > 1  3  1  13
DATA > 1  3  2  34
DATA > 1  3  2  23
DATA > 2  1  1  15
DATA > 2  1  1   7
DATA > 2  1  2   4
DATA > 2  1  2   9
DATA > 2  2  1  16
DATA > 2  2  1  22
DATA > 2  2  2   5
DATA > 2  2  2   7
DATA > 2  3  1  12
DATA > 2  3  1   6
DATA > 2  3  2   9
DATA > 2  3  2   9
DATA > 3  1  1   3
DATA > 3  1  1  11
DATA > 3  1  2   7
DATA > 3  1  2   1
DATA > 3  2  1  11
DATA > 3  2  1   7
DATA > 3  2  2  15
DATA > 3  2  2   6
DATA > 3  3  1   9
DATA > 3  3  1   2
DATA > 3  3  2   2
DATA > 3  3  2  12
DATA > 4  1  1  18
DATA > 4  1  1  23
DATA > 4  1  2   8
DATA > 4  1  2   7
DATA > 4  2  1  11
DATA > 4  2  1  25
DATA > 4  2  2  11
DATA > 4  2  2  15
DATA > 4  3  1  19
```

```
DATA > 4  3  1  19
DATA > 4  3  2   9
DATA > 4  3  2  15
DATA > END
     48 ROWS READ

MTB > ANOVA MILKQUAL = BREED DIET MACHINE BREED*DIET &
CONT > BREED*MACHINE DIET*MACHINE BREED*DIET*MACHINE;
SUBC > MEANS BREED DIET MACHINE BREED*DIET BREED*MACHINE &
CONT > DIET*MACHINE BREED*DIET*MACHINE.

Factor         Type    Levels  Values
BREED         fixed       4       1       2       3       4
DIET          fixed       3       1       2       3
MACHINE       fixed       2       1       2

Analysis of Variance for MILKQUAL

Source                  DF        SS        MS       F       P
BREED                    3   2312.17    770.72   29.27   0.000
DIET                     2    260.17    130.08    4.94   0.016
MACHINE                  1      0.75      0.75    0.03   0.867
BREED*DIET               6    162.83     27.14    1.03   0.430
BREED*MACHINE            3    829.75    276.58   10.50   0.000
DIET*MACHINE             2      6.50      3.25    0.12   0.884
BREED*DIET*MACHINE       6    177.50     29.58    1.12   0.379
Error                   24    632.00     26.33
Total                   47   4381.67

     MEANS

BREED    N   MILKQUAL
    1   12     25.417
    2   12     10.083
    3   12      7.167
    4   12     15.000

DIET     N   MILKQUAL
   1    16     11.875
   2    16     17.500
   3    16     13.875
```

MACHINE	N	MILKQUAL
1	24	14.542
2	24	14.292

BREED	DIET	N	MILKQUAL
1	1	4	19.250
1	2	4	32.250
1	3	4	24.750
2	1	4	8.750
2	2	4	12.500
2	3	4	9.000
3	1	4	5.500
3	2	4	9.750
3	3	4	6.250
4	1	4	14.000
4	2	4	15.500
4	3	4	15.500

BREED	MACHINE	N	MILKQUAL
1	1	6	18.833
1	2	6	32.000
2	1	6	13.000
2	2	6	7.167
3	1	6	7.167
3	2	6	7.167
4	1	6	19.167
4	2	6	10.833

DIET	MACHINE	N	MILKQUAL
1	1	8	12.500
1	2	8	11.250
2	1	8	17.500
2	2	8	17.500
3	1	8	13.625
3	2	8	14.125

BREED	DIET	MACHINE	N	MILKQUAL
1	1	1	2	11.500
1	1	2	2	27.000
1	2	1	2	24.000
1	2	2	2	40.500

```
     1         3           1    2       21.000
     1         3           2    2       28.500
     2         1           1    2       11.000
     2         1           2    2        6.500
     2         2           1    2       19.000
     2         2           2    2        6.000
     2         3           1    2        9.000
     2         3           2    2        9.000
     3         1           1    2        7.000
     3         1           2    2        4.000
     3         2           1    2        9.000
     3         2           2    2       10.500
     3         3           1    2        5.500
     3         3           2    2        7.000
     4         1           1    2       20.500
     4         1           2    2        7.500
     4         2           1    2       18.000
     4         2           2    2       13.000
     4         3           1    2       19.000
     4         3           2    2       12.000
MTB > STOP
```

Discussion of MINITAB Transcript

For each observation, values of BREED, DIET, MACHINE, and MILKQUAL are entered in columns C1-C4. ANOVA is requested on MILKQUAL, with a model that includes all sources of variability in our experiment. Each interaction is specified with asterisks between the factors involved. The ampersand at the end of the ANOVA line indicates continuation of this command on the next line. A semicolon is used to indicate that a subcommand follows. The MEANS subcommand requests means for each effect in the model, and is also continued on a second line. The subcommand ends in a period, denoting completion of the ANOVA command.

Output includes the analysis of variance table and means for each main effect, for each pair of factors averaged over levels of the third factor, and for individual cells.

6.12 HIGHER ORDER DESIGNS

The logic and procedure of two-way and three-way anova generalize to any number of factors. A four-way design, for example, involves four main effects, six two-way interactions, four three-way interactions, and a four-way interaction. Sums of squares and degrees of freedom are obtained in the same two-step partition, first by stringing out all the cells in one-way fashion and then by breaking down the between groups term.

In general, we always examine the highest order interaction in a design first, and move sequentially downward toward main effects. Existence of interaction at any level may imply masking of lower order terms, and obscure interpretations of lower effects. To begin an examination of results at the lowest level, as uninformed researchers often do, is to invite disappointment when expected results are not obtained, due to masking by interactions. In most practical applications, however, it is difficult or impossible to use, in a meaningful way, the knowledge that a high order interaction exists. Typically, we hope that interactions are absent, so tests of main effects are unambiguous. But we must start by studying the most complex term in any design.

One of the most fundamental questions in planning an experiment is the level of complexity to investigate. How many factors should be included? Some researchers plan extremely complex studies, in an effort to model the complex real world. Others, fearful of obtaining interactions they cannot interpret, perform research one factor at a time. Clearly, any extreme strategy is fraught with difficulties.

If we fail to study sources having substantial impact on observed measurements, their influences are still present. But we will not be able to assess their strength. Variability due to unmeasured sources is automatically pooled into the error term. In addition to being unable to test omitted sources, all tests of effects *in the model* are less powerful due to inflation of the denominator. We do not want to err on the side of an overly simple experiment.

On the other hand, unduly complex experimentation is also not advisable. Inclusion of extraneous factors—classification variables that have no real effect—reveals no new information. But the cost is tremendous. Adding a factor with only two levels doubles the size of the experiment! Obviously, we do not want to design overly complex experiments.

Logically, we should include in our design all factors that truly influence the dependent variable, but no more. Of course, this simple statement begs the question. The whole reason for performing research is to learn which factors affect the results and their relative strengths. The answer to this dilemma is to apply your best professional judgement and common sense. You must know something about the phenomena under study to be able to plan an experiment intelligently. A statistician cannot plan research in your field if he or she knows nothing about it. In experimental design, as in most other aspects of life, there is no substitute for knowing what you are doing.

One strategy frequently used in industrial settings is to perform preliminary investigations using several factors, each at only two levels. Such 2^K experiments reveal which influences are worthy of detailed study in subsequent research. If, as often happens, high order interactions are insignificant, we can pool them into the error term, to obtain more degrees of freedom and, therefore, more power for other tests. These techniques are demonstrated for a four-way design in Example 6.5.

The anova summary table for this example is shown in Table 6.13. Beginning our examination at the bottom of this table, we see that the four-way interaction and all four three-way interactions are clearly nonsignificant. At this point, we might decide to pool these high order interactions into the error term.

After pooling the four-way interaction and all three-way interactions into the error term, the anova summary table for remaining effects is shown in Table 6.14. Note that p-values for main effects and two-way interactions in this table are all smaller than in the full four-way analysis, reflecting greater power as a result of pooling. To complete the analysis of this example, we would look at two-way interactions and then main effects, as was demonstrated in the previous section.

High order anova involves a large number of statistical tests. A full four-way design, for example, includes tests of 15 effects. When many tests are performed,

---

Example 6.5

In a study of processes used to remove impurities from cellulose goods, the following data resulted from a 2^4 design that involved the desizing process. Two observations of starch percentage by weight were obtained for each combination of the four factors: enzyme concentration, pH, temperature, and time. Each row of the table below shows the two observations for a single cell.

Enzyme (g/l)	pH	Temp (°C)	Time (hrs.)	Obs. 1	Obs. 2
.50	6	60	6	9.72	13.50
.75	6	60	6	9.80	14.04
.50	7	60	6	10.13	11.27
.75	7	60	6	11.80	11.30
.50	6	70	6	12.70	11.37
.75	6	70	6	11.96	12.05
.50	7	70	6	11.38	9.92
.75	7	70	6	11.80	11.10
.50	6	60	8	13.15	13.00
.75	6	60	8	10.60	12.37
.50	7	60	8	10.37	12.00
.75	7	60	8	11.30	11.64
.50	6	70	8	13.05	14.55
.75	6	70	8	11.15	15.00
.50	7	70	8	12.70	14.10
.75	7	70	8	13.20	16.12

TABLE 6.13 Anova Summary Table for Starch Percentage

Source	df	SS	MS	F	P
Enzyme	1	0.168	0.168	0.08	0.7876
pH	1	1.940	1.940	0.87	0.3659
Temp	1	8.161	8.161	3.64	0.0744
Time	1	13.082	13.082	5.84	0.0280
Enz × pH	1	3.419	3.419	1.53	0.2346
Enz × temp	1	0.263	0.263	0.12	0.7364
Enz × time	1	0.911	0.911	0.41	0.5327
pH × temp	1	0.738	0.738	0.33	0.5740
Temp × time	1	6.771	6.771	3.02	0.1013
pH × time	1	0.781	0.781	0.35	0.5631
Enz × pH × temp	1	0.020	0.020	0.01	0.9259
Enz × pH × time	1	0.775	0.775	0.35	0.5647
Enz × temp × time	1	0.622	0.622	0.28	0.6056
pH × temp × time	1	1.758	1.758	0.78	0.3889
Enz × pH × temp × time	1	0.004	0.004	0.00	0.9666
Within	16	35.851	2.241		
Total	31	75.264			

TABLE 6.14 Anova Summary Table for Starch Percentage After Pooling Three-way and Four-way Interactions into Error Term

Source	df	SS	MS	F	P
Enzyme	1	0.168	0.168	0.09	0.7665
pH	1	1.940	1.940	1.04	0.3185
Temp	1	8.161	8.161	4.39	0.0484
Time	1	13.082	13.082	7.04	0.0149
Enz × pH	1	3.419	3.419	1.84	0.1894
Enz × temp	1	0.263	0.263	0.14	0.7107
Enz × time	1	0.911	0.911	0.49	0.4915
pH × temp	1	0.738	0.738	0.40	0.5354
pH × time	1	0.781	0.781	0.42	0.5238
Temp × time	1	6.771	6.771	3.64	0.0701
Error	21	39.029	1.859		
Total	31	75.264			

each allowing a 5-percent chance of type one error (i.e., rejecting H_0 when it is true), there is substantial probability of making one or more type one errors in the entire analysis. This is the same problem discussed in Section 5.6 in connection with multiple range tests. In that context, Tukey's Honestly Significant Difference approach was used to control type one error rate on an experimentwise basis.

A similar technique exists for experimentwise control of type one error rate in more general settings, such as high order anova designs. If a total of T tests are performed, the **Bonferroni procedure** requires each test to be significant at the 0.05/T level. This approach controls experimentwise error rate at 5 percent for the set of T tests. To apply the Bonferroni method in a four-way design, for instance, each test would be deemed significant if the p-value is less than 0.05/15 = 0.003, since there are 15 tests involved in the complete analysis. Experimentwise control of type one error rate is rather stringent, so researchers sometimes apply the Bonferroni method separately to various subgroups of tests. For example, we might use the 0.05/5 = 0.01 level of significance for testing the set of four-way and three-way interactions (a total of five tests).

By including all possible combinations of factor levels, factorial designs enable study of all main effects and interactions. But even with a modest number of factors, such experiments quickly become large and expensive. Alternative designs, requiring fewer observations, are available when some interactions are known or suspected to be absent. Designs such as Latin squares, split plots, incomplete blocks, and fractional factorials allow study of a large number of factors with fewer cells than full factorial designs. All such designs deliberately confound certain high order interactions with low order effects. As long as these high order interactions do not exist, this strategy provides valid tests of low order terms. Researchers interested in such approaches should further investigate the topic of experimental design.

6.13 FOUR-WAY ANOVA IN COMPUTER PROGRAMS

Four-way anova is demonstrated on SAS, SPSS, and MINITAB for the data of Example 6.5.

SAS Input

```
OPTIONS LS = 80;
TITLE 'EXAMPLE 6.5 - CELLULOSE IMPURITIES';
DATA;
INPUT ENZYME PH TEMP TIME STARCH;
LINES;
.50   6   60   6    9.72
.50   6   60   6   13.50
.75   6   60   6    9.80
.75   6   60   6   14.04
.50   7   60   6   10.13
```

```
. 50   7   60   6   11.27
. 75   7   60   6   11.80
. 75   7   60   6   11.30
. 50   6   70   6   12.70
. 50   6   70   6   11.37
. 75   6   70   6   11.96
. 75   6   70   6   12.05
. 50   7   70   6   11.38
. 50   7   70   6    9.92
. 75   7   70   6   11.80
. 75   7   70   6   11.10
. 50   6   60   8   13.15
. 50   6   60   8   13.00
. 75   6   60   8   10.60
. 75   6   60   8   12.37
. 50   7   60   8   10.37
. 50   7   60   8   12.00
. 75   7   60   8   11.30
. 75   7   60   8   11.64
. 50   6   70   8   13.05
. 50   6   70   8   14.55
. 75   6   70   8   11.15
. 75   6   70   8   15.00
. 50   7   70   8   12.70
. 50   7   70   8   14.10
. 75   7   70   8   13.20
. 75   7   70   8   16.12
PROC ANOVA;
CLASS ENZYME PH TEMP TIME;
MODEL STARCH = ENZYME | PH | TEMP | TIME;
PROC ANOVA;
CLASS ENZYME PH TEMP TIME;
MODEL STARCH = ENZYME PH TEMP TIME ENZYME*PH ENZYME*TEMP
   ENZYME*TIME PH*TEMP PH*TIME TEMP*TIME;
```

Discussion of SAS Input

ENZYME, PH, TEMP, TIME, and STARCH are entered for each of the 32 observations. In the MODEL statement of PROC ANOVA, vertical bars separating factors automatically produce the full factorial design. After examining output from the full design, and noting lack of significance for the four-way and all three-way interactions, these terms are pooled into error by omitting them from the model statement. Although in reality this was performed in a separate computer run, the second call to PROC ANOVA and its reduced MODEL statement are appended to the input shown here, to save space and eliminate duplication in presentation.

SAS Output

```
               EXAMPLE 6.5 - CELLULOSE IMPURITIES                 1
                 Analysis of Variance Procedure
                    Class Level Information

                Class       Levels      Values
                ENZYME         2        0.5  0.75
                PH             2        6  7
                TEMP           2        60  70
                TIME           2        6  8

            Number of observations in data set = 32
```

```
                EXAMPLE 6.5 - CELLULOSE IMPURITIES                2
                  Analysis of Variance Procedure

Dependent Variable: STARCH

                                 Sum of          Mean
Source                 DF       Squares        Square   F Value    Pr > F
Model                  15   39.41328750    2.62755250      1.17    0.3768
Error                  16   35.85050000    2.24065625
Corrected Total        31   75.26378750

R-Square        C.V.      Root MSE   STARCH Mean
0.523669    12.34097     1.4968822     12.129375

Source                  DF     Anova SS   Mean Square  F Value    Pr > F
ENZYME                   1   0.16820000    0.16820000     0.08    0.7876
PH                       1   1.94045000    1.94045000     0.87    0.3659
ENZYME*PH                1   3.41911250    3.41911250     1.53    0.2346
TEMP                     1   8.16080000    8.16080000     3.64    0.0744
ENZYME*TEMP              1   0.26281250    0.26281250     0.12    0.7364
PH*TEMP                  1   0.73811250    0.73811250     0.33    0.5740
ENZYME*PH*TEMP           1   0.02000000    0.02000000     0.01    0.9259
TIME                     1  13.08161250   13.08161250     5.84    0.0280
```

ENZYME*TIME	1	0.91125000	0.91125000	0.41	0.5327
PH*TIME	1	0.78125000	0.78125000	0.35	0.5631
ENZYME*PH*TIME	1	0.77501250	0.77501250	0.35	0.5647
TEMP*TIME	1	6.77120000	6.77120000	3.02	0.1013
ENZYME*TEMP*TIME	1	0.62161250	0.62161250	0.28	0.6056
PH*TEMP*TIME	1	1.75781250	1.75781250	0.78	0.3889
ENZYME*PH*TEMP*TIME	1	0.00405000	0.00405000	0.00	0.9666

EXAMPLE 6.5 - CELLULOSE IMPURITIES 3
Analysis of Variance Procedure
Class Level Information

Class	Levels	Values
ENZYME	2	0.5 0.75
PH	2	6 7
TEMP	2	60 70
TIME	2	6 8

Number of observations in data set = 32

EXAMPLE 6.5 - CELLULOSE IMPURITIES 4
Analysis of Variance Procedure

Dependent Variable: STARCH

Source	DF	Sum of Squares	Mean Square	F Value	Pr > F
Model	10	36.23480000	3.62348000	1.95	0.0949
Error	21	39.02898750	1.85852321		
Corrected Total	31	75.26378750			

R-Square	C.V.	Root MSE	STARCH Mean
0.481437	11.23946	1.3632766	12.129375

Source	DF	Anova SS	Mean Square	F Value	Pr > F
ENZYME	1	0.16820000	0.16820000	0.09	0.7665
PH	1	1.94045000	1.94045000	1.04	0.3185
TEMP	1	8.16080000	8.16080000	4.39	0.0484
TIME	1	13.08161250	13.08161250	7.04	0.0149
ENZYME*PH	1	3.41911250	3.41911250	1.84	0.1894
ENZYME*TEMP	1	0.26281250	0.26281250	0.14	0.7107
ENZYME*TIME	1	0.91125000	0.91125000	0.49	0.4915

```
PH*TEMP          1     0.73811250      0.73811250        0.40     0.5354
PH*TIME          1     0.78125000      0.78125000        0.42     0.5238
TEMP*TIME        1     6.77120000      6.77120000        3.64     0.0701
```

Discussion of SAS Output

The anova table from the full design on the second page of output reveals clearly nonsignificant four-way and three-way interactions. As shown on page 4, for the reduced model without these high order interactions, p-values for all remaining effects are lower than in tests with the full model, reflecting the increase in power due to pooling.

SPSS Input

```
SET WIDTH = 80
TITLE 'EXAMPLE 6.5 - CELLULOSE IMPURITIES'
DATA LIST FREE / ENZYME PH TEMP TIME STARCH
COMPUTE ENZ4 = ENZYME*4
COMPUTE TEMP10 = TEMP/10
ANOVA VARIABLES = STARCH BY ENZ4 (2,3) PH (6,7)
      TEMP10 (6,7) TIME (6,8)
BEGIN DATA
.50   6   60   6    9.72
.50   6   60   6   13.50
.75   6   60   6    9.80
.75   6   60   6   14.04
.50   7   60   6   10.13
.50   7   60   6   11.27
.75   7   60   6   11.80
.75   7   60   6   11.30
.50   6   70   6   12.70
.50   6   70   6   11.37
.75   6   70   6   11.96
.75   6   70   6   12.05
.50   7   70   6   11.38
.50   7   70   6    9.92
.75   7   70   6   11.80
.75   7   70   6   11.10
.50   6   60   8   13.15
.50   6   60   8   13.00
.75   6   60   8   10.60
```

```
.75   6   60   8   12.37
.50   7   60   8   10.37
.50   7   60   8   12.00
.75   7   60   8   11.30
.75   7   60   8   11.64
.50   6   70   8   13.05
.50   6   70   8   14.55
.75   6   70   8   11.15
.75   6   70   8   15.00
.50   7   70   8   12.70
.50   7   70   8   14.10
.75   7   70   8   13.20
.75   7   70   8   16.12
END DATA
ANOVA VARIABLES = STARCH BY ENZ4 (2,3) PH (6,7)
      TEMP10 (6,7) TIME (6,8)
      / MAXORDERS = 2
FINISH
```

Discussion of SPSS Input

ENZYME, PH, TEMP, TIME, and STARCH are entered in free format for each of the 32 observations. Factor levels must be integers for the SPSS ANOVA procedure. Since ENZYME levels are not integers, a new variable called ENZ4 is defined by a COMPUTE command to be four times ENZYME. The values 0.50 and 0.75 on ENZYME equate to values of 2 and 3 on ENZ4. The ANOVA routine uses computer space for the maximum possible design, according to limits on factor levels. Using TEMP as a factor, with limits 60 and 70, would result in allocating space for 11 possible levels. To conserve space, a new variable called TEMP10 is defined by a COMPUTE command to be TEMP divided by 10. Values of 60 and 70 on TEMP yield values of 6 and 7 on TEMP10. ENZ4 and TEMP10 are computed for each observation and added to the data, as if they had been input with other variables.

Factors ENZ4, PH, TEMP10, and TIME are listed in the ANOVA command. Since SPSS automatically includes all interactions up to fifth order, this analysis corresponds to the full factorial design for our four factors. After examining output from the full design, and noting lack of significance for the four-way and all three-way interactions, these terms are pooled into error by setting MAXORDERS equal to 2. Although in reality this was performed in a separate computer run, the second ANOVA command is appended to the input shown here, to save space and eliminate duplication in presentation.

SPSS Output

```
                EXAMPLE 6.5 - CELLULOSE IMPURITIES          Page 1
          * * * A N A L Y S I S   O F   V A R I A N C E * * *

             STARCH
        by   ENZ4
             PH
             TEMP10
             TIME

                                     Sum of             Mean               Sig
Source of Variation                 Squares    DF     Square       F      of F
Main Effects                         23.351     4      5.838   2.605      .075
   ENZ4                                .168     1       .168    .075      .788
   PH                                 1.940     1      1.940    .866      .366
   TEMP10                             8.161     1      8.161   3.642      .074
   TIME                              13.082     1     13.082   5.838      .028
2-Way Interactions                   12.884     6      2.147    .958      .483
   ENZ4    PH                         3.419     1      3.419   1.526      .235
   ENZ4    TEMP10                      .263     1       .263    .117      .736
   ENZ4    TIME                        .911     1       .911    .407      .533
   PH      TEMP10                      .738     1       .738    .329      .574
   PH      TIME                        .781     1       .781    .349      .563
   TEMP10  TIME                       6.771     1      6.771   3.022      .101
3-Way Interactions                    3.174     4       .794    .354      .837
   ENZ4    PH        TEMP10            .020     1       .020    .009      .926
   ENZ4    PH        TIME              .775     1       .775    .346      .565
   ENZ4    TEMP10    TIME              .622     1       .622    .277      .606
   PH      TEMP10    TIME             1.758     1      1.758    .785      .389
4-Way Interactions                     .004     1       .004    .002      .967
   ENZ4    PH    TEMP10    TIME        .004     1       .004    .002      .967
Explained                            39.413    15      2.628   1.173      .377
Residual                             35.850    16      2.241
Total                                75.264    31      2.428
32 cases were processed.
0 cases (.0 pct) were missing.
```

```
             EXAMPLE 6.5 - CELLULOSE IMPURITIES          Page 2
     * * * A N A L Y S I S   O F   V A R I A N C E * * *

             STARCH
        by   ENZ4
             PH
             TEMP10
             TIME
```

Source of Variation		Sum of Squares	DF	Mean Square	F	Sig of F
Main Effects		23.351	4	5.838	3.141	.036
ENZ4		.168	1	.168	.091	.766
PH		1.940	1	1.940	1.044	.319
TEMP10		8.161	1	8.161	4.391	.048
TIME		13.082	1	13.082	7.039	.015
2-Way Interactions		12.884	6	2.147	1.155	.366
ENZ4	PH	3.419	1	3.419	1.840	.189
ENZ4	TEMP10	.263	1	.263	.141	.711
ENZ4	TIME	.911	1	.911	.490	.491
PH	TEMP10	.738	1	.738	.397	.535
PH	TIME	.781	1	.781	.420	.524
TEMP10	TIME	6.771	1	6.771	3.643	.070
Explained		36.235	10	3.623	1.950	.095
Residual		39.029	21	1.859		
Total		75.264	31	2.428		

```
32 cases were processed.
0 cases (.0 pct) were missing.
```

Discussion of SPSS Output

The anova table from the full design on the first page of output reveals clearly nonsignificant four-way and three-way interactions. As shown on page 2, for the reduced model without these high order interactions, p-values for all remaining effects are lower than in tests with the full model, reflecting the increase in power due to pooling.

MINITAB Transcript

```
MTB > NAME C1 = 'ENZYME'
MTB > NAME C2 = 'PH'
MTB > NAME C3 = 'TEMP'
MTB > NAME C4 = 'TIME'
MTB > NAME C5 = 'STARCH'
MTB > READ C1-C5
DATA > .50   6   60   6    9.72
DATA > .50   6   60   6   13.50
DATA > .75   6   60   6    9.80
DATA > .75   6   60   6   14.04
DATA > .50   7   60   6   10.13
```

```
DATA > .50   7   60   6   11.27
DATA > .75   7   60   6   11.80
DATA > .75   7   60   6   11.30
DATA > .50   6   70   6   12.70
DATA > .50   6   70   6   11.37
DATA > .75   6   70   6   11.96
DATA > .75   6   70   6   12.05
DATA > .50   7   70   6   11.38
DATA > .50   7   70   6    9.92
DATA > .75   7   70   6   11.80
DATA > .75   7   70   6   11.10
DATA > .50   6   60   8   13.15
DATA > .50   6   60   8   13.00
DATA > .75   6   60   8   10.60
DATA > .75   6   60   8   12.37
DATA > .50   7   60   8   10.37
DATA > .50   7   60   8   12.00
DATA > .75   7   60   8   11.30
DATA > .75   7   60   8   11.64
DATA > .50   6   70   8   13.05
DATA > .50   6   70   8   14.55
DATA > .75   6   70   8   11.15
DATA > .75   6   70   8   15.00
DATA > .50   7   70   8   12.70
DATA > .50   7   70   8   14.10
DATA > .75   7   70   8   13.20
DATA > .75   7   70   8   16.12
DATA > END
     32 ROWS READ

MTB > NAME C6 = 'ENZ4'
MTB > LET 'ENZ4' = 'ENZYME'*4
MTB > ANOVA STARCH = ENZ4 | PH | TEMP | TIME

Factor   Type    Levels   Values
ENZ4     fixed     2         2    3
PH       fixed     2         6    7
TEMP     fixed     2        60   70
TIME     fixed     2         6    8

Analysis of Variance for STARCH
```

Source	DF	SS	MS	F	P
ENZ4	1	0.168	0.168	0.08	0.788
PH	1	1.940	1.940	0.87	0.366
TEMP	1	8.161	8.161	3.64	0.074
TIME	1	13.082	13.082	5.84	0.028
ENZ4*PH	1	3.419	3.419	1.53	0.235
ENZ4*TEMP	1	0.263	0.263	0.12	0.736
ENZ4*TIME	1	0.911	0.911	0.41	0.533
PH*TEMP	1	0.738	0.738	0.33	0.574
PH*TIME	1	0.781	0.781	0.35	0.563
TEMP*TIME	1	6.771	6.771	3.02	0.101
ENZ4*PH*TEMP	1	0.020	0.020	0.01	0.926
ENZ4*PH*TIME	1	0.775	0.775	0.35	0.565
ENZ4*TEMP*TIME	1	0.622	0.622	0.28	0.606
PH*TEMP*TIME	1	1.758	1.758	0.78	0.389
ENZ4*PH*TEMP*TIME	1	0.004	0.004	0.00	0.967
Error	16	35.851	2.241		
Total	31	75.264			

```
MTB > ANOVA STARCH = ENZ4 PH TEMP TIME ENZ4*PH ENZ4*TEMP &
CONT > ENZ4*TIME PH*TEMP PH*TIME TEMP*TIME
```

Factor	Type	Levels	Values	
ENZ4	fixed	2	2	3
PH	fixed	2	6	7
TEMP	fixed	2	60	70
TIME	fixed	2	6	8

Analysis of Variance for STARCH

Source	DF	SS	MS	F	P
ENZ4	1	0.168	0.168	0.09	0.766
PH	1	1.940	1.940	1.04	0.319
TEMP	1	8.161	8.161	4.39	0.048
TIME	1	13.082	13.082	7.04	0.015
ENZ4*PH	1	3.419	3.419	1.84	0.189
ENZ4*TEMP	1	0.263	0.263	0.14	0.711
ENZ4*TIME	1	0.911	0.911	0.49	0.491
PH*TEMP	1	0.738	0.738	0.40	0.535
PH*TIME	1	0.781	0.781	0.42	0.524
TEMP*TIME	1	6.771	6.771	3.64	0.070

```
Error              21      39.029        1.859
Total              31      75.264

MTB > STOP
```

Discussion of MINITAB Transcript

ENZYME, PH, TEMP, TIME, and STARCH are read into columns C1-C5 of the worksheet. Factor levels must be integers for the MINITAB ANOVA procedure. Since ENZYME levels are not integers, a new variable named ENZ4 is defined by a LET command to be four times ENZYME. The values 0.50 and 0.75 on ENZYME equate to values of 2 and 3 on ENZ4. ENZ4 is computed for each observation and stored in column C6, as if it had been input with other variables.

In the model portion of the ANOVA command, vertical bars separating factors automatically produce the full factorial design. Output from this analysis reveals lack of significance for the four-way and all three-way interactions. In the second ANOVA command, these terms are pooled into error by omitting them from the model. In the reduced model without these high order interactions, p-values for all remaining effects are lower than in tests with the full model, reflecting the increase in power due to pooling.

6.14 SUMMARY

This chapter has discussed two-way and higher order designs for analysis of variance. These factorial experiments allow for examination of effects associated with each factor and interactions among factors. Separate partitions of sums of squares and degrees of freedom are employed. Each partition is a two-step process, with the first step representing the breakdown of total into between and within groups components, as in one-way anova. In the second step, the between groups term is divided into terms for each main effect and interaction in the design. Each sum of squares is converted to a mean square by dividing by degrees of freedom. For each effect, an F statistic is constructed as the ratio of its mean square to mean square within groups.

Interaction between two factors is represented graphically as nonparallel lines. Three-way interaction occurs when the pattern of a two-way interaction varies over levels of a third factor. Interaction at any level may mask lower order terms. For example, two-way interaction may mask main effects, which are comparisons among levels of one factor averaged over levels of all other factors. In the presence of interaction, we should look at main effect tests only if we are interested in such averages. For significant main effects, multiple range tests can be employed to determine pairwise differences among levels. When interactions are absent, their sums of squares and degrees of freedom may be pooled into the error term to increase power for tests of remaining effects.

Unequal sample sizes cause problems because differences between cells contribute to more than one effect. Tests of unique contributions to total variance, as demonstrated in this chapter, are one way of dealing with this confounding. Examples of two-way, three-way, and four-way anovas were presented on SAS, SPSS, and MINITAB computer programs.

7

Correlation

The last several chapters have presented techniques for analyzing observations on a single variable. Now we consider two variables and methods for examining the relationship between them. Different approaches are used, depending on the level of measurement of each variable. Section 1.2 discussed variables and defined the measurement levels categorical, ordered categories, and numeric. Subsequent sections of this chapter address relationships between variables according to this classification scheme. Regardless of the level of measurement, however, we can only assess the relationship between two variables when each observation is measured on both variables.

7.1 TEST FOR ASSOCIATION WITH CATEGORICAL DATA

For a categorical variable, observations are sorted into categories, and we analyze the frequencies or counts in the various categories. When observations are cross classified according to two categorical variables, counts are obtained for each combination of categories on the two variables. In Example 7.1, the two categorical variables are meal choice (row variable) and beverage selection (column variable). Each **observed count** is the number of observations for a combination of categories on the two variables. For instance, 79 people ordered a hamburger and a soft drink. We are interested in determining if the two categorical variables are related in the population of all customers. Our null hypothesis states that there is no relationship, and the alternative indicates that a relationship exists:

H_0: meal and beverage are not related
H_1: meal and beverage are related.

Example 7.1

A major fast food chain is contemplating an advertising campaign based on a special sale price for purchase of certain combinations of meals and beverages. To determine if any relationship exists between meal choice and beverage selection, 200 combined meal and beverage purchases were randomly sampled from recent sales. Each sale was cross classified according to meal and beverage, with results below.

Observed Count	Soft Drink	Milk	Coffee/Tea	Row Total
Hamburger	79	6	17	102
Roast beef	25	6	15	46
Fish sandwich	9	4	18	31
Salad	9	8	4	21
Column Total	122	24	54	200

To test these hypotheses, we compare the observed counts with counts that would be expected if no relationship existed between meal and beverage. **Expected counts** are determined by computing overall proportions for each row and applying them separately to each column.

Overall, the proportion of hamburgers ordered was 102/200, or 51 percent. If no association existed, we would expect to find this same proportion of hamburgers within each beverage category. Since 122 people ordered soft drinks, we would expect 51 percent of these, or $0.51(122) = 62.22$ of the soft drink orders, to include hamburgers. Similarly, for milk selections, 51 percent, or $0.51(24) = 12.24$, would include hamburgers, and $0.51(54) = 27.54$ of coffee or tea orders would occur in connection with hamburgers. These numbers are shown in the first row of Table 7.1. Subsequent rows are obtained in the same fashion. For example, the overall proportion of roast beef orders is $46/200 = 23$ percent. Applying this percentage to the column total for each beverage yields $0.23(122) = 28.06$ roast beef orders with soft drinks, $0.23(24) = 5.52$ with milk, and $0.23(54) = 12.42$ with coffee or tea.

Clearly, we recognize that actual counts must be whole numbers. It is impossible to find 62.22 orders consisting of a hamburger and soft drink. To interpret expected counts, we imagine repeating this experiment many times, with each replication consisting of a random sample of 200 orders. In the long run, assuming there is no relationship between meal choice and beverage selection, the *average* number of orders including a hamburger and soft drink would be 62.22. Each number in Table 7.1 is a long-run average, or expected count, assuming the null hypothesis is true.

TABLE 7.1 Expected cell counts when meal and beverage not related

Expected Count	Soft drink	Milk	Coffee/Tea	Row Total
Hamburger	62.22	12.24	27.54	102.00
Roast beef	28.06	5.52	12.42	46.00
Fish sandwich	18.91	3.72	8.37	31.00
Salad	12.81	2.52	5.67	21.00
Column Total	122.00	24.00	54.00	200.00

Row and column totals in Table 7.1 are identical to those of the original sample. The null hypothesis does not suppose that preferences for meals are affected or that choices among beverages are altered. The expected counts simply indicate how these totals would be distributed in the body of the table if there were no relationship between meal choice and beverage selection.

To test for association between meal and beverage, we compare observed counts with expected counts. Do the two tables differ by more than chance? Even if there is no relationship between meal and beverage in the population of all customers, we are likely to find observed counts somewhat different from expected counts in our sample. We ask whether the observed counts differ from the expected counts by more than random sampling error. (This logic is analogous to other hypothesis tests. For example, when testing whether the population mean, μ, is equal to 170, we do not expect to find $\overline{X}$ exactly equal to 170 in a sample. The difference between $\overline{X}$ and 170 is considered relative to random variability of the sample average.)

To assess the degree of relationship in our sample, we need a single measure that compares observed counts with expected counts across all cells of the table. One approach might be to subtract expected from observed for each cell, and sum over all cells:

$$\Sigma(\text{observed count} - \text{expected count}).$$

However, this simple strategy does not yield a satisfactory solution. In some cells, the observed count is higher than the expected count, producing a positive difference, while in others it is lower and results in a negative difference. Since marginal totals are identical in the two tables, positives and negatives exactly cancel, yielding a sum of 0.

To avoid cancellation of positives and negatives, we might square each difference before summing, resulting in the formula:

$$\Sigma(\text{observed count} - \text{expected count})^2.$$

This approach also has drawbacks, as can be seen by considering the two hypothetical cells shown below:

Cell 1	Cell 2
observed = 10 expected = 5	observed = 100 expected = 95

In cell 1, the observed count is twice the expected count, and suggests a serious departure from the null hypothesis. In cell 2, observed and expected counts are almost identical, in accordance with the null hypothesis. Yet the squared difference between observed and expected counts is 25 in both cells.

To eliminate this problem, we compute the squared difference between observed and expected counts as a proportion of the expected count in each cell. The sum of these quantities has approximately a chi-square distribution:

$$\chi^2 = \sum \left(\frac{(\text{observed count} - \text{expected count})^2}{\text{expected count}} \right). \tag{7.1.1}$$

Table 7.2 displays the contributions to this sum for each cell. For example, in the first cell,

$$\frac{(\text{observed count} - \text{expected count})^2}{\text{expected count}} = \frac{(79 - 62.22)^2}{62.22} = 4.53.$$

Summing over all cells of this table, we obtain $\chi^2 = 42.49$ for these data.

Degrees of freedom for this chi-square statistic are $(r - 1)(c - 1)$, where r is the number of rows in the table, and c is the number of columns (excluding margins).

TABLE 7.2 Cell Contributions to Chi-square Value

Cell Contribution	Soft Drink	Milk	Coffee/Tea	Row Total
Hamburger	4.53	3.18	4.03	11.74
Roast beef	0.33	0.04	0.54	0.91
Fish sandwich	5.19	0.02	11.08	16.29
Salad	1.13	11.92	0.49	13.54
Column Total	11.18	15.16	16.14	42.49

Example 7.1 has $r = 4$ rows and $c = 3$ columns, resulting in $(r - 1)(c - 1) = (4 - 1)(3 - 1) = 6$ degrees of freedom.

When the null hypothesis is true, meal and beverage are unrelated in the population of all customers, and relatively little relationship should occur in a random sample. Observed counts should be close to expected counts (which are computed assuming no relationship), and chi-square should be small. However, if the alternative hypothesis is true, observed counts should differ substantially from expected counts, and chi-square should be large. Consequently, if we find a large chi-square value, we are inclined to believe that the alternative hypothesis is true and will reject the null. By this logic, we locate our 5% rejection region in the upper tail of the chi-square distribution. From Table 3b in the Appendix, we find the critical chi-square value of 12.59 for six degrees of freedom. Our computed chi-square value of 42.49 exceeds this critical value, so we reject the null hypothesis. We conclude that meal choice and beverage selection are related in the population of all customers.

Cells with large entries in Table 7.2 contribute to the relationship between meal and beverage. By comparing observed and expected counts for these cells, we can characterize the relationship more precisely. For example, all entries in the first row of Table 7.2 are sizeable, indicating that hamburgers are not paired with beverage selections in accordance with expected counts. Hamburgers are more frequently ordered with soft drinks than would be expected if there were no relationship between meal and beverage (79 observed, 62.22 expected), and are less commonly ordered with milk, coffee, or tea than expected. Since all entries in the second row of Table 7.2 are small, roast beef purchasers tend to order beverages roughly in accordance with expected counts. Fish purchasers buy fewer soft drinks and more coffee or tea than would be expected, and salads are more commonly ordered with milk than would be expected.

The statistic shown in (7.1.1) is approximately distributed as chi-square with $(r - 1)(c - 1)$ degrees of freedom. This approximation is adequate if each expected count is greater than or equal to 1, and at least 75 percent of the expected counts are greater than or equal to 5. In our example, all expected counts are greater than 1, and 83 percent (10 cells out of 12) have expected counts greater than 5. The chi-square approximation is therefore satisfactory.

If these rules of thumb for the approximation are not met, certain rows and/or columns may be combined to increase expected counts. For example, we might combine fish and salad into a single category (health food?). Such pooling alters the definition of the variable "meal" and should only be performed if the new variable makes logical sense.

Pooling is impossible when each categorical variable has only two categories—that is, when the table has two rows and two columns. For this special case, a continuity correction is available to improve the approximation. Most computer programs automatically print the corrected chi-square for 2×2 tables. As mentioned at the end of Section 4.13, testing for relationship in a 2×2 table is identical to comparing two proportions. The null hypothesis of no relationship between row and column variables is equivalent to hypothesizing that the proportion of observations falling in the first row is the same in the two columns.

7.2 STRENGTH OF RELATIONSHIP FOR CATEGORICAL DATA

In Section 3.10, we discussed the distinction between statistical significance and practical importance. Statistical significance addresses the question, "How *sure* are we that H_0 is wrong?" The answer to this question is measured by the p-value. The smaller the p-value, the more convinced we are that the null hypothesis is false. Practical importance, on the other hand, addresses the question, "How *wrong* is H_0?" In the present context, the null hypothesis says that row and column variables are not related. The stronger the relationship, the more wrong H_0 is. To assess practical importance, we need a measure of the strength of the relationship between row and column variables.

At first thought it might seem we can use the observed value of chi-square as a measure of strength of relationship. The larger chi-square is, the stronger the relationship should be between row and column variables. However, chi-square is not an adequate measure of strength of relationship because it depends on sample size. A simple example will demonstrate this point.

Consider the hypothetical data shown in Table 7.3, relating sex and response to a yes/no question for a sample of size 100. The 5% critical value for $(r - 1)(c - 1) = (2 - 1)(2 - 1) = 1$ degree of freedom is 3.841. Chi-square for the data in this table is 0.16, and is not significant. For females, 26 out of 50, or 52 percent, responded yes, while for males, 24 out of 50, or 48 percent, responded yes. Since these percentages are very close, there is little relationship between sex and response.

Now consider the hypothetical data in Table 7.4 for 10,000 observations on theses two variables. Since the table is still 2×2, the critical value is unchanged. The computed chi-square for these data is 16.00, and is highly significant. We are very

TABLE 7.3 Hypothetical Sample Of 100 on Sex and Yes/No Response

Observed Count	Yes	No	Row Total
Female	26	24	50
Male	24	26	50
Column Total	50	50	100

TABLE 7.4 Hypothetical Sample of 10,000 on Sex and Yes/No Response

Observed Count	Yes	No	Row Total
Female	2,600	2,400	5,000
Male	2,400	2,600	5,000
Column Total	5,000	5,000	10,000

sure that H_0 is wrong. But the strength of relationship between sex and response is identical to the earlier sample. We still have 52 percent of females responding yes and 48 percent of males. The null hypothesis appears to be only slightly wrong—there is a weak relationship between sex and response. The additional evidence provided by the larger sample convinces us that this weak relationship exists, but does not make the relationship any stronger. Chi-square is sensitive to sample size, and is therefore not appropriate as a measure of strength of relationship.

To measure strength of relationship, the dependence of chi-square on sample size must be eliminated. Three methods of accomplishing this are commonly used. The first is the **phi coefficient**, given by:

$$\varphi = \sqrt{\chi^2/n}\,.$$

For the data of Example 7.1, we find

$$\varphi = \sqrt{\chi^2/n} = \sqrt{42.49/200} = 0.461.$$

The phi coefficient has a lower bound of 0, attained when each observed count equals its expected count exactly, and chi-square is 0. The upper bound of phi is $\sqrt{\min(r-1,c-1)}$, where min(r – 1,c – 1) stands for the smaller of (r – 1) and (c – 1). For a table with either two rows or two columns, the upper bound of phi is 1, but larger tables can have phi greater than 1. Thus, phi is generally only used for tables with two rows and/or two columns.

For tables with more than two rows and two columns, **Cramer's V** is often used to measure the strength of relationship. This statistic, given by

$$V = \frac{\varphi}{\sqrt{\min(r-1,\,c-1)}},$$

eliminates the upper bound problem, and has a range of 0 to 1. For Example 7.1, Cramer's V is

$$V = \frac{\varphi}{\sqrt{\min(r-1,\,c-1)}} = \frac{0.461}{\sqrt{\min(4-1,\,3-1)}} = \frac{0.461}{\sqrt{2}} = 0.326.$$

A third measure, the **contingency coefficient**, is given by

$$CC = \sqrt{\frac{\chi^2}{\chi^2 + n}}.$$

For the data of Example 7.1, we obtain

$$CC = \sqrt{\frac{\chi^2}{\chi^2 + n}} = \sqrt{\frac{42.49}{42.49 + 200}} = 0.419.$$

The contingency coefficient has a lower bound of 0, attained when chi-square is 0. It is always less than 1, but its upper bound approaches 1 as the size of the table increases. Some or all of these measures are frequently printed automatically by computer programs whenever a chi-square test is performed.

7.3 CATEGORICAL DATA IN COMPUTER PROGRAMS

The chi-square test for relationship between two categorical variables can be performed with each of the programs, SAS, SPSS, and MINITAB. The data of Example 7.1 are used to demonstrate this test in each package.

In each program, two different methods are available for data entry, depending on whether or not observed cell counts are known. If they are not known, data is entered one line per observation, and the program tallies observed counts as it sorts observations into cells. Section 4.14, which describes computer analysis for comparing two proportions, demonstrates this approach to data input for each program. Alternatively, when observed cell counts are known, only these counts need be entered. This second approach to data entry is illustrated in this section for each program.

SAS Input

```
OPTIONS LS = 80;
TITLE 'EXAMPLE 7.1 - FAST FOOD PREFERENCES';
DATA;
INPUT MEAL BEVERAGE NUMBER;
LINES;
1  1   79
1  2    6
1  3   17
2  1   25
2  2    6
2  3   15
3  1    9
3  2    4
```

```
3  3   18
4  1    9
4  2    8
4  3    4
PROC FREQ;
TABLES MEAL*BEVERAGE / CHISQ EXPECTED CELLCHI2;
WEIGHT NUMBER;
```

Discussion of SAS Input

Three values are entered for each cell—MEAL (row number), BEVERAGE (column number), and NUMBER (observed cell count). Twelve lines of input are required, one line per cell of the table. The SAS procedure FREQ (frequencies) and its TABLES statement request a table of the observations sorted by the two variables MEAL and BEVERAGE. The CHISQ option asks for a chi-square test to be performed on this table. EXPECTED requests display of expected counts in this table, and CELLCHI2 requests cell contributions to chi-square. The WEIGHT statement indicates that NUMBER is the observed count in each cell.

SAS Output

```
                EXAMPLE 7.1 - FAST FOOD PREFERENCES                    1
                     TABLE OF MEAL BY BEVERAGE

MEAL                   BEVERAGE

Frequency          |
Expected           |
Cell Chi-Square    |
Percent            |
Row Pct            |
Col Pct            |         1 |          2 |          3 |      Total
-------------------|-----------|------------|------------|
                 1 |      79   |       6    |      17    |       102
                   |    62.22  |    12.24   |    27.54   |
                   |   4.5254  |   3.1812   |   4.0338   |
                   |    39.50  |     3.00   |     8.50   |     51.00
                   |    77.45  |     5.88   |    16.67   |
                   |    64.75  |    25.00   |    31.48   |
-------------------|-----------|------------|------------|
                 2 |      25   |       6    |      15    |        46
                   |    28.06  |     5.52   |    12.42   |
```

```
           | 0.3337  | 0.0417  | 0.5359  |
           |  12.50  |   3.00  |   7.50  |   23.00
           |  54.35  |  13.04  |  32.61  |
           |  20.49  |  25.00  |  27.78  |
-----------+---------+---------+---------+
         3 |      9  |      4  |     18  |     31
           |  18.91  |   3.72  |   8.37  |
           | 5.1934  | 0.0211  |  11.08  |
           |   4.50  |   2.00  |   9.00  |   15.50
           |  29.03  |  12.90  |  58.06  |
           |   7.38  |  16.67  |  33.33  |
-----------+---------+---------+---------+
         4 |      9  |      8  |      4  |     21
           |  12.81  |   2.52  |   5.67  |
           | 1.1332  | 11.917  | 0.4919  |
           |   4.50  |   4.00  |   2.00  |   10.50
           |  42.86  |  38.10  |  19.05  |
           |   7.38  |  33.33  |   7.41  |
-----------+---------+---------+---------+
Total            122        24       54       200
               61.00     12.00    27.00    100.00
```

```
              EXAMPLE 7.1 - FAST FOOD PREFERENCES               2
           STATISTICS FOR TABLE OF MEAL BY BEVERAGE

Statistic                          DF        Value         Prob
---------------------------------------------------------------
Chi-Square                          6       42.488        0.000
Likelihood Ratio Chi-Square         6       37.777        0.000
Mantel-Haenszel Chi-Square          1       15.998        0.000
Phi Coefficient                              0.461
Contingency Coefficient                      0.419
Cramer's V                                   0.326

Sample Size = 200
```

Discussion of SAS Output

In the upper left corner of the table on page 1, SAS identifies the entries in each cell. The first entry (Frequency) is the observed count, followed by the expected count (Expected), the cell contribution to chi-square (Cell Chi-Square), the observed count as a percentage of overall sample size (Percent), and as a percentage of row total (Row Pct) and column total (Col Pct).

The second page of output shows the degrees of freedom (6), chi-square (42.488), and p-value (0.000). Clearly, there is *some* tail area beyond the chi-square value, but SAS rounds the area to three decimal places, and it is closer to 0.000 than to 0.001. Since p is less than 5 percent, we reject the null hypothesis. We are very sure that there is some relationship between meal choice and beverage selection. SAS automatically shows the values of phi (0.461), the contingency coefficient (0.419), and Cramer's V (0.326) whenever a chi-square test is performed.

SPSS Input

```
SET WIDTH = 80
TITLE 'EXAMPLE 7.1 - FAST FOOD PREFERENCES'
DATA LIST FREE / MEAL BEVERAGE NUMBER
WEIGHT BY NUMBER
CROSSTABS TABLES = MEAL BY BEVERAGE
     / CELLS = COUNT ROW COLUMN TOTAL EXPECTED SRESID
     / STATISTICS = CHISQ PHI CC
BEGIN DATA
1  1   79
1  2    6
1  3   17
2  1   25
2  2    6
2  3   15
3  1    9
3  2    4
3  3   18
4  1    9
4  2    8
4  3    4
END DATA
FINISH
```

Discussion of SPSS Input

Three values are entered in free format for each cell—MEAL (row number), BEVERAGE (column number), and NUMBER (observed cell count). Twelve lines of input are required, one line per cell of the table. The WEIGHT statement indicates that NUMBER is the observed count in each cell. The SPSS procedure

CROSSTABS requests a table of observations sorted by the two variables MEAL and BEVERAGE. Optional information requested in each cell includes the observed count (COUNT) and observed count as a percentage of row total (ROW), as a percentage of column total (COLUMN), and as a percentage of total sample size (TOTAL). Also requested are the expected count in each cell (EXPECTED) and the standardized residual (SRESID), which is the square root of the cell contribution to chi-square, multiplied by the sign of (observed count - expected count). The CHISQ statistic asks for a chi-square test to be performed on this table. PHI produces the phi coefficient and Cramer's V, while CC stands for contingency coefficient.

SPSS Output

```
             EXAMPLE 7.1 - FAST FOOD PREFERENCES          Page 1

MEAL by BEVERAGE
                      BEVERAGE
        Count    |
        Exp Val  |
        Row Pct  |
        Col Pct  |
        Tot Pct  |                                             Row
        Std Res  |      1.00 |       2.00 |       3.00 |   Total
MEAL    - - - - -| - - - - - |- - - - - - |- - - - - - |
                 |           |            |            |
         1.00    |      79   |       6    |      17    |     102
                 |    62.2   |    12.2    |    27.5    |   51.0%
                 |   77.5%   |    5.9%    |   16.7%    |
                 |   64.8%   |   25.0%    |   31.5%    |
                 |   39.5%   |    3.0%    |    8.5%    |
                 |     2.1   |    -1.8    |    -2.0    |
                 |- - - - - -| - - - - - -| - - - - - -|
         2.00    |      25   |       6    |      15    |      46
                 |    28.1   |     5.5    |    12.4    |   23.0%
                 |   54.3%   |   13.0%    |   32.6%    |
                 |   20.5%   |   25.0%    |   27.8%    |
                 |   12.5%   |    3.0%    |    7.5%    |
                 |     -.6   |      .2    |      .7    |
                 |- - - - - -| - - - - - -| - - - - - -|
         3.00    |       9   |       4    |      18    |      31
                 |    18.9   |     3.7    |     8.4    |   15.5%
```

```
                    |  29.0%     | 12.9%     | 58.1%     |
                    |   7.4%     | 16.7%     | 33.3%     |
                    |   4.5%     |   2.0%    |   9.0%    |
                    |   -2.3     |     .1    |    3.3    |
                    |- - - - - - | - - - - - | - - - - - |
       4.00         |      9     |      8    |      4    |      21
                    |   12.8     |    2.5    |    5.7    |   10.5%
                    |  42.9%     | 38.1%     | 19.0%     |
                    |   7.4%     | 33.3%     |   7.4%    |
                    |   4.5%     |   4.0%    |   2.0%    |
                    |   -1.1     |    3.5    |    -.7    |
                    | - - - - - -|- - - - - -|- - - - - -|
     Column            122          24          54           200
      Total          61.0%       12.0%       27.0%        100.0%
```

```
          EXAMPLE 7.1 - FAST FOOD PREFERENCES          Page 2

   Chi-Square              Value              DF      Significance
-------------------      --------            ----     ------------
Pearson                  42.48782             6          .00000
Likelihood Ratio         37.77706             6          .00000
Mantel-Haenszel          15.99788             1          .00006

Minimum Expected Frequency-      2.520
Cells with Expected Frequency  < 5 -   2 OF  12 (16.7%)

                                                      Approximate
       Statistic              Value   ASE1   T-value  Significance
---------------------        ------   ----   -------  ------------
Phi                          .46091                      .00000 *1
Cramer's V                   .32591                      .00000 *1
Contingency Coefficient      .41859                      .00000 *1

*1 Pearson chi-square probability
Number of Missing Observations:     0
```

Discussion of SPSS Output

In the upper left corner of the table on page 1, SPSS identifies the entries in each cell. The first entry (Count) is the observed count, followed by the expected count (Exp Val) and the observed count as a percentage of row total (Row Pct), as a percentage of column total (Col Pct), and as a percentage of overall sample size (Tot Pct). The last entry in each cell is the standardized residual (Std Res).

Shown on the second page are the chi-square value of 42.48782 (labeled Pearson), its degrees of freedom (6), and p-value (0.00000). Clearly, there is *some* tail area beyond the chi-square value, but SPSS rounds the area to five decimal places, and it is closer to 0.00000 than to 0.00001. Since p is less than 5 percent, we reject the null hypothesis. We are very sure that there is some relationship between meal choice and beverage selection. Also shown are the smallest expected count (2.520) and the number of cells with expected counts less than 5 (2 of 12). Below this table, SPSS prints the values of phi, Cramer's V, and the contingency coefficient.

MINITAB Transcript

```
MTB > READ C1-C3
DATA > 79 6 17
DATA > 25 6 15
DATA >  9 4 18
DATA >  9 8  4
DATA > END
       4 ROWS READ

MTB > CHISQUARE C1-C3

Expected counts are printed below observed counts

               C1         C2         C3      Total
    1          79          6         17        102
            62.22      12.24      27.54

    2          25          6         15         46
            28.06       5.52      12.42

    3           9          4         18         31
            18.91       3.72       8.37

    4           9          8          4         21
            12.81       2.52       5.67

Total         122         24         54        200

ChiSq =     4.525  +  3.181  +  4.034  +
            0.334  +  0.042  +  0.536  +
            5.193  +  0.021  + 11.080  +
            1.133  + 11.917  +  0.492  =  42.488
```

```
df = 6
2 cells with expected counts less than 5.0

MTB > CDF 42.488 K1;
SUBC > CHISQUARE 6.
MTB > LET K1 = 1-K1
MTB > PRINT K1
K1        0.000000119

MTB > STOP
```

Discussion of MINITAB Transcript

The first six lines of this transcript read the observed counts into columns C1 through C3, each row of data corresponding to a row of the table. The CHISQUARE command requests a chi-square statistic for the data stored in columns C1 through C3. MINITAB prints a table containing the observed count and expected count in each cell. Below the table, cell contributions to chi-square are shown, summing to the chi-square value of 42.488.

To compute the p-value, we use MINITAB's built-in CDF function, storing the area below 42.488 in K1. The area *above* 42.488 is the p-value of 0.00000019. Since this is less than 5 percent, we reject the null hypothesis. We are very sure that there is some relationship between meal choice and beverage selection.

7.4 ORDERED CATEGORIES

For some categorical variables, there is a natural ordering associated with the categories. Residential location, for example, might be classified according to the ordered categories: urban, suburban, and rural. Variables with ordered categories are frequently included in surveys to obtain information about sensitive issues, such as income and age. People may be reluctant to reveal such information precisely, but are usually less hesitant to check a category describing their income or age. Also common are opinion items in which respondents are asked to react to a statement on a strongly agree/agree/disagree/strongly disagree basis. Example 7.2 involves two such variables.

The two variables in this example are level of education (row variable) and rating of President's performance (column variable). Each of these variables is measured according to ordered categories. Level of education increases as we move down the rows, and Presidential rating increases across the columns. Based on this sample, we might be interested in testing for relationship between education and rating in the population surveyed. If a relationship exists, it might be characterized as positive or negative. The relationship is positive if those with a low level

Example 7.2

Included among the items in a recent survey were the two questions:

What is your level of education? (check one)
______ did not finish high school
______ high school degree
______ college degree
______ post graduate or professional degree

Overall, how would you rate the President's performance? (circle one)

POOR FAIR GOOD EXCELLENT

Responses of 180 people to these two items are summarized in the table of observed counts below.

	Rating			
Education	Poor	Fair	Good	Excellent
Less than HS	4	9	10	7
HS degree	5	8	22	14
College	20	31	11	12
Post graduate	9	6	8	4

of education tend to give the President a low rating, and those with a higher level of education tend to rate the President higher. The relationship is negative if those with low education tend to give high ratings, while those with high education tend to give low ratings.

When categories are ordered on both variables, direction of association is relevant, and the relationship between the variables is called **correlation**. This term is also used to characterize relationships between numeric variables, since numbers are inherently ordered. For categorical variables that are unordered, the concept of positive or negative relationship is meaningless. Association may exist between such variables, but it is not called correlation.

The symbol ρ (rho) represents population correlation. Correlation is standardized to range from -1 to $+1$, where $\rho = -1$ for perfect negative association, $\rho = +1$ for perfect positive correlation, and $\rho = 0$ when no relationship exists between the two variables. Typically, we test the null hypothesis that two variables are unrelated:

$$H_0: \rho = 0.$$

Depending on the research setting, the alternative hypothesis may be directional (one-sided) or nondirectional (two-sided). The alternative hypothesis is either

$$H_1: \rho < 0,$$
$$H_1: \rho > 0, \text{ or}$$
$$H_1: \rho \neq 0.$$

Since observations are cross classified according to education and rating in Example 7.2, we could test for association with a chi-square test. However, chi-square would not allow us to determine the *direction* of relationship. A different approach is therefore used for assessing relationship between two categorical variables when both are ordered.

To measure correlation in our sample, we examine each pair of observations and classify it as **concordant** or **discordant.** A pair of observations is concordant if the observation in the *higher* category on one variable is also in a *higher* category on the other variable. A pair is called discordant if the observation in the *higher* category on one variable is in a *lower* category on the other variable.

To demonstrate, consider picking two observations from the 180 people surveyed in Example 7.2. Suppose one of the people selected has a high school degree and gave the President a FAIR rating, and the other person has a college degree and rated the President's performance as GOOD. The second person is in a higher category than the first on level of education (college degree as opposed to high school degree), and is also in a higher category on Presidential rating (GOOD as opposed to FAIR). Since the person in the *higher* category on education is also in a *higher* category on rating, this pair of observations is concordant. A large number of concordant pairs would be expected in the sample if the relationship between education and rating were positive in the population.

Now suppose that a different pair of observations is selected—one person with less than high school education who rated the President as GOOD, and one with post graduate education who gave a POOR rating. The second person is in a *higher* category than the first on education (post graduate as opposed to less than high school degree), but in a *lower* category on Presidential rating (POOR as opposed to FAIR). This pair of observations is discordant. A large number of discordant pairs would be expected in the sample if the relationship between education and rating were negative in the population.

Clearly, there are many pairs of observations to examine. Computers, however, perform such tedious tasks quite efficiently. Let the total number of concordant pairs be denoted by C, and the total number of discordant pairs by D. If concordant pairs predominate (C greater than D), the relationship between education and rating is positive. If discordant pairs predominate (C less than D), the relationship is negative. The difference, C–D, measures the direction of association in the sample. We should also note that many pairs of observations are neither concordant nor discordant. If two people are selected in the same category on either (or both) variables, they do not contribute to C or D. Thus, the difference between C and D should be considered relative to the total number of pairs that are either concordant or discordant, C+D. The ratio of C–D to C+D is a measure of correlation in the sample between two ordered categorical variables known as **gamma** (γ):

$$\gamma = \frac{C-D}{C+D}. \tag{7.4.1}$$

If there is no relationship between two ordered categorical variables in the population ($\rho = 0$), we should find about as many concordant as discordant pairs in the sample, and a value of gamma near 0. If the population correlation is positive, gamma should also be positive. The maximum value for gamma is +1, obtained only if all pairs of observations in the sample are concordant (and thus D = 0). If the population correlation is negative, gamma should be negative. The minimum value for gamma is -1, when all pairs of observations in the sample are discordant (and thus C = 0). Gamma serves as an estimate of population correlation, ρ. Just as ρ is between −1 and +1, so is gamma.

For Example 7.2, the number of concordant pairs of observations is C = 3171, and the number of discordant pairs is D = 5530. From (7.4.1), the value of gamma is

$$\gamma = \frac{C-D}{C+D} = \frac{3171 - 5530}{3171 + 5530} = -0.27.$$

In the sample, education and rating are negatively related. Those with a higher level of education tend to give the President lower ratings than those with less education.

To test for correlation in the population, we use the fact that gamma is approximately normally distributed. When there is no correlation in the population (our null hypothesis), gamma has a mean value of 0. The variance of gamma can be estimated from the sample. Taking the square root of this estimate produces the estimated standard deviation, or standard error, of gamma. Since gamma is approximately normal, we standardize it by subtracting its mean (which is 0), and dividing by its standard error, to obtain an approximate standard normal statistic

$$Z = \frac{\gamma}{SE(\gamma)}.$$

This ratio is approximately normal, not t, even though the standard deviation of gamma is estimated from the sample.

For Example 7.2, the standard error of gamma is $SE(\gamma) = 0.082$, yielding

$$Z = \frac{\gamma}{SE(\gamma)} = \frac{-0.27}{0.082} = -3.29.$$

For a two-sided test (H_1: $\rho \neq 0$), we reject the null hypothesis, since the absolute value of Z is greater than 1.96. Based on the negative value of gamma, and therefore of Z, we conclude that education and rating of the President are negatively related in the population surveyed.

7.5 ORDERED CATEGORIES IN COMPUTER PROGRAMS

The gamma estimate of correlation between two ordered categorical variables can be calculated on SAS and SPSS, but is not available on MINITAB. The data of Example 7.2 are used for demonstration.

In both SAS and SPSS, the procedure that calculates gamma is the same procedure that performs the chi-square test. Thus, as with the chi-square test, data may be entered either one line per observation (when cell counts are not known), or one line per cell (when cell counts are known). The latter approach is employed here.

SAS Input

```
OPTIONS LS = 80;
TITLE 'EXAMPLE 7.2 - PRESIDENT RATING BY EDUCATION';
DATA;
INPUT EDUC RATING COUNT;
LINES;
1  1    4
1  2    9
1  3   10
1  4    7
2  1    5
2  2    8
2  3   22
2  4   14
3  1   20
```

```
3  2   31
3  3   11
3  4   12
4  1    9
4  2    6
4  3    8
4  4    4
PROC FREQ;
TABLES EDUC*RATING / MEASURES;
WEIGHT COUNT;
```

Discussion of SAS Input

Three values are entered for each cell—EDUC (row number), RATING (column number), and COUNT (observed cell count). Sixteen lines of input are required, one line per cell of the table. PROC FREQ (frequencies) and its TABLES statement request a table of the observations sorted by the two variables EDUC and RATING. The MEASURES option requests a number of sample estimates of correlation and the standard error of each. The WEIGHT statement indicates that COUNT is the observed count in each cell.

SAS Output

```
          EXAMPLE 7.2 - PRESIDENT RATING BY EDUCATION          1
                   TABLE OF EDUC BY RATING

EDUC         RATING

Frequency |
Percent   |
Row Pct   |
Col Pct   |       1 |       2 |       3 |       4 |     Total
----------|---------|---------|---------|---------|
        1 |       4 |       9 |      10 |       7 |        30
          |    2.22 |    5.00 |    5.56 |    3.89 |     16.67
          |   13.33 |   30.00 |   33.33 |   23.33 |
          |   10.53 |   16.67 |   19.61 |   18.92 |
----------|---------|---------|---------|---------|
        2 |       5 |       8 |      22 |      14 |        49
          |    2.78 |    4.44 |   12.22 |    7.78 |     27.22
          |   10.20 |   16.33 |   44.90 |   28.57 |
          |   13.16 |   14.81 |   43.14 |   37.84 |
```

```
------ |------| ----- |------| ----- |
   3   |    20 |    31 |    11 |    12 |      74
       | 11.11 | 17.22 |  6.11 |  6.67 |   41.11
       | 27.03 | 41.89 | 14.86 | 16.22 |
       | 52.63 | 57.41 | 21.57 | 32.43 |
------ |------| ----- |------| ----- |
   4   |     9 |     6 |     8 |     4 |      27
       |  5.00 |  3.33 |  4.44 |  2.22 |   15.00
       | 33.33 | 22.22 | 29.63 | 14.81 |
       | 23.68 | 11.11 | 15.69 | 10.81 |
------ |------| ----- |------| ----- |
Total       38      54      51      37      180
         21.11   30.00   28.33   20.56   100.00
```

```
        EXAMPLE 7.2 - PRESIDENT RATING BY EDUCATION           2
           STATISTICS FOR TABLE OF EDUC BY RATING

Statistic                                   Value         ASE
--------------------------------------------------------------
Gamma                                      -0.271       0.082
Kendall's Tau-b                            -0.201       0.062
Stuart's Tau-c                             -0.194       0.059

Somers' D C|R                              -0.206       0.063
Somers' D R|C                              -0.196       0.060

Pearson Correlation                        -0.218       0.072
Spearman Correlation                       -0.237       0.072

Lambda Asymmetric C|R                       0.143       0.059
Lambda Asymmetric R|C                       0.123       0.068
Lambda Symmetric                            0.134       0.052

Uncertainty Coefficient C|R                 0.055       0.020
Uncertainty Coefficient R|C                 0.058       0.021
Uncertainty Coefficient Symmetric           0.057       0.021

Sample Size = 180
```

Discussion of SAS Output

As indicated in the upper left corner of the table on page 1, each cell contains the observed count (Frequency), the observed count as a percentage of overall sample size (Percent), and as a percentage of row total (Row Pct) and column total (Col

Pct). Listed on the second page of output are a large number of measures of sample correlation, along with the standard error of each (identified by SAS as ASE, for asymptotic standard error). Gamma is -0.271, as computed in the previous section. SAS does not compute the Z value by dividing gamma by its standard error, and consequently does not print the obtained p-value.

SPSS Input

```
SET WIDTH = 80
TITLE 'EXAMPLE 7.2 - PRESIDENT RATING BY EDUCATION'
DATA LIST FREE / EDUC RATING COUNT
WEIGHT BY COUNT
CROSSTABS TABLES = EDUC BY RATING
     / STATISTICS = GAMMA
BEGIN DATA
1  1   4
1  2   9
1  3  10
1  4   7
2  1   5
2  2   8
2  3  22
2  4  14
3  1  20
3  2  31
3  3  11
3  4  12
4  1   9
4  2   6
4  3   8
4  4   4
END DATA
FINISH
```

Discussion of SPSS Input

Three values are entered in free format for each cell, EDUC (row number), RATING (column number), and COUNT (observed cell count). Sixteen lines of input are required—one line per cell of the table. The WEIGHT statement indicates that COUNT is the observed count in each cell. CROSSTABS requests a table of observations sorted by the two variables EDUC and RATING. The gamma statistic is requested as an option.

SPSS Output

```
     EXAMPLE 7.2 - PRESIDENT RATING BY EDUCATION          Page 1

EDUC by RATING
                    RATING
          Count    |
                   |                                        Row
                   |   1.00  |   2.00  |   3.00  |   4.00  |   Total
EDUC       ----    |----     |----     |----     |----     |
           1.00    |    4    |    9    |   10    |    7    |     30
                   |         |         |         |         |   16.7
                   |----     |----     |----     |----     |
           2.00    |    5    |    8    |   22    |   14    |     49
                   |         |         |         |         |   27.2
                   |----     |----     |----     |----     |
           3.00    |   20    |   31    |   11    |   12    |     74
                   |         |         |         |         |   41.1
                   |----     |----     |----     |----     |
           4.00    |    9    |    6    |    8    |    4    |     27
                   |         |         |         |         |   15.0
                   |----     |----     |----     |----     |
         Column        38        54        51        37          180
          Total      21.1      30.0      28.3      20.6        100.0

                                                         Approximate
Statistic         Value         ASE1         T-value    Significance
--------------------------------------------------------------------
Gamma           -.27112       .08218      -3.26919

Number of Missing Observations:    0
```

Discussion of SPSS Output

Below the table of observed cell counts, SPSS prints the value of gamma (-0.27112), its standard error, (identified as ASE1 for asymptotic standard error), and the Z value computed by dividing gamma by its standard error (but labeled as a T-value). Unfortunately, the obtained p-value is not shown.

7.6 NUMERIC DATA

Although some measurement processes involve categorical data, numeric variables are far more common. Most of the quantities we deal with in life, such as distance, time, cost, temperature, and so forth, are numeric in nature. In many practical situations, we wish to examine the relationship between two numeric variables. Example 7.3 provides a typical illustration.

Numeric variables contain more information than ordered categories, and a different estimate of correlation is employed. For each variable separately, the sample standard deviation measures the spread of observations. In Example 7.3, $S_X = 8.96$ measures the spread of the X values, and $S_Y = 2.62$ gives the spread of the Y values. The **sample covariance**, denoted by S_{XY}, indicates how the spread among X values *relates* to the spread among Y values. (The letter "S" has two subscripts when representing covariance between two variables, but only one when referring to standard deviation of a single variable.) The formula for sample covariance is shown in Table 7.5, which illustrates computation of the sample covariance for the data of Example 7.3. In this case, the sample covariance, $S_{XY} = 21.56$, is positive, indicating a positive relationship between coal consumption and sulphur in tree bark.

While the sample covariance reflects the *direction* of association between two variables, it has no upper or lower limit. Thus, it is impossible to assess the *strength* of relationship from its magnitude. However, dividing the sample covariance by

---

Example 7.3

Although power companies are monitored and regulated with respect to environmental impact, some pollution seems inevitable in generating power from fossil fuels. Effects of pollution are found in plants as well as animals. Data below show yearly average coal consumption (X) for ten coal burning power facilities, and sulphur concentration (Y) in the bark of trees near power generating operations. Is sulphur concentration in tree bark greater near plants with higher coal consumption?

X=coal	Y=sulphur
17	3.9
32	7.8
6	1.1
25	6.6
23	3.4
18	2.5
30	6.0
33	9.1
25	4.6
11	2.0

TABLE 7.5 Computation of Sample Covariance Between Coal and Sulphur

X–coal	Y–sulphur	$X-\bar{X}$	$Y-\bar{Y}$	$(X-\bar{X})(Y-\bar{Y})$
17	3.9	-5	-0.8	4.0
32	7.8	10	3.1	31.0
6	1.1	-16	-3.6	57.6
25	6.6	3	1.9	5.7
23	3.4	1	-1.3	-1.3
18	2.5	-4	-2.2	8.8
30	6.0	8	1.3	10.4
33	9.1	11	4.4	48.4
25	4.6	3	-0.1	-0.3
11	2.0	-11	-2.7	29.7
220	47.0	0	0	194.0

$$\bar{X} = \frac{\Sigma X}{n} = \frac{220}{10} = 22 \qquad \bar{Y} = \frac{\Sigma Y}{n} = \frac{47.0}{10} = 4.7$$

$$S_{XY} = \frac{\Sigma(X - \bar{X})(Y - \bar{Y})}{n - 1} = \frac{194.0}{9} = 21.56$$

the standard deviation of each variable scales the result to the usual −1 to +1 range. This ratio is referred to as **Pearson's r** or, simply, the **sample correlation**:

$$r = \frac{S_{XY}}{S_X S_Y}. \tag{7.6.1}$$

For Example 7.3, we obtain

$$r = \frac{S_{XY}}{S_X S_Y} = \frac{21.56}{(8.96)(2.62)} = 0.92.$$

Since r is near its maximum value of +1, a strong positive relationship exists in the sample between coal consumption and sulphur content in tree bark. The sample correlation, r, is the most commonly used estimate of population correlation, ρ.

To test the null hypothesis of no correlation between two variables in the population, a t test is performed:

$$t = (r)\sqrt{\frac{n-2}{1-r^2}} = (0.92)\sqrt{\frac{10-2}{1-(0.92)^2}} = 6.60.$$

The degrees of freedom for this t ratio are (n - 2) = (10 - 2) = 8. In Example 7.3, we are interested in whether sulphur concentration in tree bark is *greater* near plants with higher coal consumption. This relationship would produce positive correlation and, accordingly, our alternative hypothesis is

$$H_1: \rho > 0.$$

Locating our 5% rejection region in the upper tail of the t distribution, we find the critical value of 1.860 from Table 2 in the Appendix. As our observed t value far exceeds this critical value, we soundly reject the null hypothesis. We are very sure that there is a positive relationship between coal consumption and sulphur in the bark of trees near power generating operations.

The test on Person's r is appropriate for normally distributed data. When either variable appears substantially non-normal, a nonparametric correlation coefficient, **Spearman's r_s**, is used instead. Spearman's r_s is obtained by converting numeric values on each variable to ranks before computing the correlation. The same t test may be applied to this rank order correlation. For the data of Example 7.3, Spearman's r_s is 0.94, and the observed t value is 7.96. Again, this far exceeds the critical value of 1.860, and we conclude that coal consumption and sulphur in tree bark are positively related.

7.7 NUMERIC DATA IN COMPUTER PROGRAMS

Each of the programs, SAS, SPSS, and MINITAB, computes both Pearson's r and Spearman's r_s. The data from Example 7.3 are used to demonstrate these correlations in each package.

SAS Input

```
OPTIONS LS = 80;
TITLE 'EXAMPLE 7.3 - SULPHUR CONTENT IN TREE BARK';
DATA;
INPUT COAL SULPHUR;
LINES;
17  3.9
32  7.8
 6  1.1
25  6.6
23  3.4
18  2.5
30  6.0
33  9.1
25  4.6
11  2.0
PROC CORR PEARSON SPEARMAN;
VAR COAL SULPHUR;
```

Discussion of SAS Input

For each observation, two variables are entered, COAL and SULPHUR. The SAS procedure CORR is used with requests for both Pearson and Spearman correlations. The accompanying variables statement (VAR) specifies which variables to use in computing the correlations.

SAS Output

```
            EXAMPLE 7.3 - SULPHUR CONTENT IN TREE BARK            1
                       Correlation Analysis

                 2 'VAR' Variables: COAL  SULPHUR

                         Simple Statistics

Variable  N       Mean   Std Dev     Median  Minimum    Maximum
COAL      10   22.0000    8.9567    24.0000   6.0000    33.0000
SULPHUR   10    4.7000    2.6183     4.2500   1.1000     9.1000

  Pearson Correlation Coefficients  /  Prob  >  |R|  under
                    Ho: Rho=0  /  N  =  10

                             COAL         SULPHUR
           COAL           1.00000         0.91916
                            0.0            0.0002
           SULPHUR        0.91916         1.00000
                           0.0002          0.0

 Spearman Correlation Coefficients  /  Prob  >  |R|  under
                    Ho: Rho=0  /  N  =  10

                             COAL         SULPHUR
           COAL           1.00000         0.94225
                            0.0            0.0001
           SULPHUR        0.94225         1.00000
                           0.0001          0.0
```

Discussion of SAS Output

At the top of the page, descriptive information is shown for each variable, including the number of observations, sample mean, standard deviation, median, and smallest and largest values. Below this, SAS prints, for each pair of variables, the Pearson correlation and associated two-tailed p-value. (The t value is computed by the program to obtain the p-value, but is not printed.) Obviously, each variable is correlated perfectly (+1) with itself. The correlation between COAL and SULPHUR is 0.91916, agreeing with computations demonstrated in the previous section. For this example, a one-sided test is desired. Halving the two-tailed p-value, we obtain the one-tailed p-value of 0.0001, far beyond the 5 percent criterion. We are very sure that there is a positive relationship between coal consumption and sulphur in tree bark.

The Spearman correlation between COAL and SULPHUR is shown as 0.94225. Its one-tailed p-value (0.00005) is also far beyond the 5 percent cutoff, again leading to rejection of the null hypothesis.

SPSS Input

```
SET WIDTH = 80
TITLE 'EXAMPLE 7.3 - SULPHUR CONTENT IN TREE BARK'
DATA LIST FREE / COAL SULPHUR
CORRELATIONS VARIABLES = COAL SULPHUR / PRINT = ONETAIL
BEGIN DATA
17  3.9
32  7.8
 6  1.1
25  6.6
23  3.4
18  2.5
30  6.0
33  9.1
25  4.6
11  2.0
END DATA
NPAR CORR VARIABLES = COAL SULPHUR
FINISH
```

Discussion of SPSS Input

For each observation, two variables are entered, COAL and SULPHUR. The CORRELATIONS command produces the Pearson correlation between variables COAL and SULPHUR. By default, CORRELATIONS performs a two-sided test. The PRINT option is used to request a one-sided test (ONETAIL), as desired in Example 7.3. After the data, NPAR CORR is used to compute the Spearman correlation. Unlike the CORRELATIONS procedure, NPAR CORR performs a one-sided test by default.

SPSS Output

```
          EXAMPLE 7.3 - SULPHUR CONTENT IN TREE BARK     Page 1
                      Correlation Coefficients

                    COAL         SULPHUR
COAL              1.0000           .9192**
SULPHUR            .9192**        1.0000
* - Signif. LE.05         ** - Signif. LE.01   (1-tailed)
" . " is printed if a coefficient cannot be computed.
```

```
          EXAMPLE 7.3 - SULPHUR CONTENT IN TREE BARK     Page 2
              SPEARMAN  CORRELATION COEFFICIENTS

SULPHUR        .9423
            N(   10)
            SIG .000

                COAL
" . " IS PRINTED IF A COEFFICIENT CANNOT BE COMPUTED.
```

Discussion of SPSS Output

For each pair of variables, SPSS prints the Pearson correlation. Obviously, each variable is correlated perfectly (+1) with itself. The correlation between COAL and SULPHUR is 0.9192, agreeing with computations demonstrated in the previous section. Although SPSS does not print the actual p-value, the double asterisk indicates that the one-tailed p-value is less than 0.01. We are very sure that there is a positive relationship between coal consumption and sulphur in tree bark.

Unlike the CORRELATIONS procedure, the NPAR CORR command does not print correlations of each variable with itself. The Spearman correlation between COAL and SULPHUR is shown as 0.9423, and the one-tailed p-value is printed as 0.000, again leading to rejection of the null hypothesis.

MINITAB Transcript

```
MTB > NAME C1 = 'COAL'
MTB > NAME C2 = 'SULPHUR'
MTB > READ C1 C2
DATA > 17 3.9
DATA > 32 7.8
```

```
DATA >  6 1.1
DATA > 25 6.6
DATA > 23 3.4
DATA > 18 2.5
DATA > 30 6.0
DATA > 33 9.1
DATA > 25 4.6
DATA > 11 2.0
DATA > END
     10 ROWS READ

MTB > CORRELATION 'COAL' 'SULPHUR'

Correlation of COAL and SULPHUR = 0.919

MTB > LET K1 = .919*SQRT (8/(1 - .919**2))
MTB > CDF K1 K2;
SUBC > T 8.
MTB > LET K2 = 1-K2
MTB > PRINT K1 K2
K1        6.59296
K2        0.000063777

MTB > RANK 'COAL' C3
MTB > RANK 'SULPHUR' C4
MTB > PRINT 'COAL' C3 'SULPHUR' C4

 ROW    COAL      C3   SULPHUR       C4
   1      17     3.0       3.9        5
   2      32     9.0       7.8        9
   3       6     1.0       1.1        1
   4      25     6.5       6.6        8
   5      23     5.0       3.4        4
   6      18     4.0       2.5        3
   7      30     8.0       6.0        7
   8      33    10.0       9.1       10
   9      25     6.5       4.6        6
  10      11     2.0       2.0        2

MTB > CORRELATION C3 C4

Correlation of C3 and C4 = 0.942

MTB > LET K1 = .942*SQRT (8/(1 - .942**2))

MTB > CDF K1 K2;
```

```
SUBC > T 8.
MTB > LET K2 = 1-K2
MTB > PRINT K1 K2
K1          7.93884
K2          0.000005662

MTB > STOP
```

Discussion of MINITAB Transcript

Variables identified as COAL and SULPHUR are read into columns C1 and C2 of the worksheet for each observation. The CORRELATION command prints Pearson's r as 0.919, agreeing with computations demonstrated in the previous section. MINITAB does not automatically perform the significance test, so the t statistic is computed and stored in K1. The CDF command and its associated subcommand store the area below t in K2. Since our rejection region is located in the *upper* tail, the area above t is obtained by subtracting from one. The PRINT command shows the obtained t value (6.59296) and its one-tailed p-value (0.000063777). As p is well beyond the 5 percent criterion, we are very sure that there is a positive relationship between coal consumption and sulphur in tree bark.

To obtain Spearman's r_s, we must first convert numeric values to ranks for each variable. The ranks for COAL are stored in column C3, and for SULPHUR in C4. On COAL, observations four and nine are tied as sixth smallest. Using midranks (see Section 3.13), each is assigned the rank 6.5, the average of ranks that would have been assigned if they were slightly different. Having now ranked the sixth and seventh observations, the next smallest value (observation number seven) is given the rank of 8.

By the same commands used for Pearson's r, we find Spearman's r_S to be 0.942, and its t value to be 7.93884. The one-tailed p-value (0.000005662) is far beyond the 5 percent cutoff, again leading to rejection of the null hypothesis.

7.8 CORRELATION AS A MEASURE OF LINEAR RELATIONSHIP

Pearson's r and, to some extent, Spearman's r_s measure the degree of linear relationship between two variables. Each is restricted to the range from -1 to +1, and is positive or negative according to the direction of association. A plot of the observations on X and Y axes reveals the direction of relationship. If the best fitting line slopes up to the right, the correlation between X and Y is positive. If the line slopes up to the left, correlation is negative. The data of Example 7.3 are plotted in

Figure 7.1. As can be seen, the best fitting line slopes up to the right, consistent with the obtained positive correlation.

Correlation is unaffected by change of scale. A change of scale is frequently called a **linear transformation**. Linear transformation involves multiplying every value by a constant, and then adding a constant. For example, a linear transformation of X is obtained if every X value is multiplied by (1/4) and then 10 is added. Denoting the transformed variable by X*, we would have $X^* = (1/4)X + 10$. The correlation between X* and Y is the same as that between X and Y. Linear transformation could also be applied to Y (using either the same or different constants) without affecting correlation.

It is reasonable that correlation is invariant to linear transformation. Suppose, for example, that a correlation is computed between temperature and activity of a certain insect species, using measurements obtained each day during the summer. The correlation does not depend on whether temperature is measured on the Fahrenheit or Celsius scale. Celsius temperature can be obtained from Fahrenheit by multiplying each value by (5/9) and then adding (-32)—that is, by a linear transformation. Common sense tells us that the relationship between temperature and insect activity should be invariant to the temperature scale used. If the insect

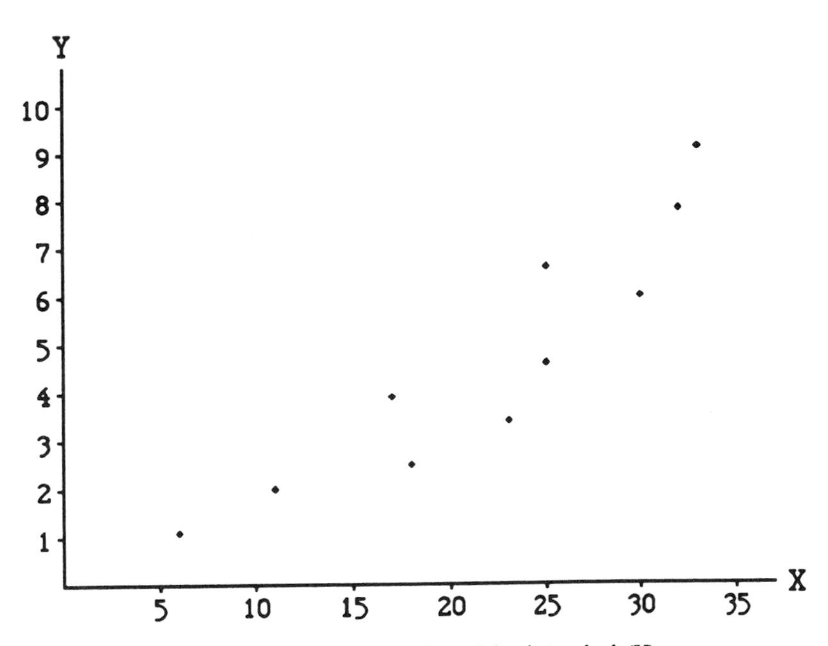

FIGURE 7.1. Plot of coal consumption (X) against sulphur in tree bark (Y).

is more active on warm days than cool, it should not matter whether warmth is measured on the Fahrenheit or Celsius scale.

While the magnitude of a correlation is unaffected by linear transformation, the sign is reversed if the multiplying constant is negative. For example, suppose insect *in*activity is defined by subtracting the activity from some standard value. Since subtracting is equivalent to multiplying by −1 and adding, the direction of correlation is reversed. Again, this makes logical sense. If temperature is positively related to insect activity, it should be negatively related to inactivity.

The invariance of correlation to linear transformation requires us to be extremely cautious about drawing conclusions from graphs. To demonstrate, suppose that coal consumption in Example 7.3 is transformed, as discussed above, to $X^* = (1/4)X + 10$. These numbers are plotted against Y in Figure 7.2. The relationship between coal and sulphur now appears to be much stronger than shown in Figure 7.1 for the original data. Yet we know that the correlation is unchanged. Such "gee whiz graphs" are used occasionally by unscrupulous advertisers to mislead unwary consumers about the effectiveness of their products.

In general, correlation measures how near the points are to lying in a line, not how steep the line is. The higher the magnitude of the correlation, the closer the

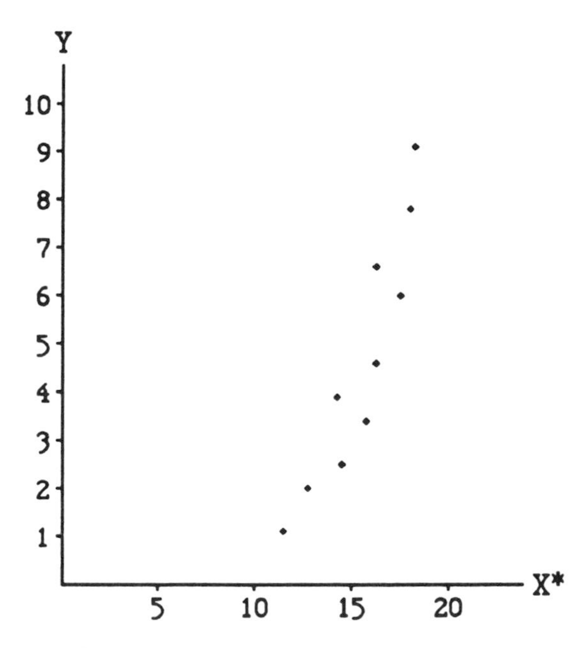

FIGURE 7.2. Plot of transformed coal consumption (X*) against sulphur in tree bark (Y).

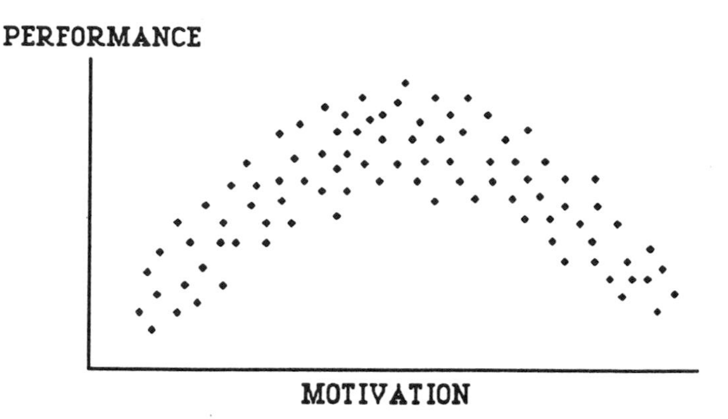

FIGURE 7.3. Plot of motivation (X) against performance (Y).

points lie to a straight line. A correlation near +1 occurs when the data lie near a line with positive slope; a correlation near -1 corresponds to near linearity in the negative direction.

Finally, we note that correlation measures only the degree of *linear* association between variables. Other relationships may exist, but they are not assessed properly by correlation. For example, motivation and performance are known to be strongly related, but in a nonlinear fashion. When motivation is low, performance tends to be low as well. As motivation increases, performance improves. But when motivation is extremely high, anxiety interferes with ability to produce results. For high levels of tension, performance actually declines. This relationship is shown for hypothetical data in Figure 7.3. Since the data do not fall near a straight line, the correlation between motivation and performance would be near 0. This does not imply that the variables are unrelated, only that they are not *linearly* related. Methods of assessing the strength of nonlinear relationships are available in multiple regression, which is discussed in Chapter 9.

7.9 OUTLIERS

Outliers are data points that do not follow the same pattern as most of the other data points. One substantial outlier can dramatically affect the correlation, weakening or even reversing the trend associated with other observations. To demonstrate, suppose that a single new observation is added to the data of Example 7.3, with coal consumption (X) of 7, and sulphur concentration (Y) of 8.9. As shown in Figure 7.4, this point clearly departs from the linear tendency that characterizes the other observations.

When Pearson's r is computed for all 11 observations, a value of 0.51 is obtained. The t test for population correlation yields a t value of 1.78, and is not

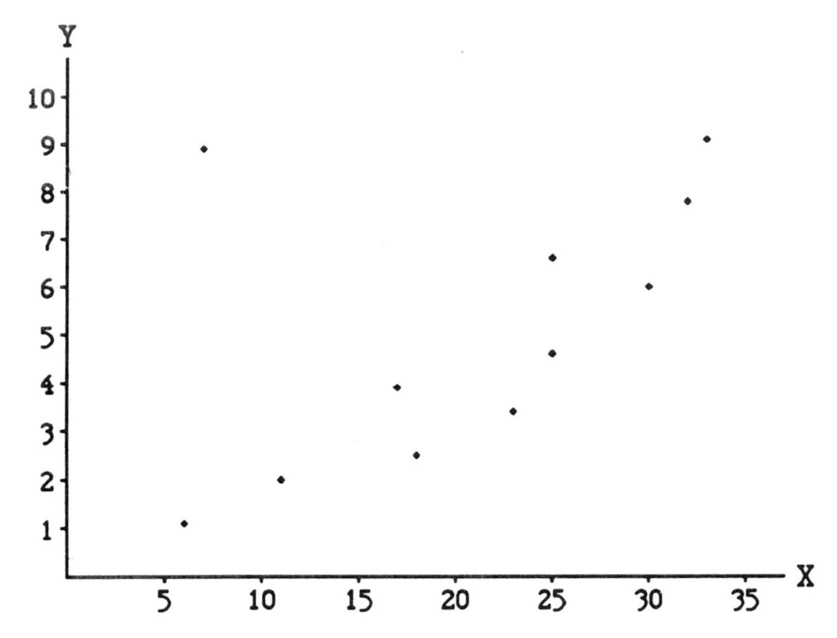

FIGURE 7.4. Plot of coal consumption (X) against sulphur in tree bark (Y) with one outlier added.

significant at the 5% level. Recall that, for the original 10 observations, we found $r = 0.92$ and $t = 6.60$, which was highly significant. The single outlier has eliminated our ability to conclude that coal consumption and sulphur in tree bark are linearly related.

Yet this cantankerous data point might have escaped our notice. It is inconsistent with the positive trend of other data points because a low value of X is paired with a high value of Y. But we might have failed to recognize it as an outlier, since its X coordinate is not the lowest of the X values, nor is its Y coordinate the highest of the Y values. In regression programs, discussed in Chapter 8, diagnostic statistics are printed to help identify outliers. But most computer programs for correlation do not include such features. Only by plotting the data can we become aware of outlying observations. Calculation of a correlation coefficient should always be accompanied by a plot of the data. In addition to revealing outliers, a plot may suggest nonlinear relationships that would be missed by linear correlation techniques.

Once we have identified an outlier, we are faced with the problem of what to do about it. Typically, we begin by trying to uncover an explanation for the aberrant value. Outliers can frequently be traced to malfunctioning equipment, inaccurate readings, unusual outside influences, recording errors, or other sources that invalidate the data point. In such cases, one is justified in simply deleting the outlier and analyzing the remaining data. But, even if the cause of an outlier cannot be determined, some remedial action must be taken. Some programs include options for weighting outliers less than other observations in calculating correlations. Often, converting to ranks is sufficient to minimize the effects of an outlier. But the main point to be stressed is that outliers cannot be simply ignored.

7.10 CORRELATION IN SUBGROUPS

To infer the nature of a population relationship accurately, estimated correlation must be based on a random sample from the population. Use of nonrandom samples, or specific subgroups from the population, can produce misleading results. Consider, for example, the relationship between motivation and performance depicted in Figure 7.3. If we only examine data points at the low end of the motivation axis, we might infer that motivation and performance are positively related. On the other hand, basing a sample estimate only on observations at the high end of the motivation axis might lead to the conclusion that motivation and performance are negatively related. To determine the proper pattern, we must examine all reasonable values of motivation.

In the situation discussed in the previous paragraph, sampling from a restricted range leads to overstating the extent of linear relationship between two variables. But the reverse can also occur. Restricting the range of a sample can produce an underestimate of true population correlation. If all observations shown in Figure 7.5(a) are included in computations, relatively strong positive correlation results. But the data points to the right of the vertical line in Figure 7.5(b) are nearly randomly scattered. A correlation computed from only these values is close to 0.

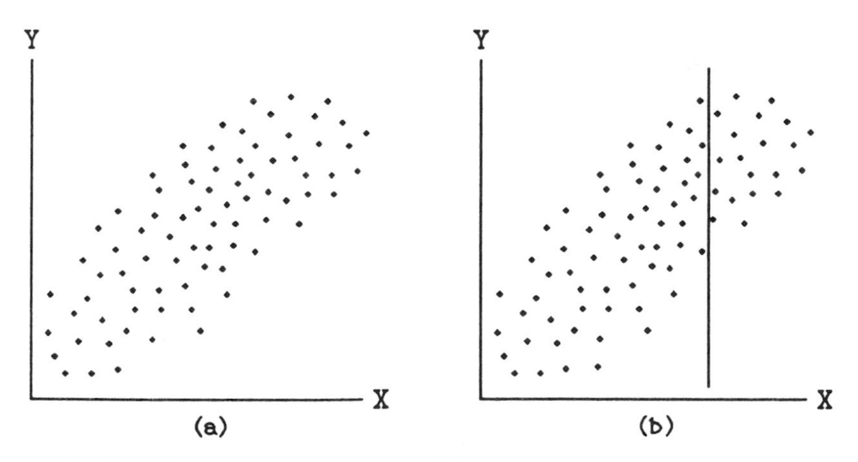

FIGURE 7.5. Effect of restricted range on correlation.

When scores on one variable determine whether the other variable is measured, a low correlation may arise due to restricted range. For example, suppose the X variable in Figure 7.5 is score on the Scholastic Aptitude Test (SAT), and Y is grade point average (GPA) after the first year of college. If all persons taking the SAT were admitted to college, a strong relationship between the SAT and GPA, such as depicted in Figure 7.5(a), might result. But, in practice, the SAT is used as an admission criterion. Those with low SAT scores are not admitted to college, and GPA is not available for them. Attempts to validate the SAT as an admission criterion based on its correlation with GPA are doomed to failure. Since we can only compute correlation when both variables are measured, we may see a nearly random pattern, as in Figure 7.5(b), instead of strong correlation.

Restricting the range of one of the variables involved in correlation can produce results uncharacteristic of the whole population. But other variables that divide the population into subgroups may also play a role in disguising true relationships. Figure 7.6 displays specific patterns that might be obtained when both females and males are measured on the same variables. In Figure 7.6(a), the two variables are positively related for females, but negatively related for males. Consequently, a single correlation computed across all observations would be nearly 0. In Figure 7.6(b), the correlation would be near 0 if computed separately for either females or males, but would show a negative relationship if based on a composite sample.

The lesson to be learned from these figures is that inferences about population correlation must be based on appropriately drawn samples. A sample correlation computed from a subgroup may not reflect the pattern of relationship in the whole population. Also, a correlation calculated from a composite sample may not portray the proper picture for specific subgroups. We should only draw conclusions

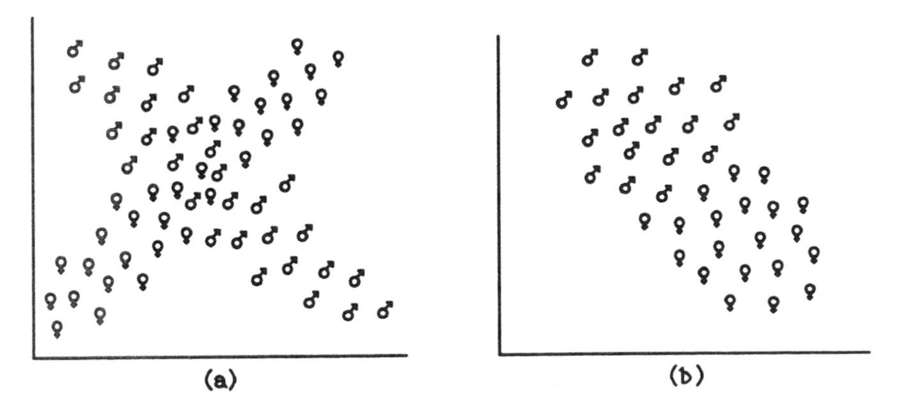

FIGURE 7.6. Effects of subgroups on correlation.

about population relationships when our estimated correlation is based on a random sample from the population of interest.

7.11 CAUSALITY

In Example 7.3, a strong positive correlation was found between coal consumption and sulphur concentration in the bark of trees near power plants. The obvious conclusion is that pollution from the burning of coal is absorbed by the trees and appears as sulphur in the bark. But the obvious conclusion may be wrong. It is possible that some power facilities burn more coal than others because they are located closer to coal deposits in the earth. Also, trees growing near underground coal may absorb sulphur through their roots, drinking ground water contaminated by flowing through the coal deposits. In short, it is possible that coal burning is not the cause of sulphur in tree bark. Both coal consumption and sulphur concentration may be influenced by a third variable, proximity to underground coal.

In general, if we find that two variables are correlated, we cannot draw the conclusion that one of them is *causing* the other. In Example 7.3, we are not permitted to conclude that coal burning causes sulphur to concentrate in tree bark. This conclusion would be no more valid than claiming that increased sulphur in tree bark causes higher levels of coal consumption in power plants! Correlation does not imply causation. Simple regression, unlike correlation, provides an opportunity to draw causal conclusions when certain experimental procedures are followed (see Section 8.11). But simply observing a correlation between two variables is not justification for concluding that a causal relationship exists.

7.12 SELECTING AN APPROPRIATE TEST FOR RELATIONSHIP

Many statistical procedures, in this and earlier chapters, can be characterized in terms of testing for relationship between two variables. In this chapter, we have specifically addressed relationships for two variables that are both categorical (chi-square test), both ordered categories (gamma), or both continuous (Pearson's r or Spearman's r_S). But a broader view of testing for relationships enables the inclusion of t tests, one-way anova, and their nonparametric equivalents.

When analyzing data, one should attempt to identify all variables involved in the research. In Section 1.2, we introduced the terms random variable and design variable. A random, or dependent, variable is measured during the experiment for each observation. A design, or independent, variable (such as a grouping variable) helps determine the structure of the experiment. Both random and design variables may be involved in relationships, and the distinction between them is not important in choosing an appropriate relational technique. What is important, however, is the level of measurement of each variable. Every variable, whether random or design, can be classified as categorical, ordered categories or numeric.

If both variables have the same level of measurement, the techniques of this chapter apply (chi-square, gamma, and Pearson and Spearman correlations). When the variables have different levels of measurement, t tests, anova, or some other procedure may be employed to assess relationship. In the next few paragraphs, we consider the possibilities for two variables whose levels of measurement are different, to help shed light on selecting an appropriate test.

When one variable is categorical and the other has ordered categories, the lack of order on *both* variables precludes a directional relationship. The concept of correlation is meaningless unless both variables are ordered. As a consequence, the chi-square test is performed, essentially ignoring the information contained in the ordering of categories on one of the variables.

If one variable is categorical and the other numeric, a two-group t test or one-way analysis of variance may be employed. The t test is used when the categorical variable has only two categories, and anova is used when there are more than two categories. The numeric variable is the one on which sample means and variances are computed. The categorical variable is the group, with two levels for a t test and more for anova. If we find a significant difference between the means, there is some relationship between the categorical variable and the numeric variable in the population. Equivalent nonparametric procedures—the rank sum, median, and Kruskal-Wallis tests—can also be used to test for relationship between categorical and numeric variables.

Testing for relationship between an ordered category variable and a numeric variable depends on how much we know or are willing to assume about the ordered categories. In some cases, we may be able to assign numeric values to the categories. Level of education, for example, might be approached by assigning the value 12 to high school degree (12 years of education), 16 to college degree (high school plus 4 years), and 19 to post graduate degree (averaging 2 years past college for a master's degree and 4 for a doctorate). Assigning a numeric value to less than high school education is more problematic, but 6 (half a high school degree) may be a reasonable solution. If numeric values can rationally be assigned to the ordered categories, as in this case, the Pearson or Spearman correlation can be computed between these values and the values of a numeric variable.

Sometimes we cannot assign specific numeric values to ordered categories, but are willing to assume equal spacing. For example, strongly agree/agree/disagree/strongly disagree might be viewed as equally spaced, the difference between each adjacent pair of responses representing approximately the same increment in opinion. In this case, the numbers 1, 2, 3, and 4 can be assigned to these responses, respectively, and this value can be related to a numeric variable using a Pearson or Spearman correlation. Since correlation is unaffected by linear transformation, it makes no difference whether we use the numbers 1 through 4, or assign values 10, 20, 30, and 40 to the four ordered categories.

However, some ordered categories are difficult or impossible to convert to numeric values. For example, urban, suburban, and rural do not lend themselves naturally to any set of numbers. Equal spacing of these levels might be considered an unreasonable assumption. For such variables, we may be safer to ignore the order of the categories and resort to one-way analysis of variance, which is valid for unordered categories.

When two variables are not assessed at the same level of measurement, selection of an appropriate test for relationship is a subjective decision. As much as possible, we should attempt to use all the information contained in our measurements. If we analyze a variable using a technique appropriate for a lower level of measurement, we are not using all the information available in the data. A numeric variable contains more information than mere ordered categories, and ordered categories provide more information than *un*ordered categories. Information helps us reach correct decisions. The more information used in making a decision, the more likely we are to find the correct answer.

7.13 SUMMARY

This chapter presented tests for relationship between variables, with different tests employed depending on the level of measurement of the variables. For two categorical variables, the chi-square test is used to compare observed counts in the cross classification table with counts that would be expected if there were no association between the variables. Characterization of the form of relationship is accomplished by examining cell contributions to chi-square. With small sample sizes, pooling of rows and/or columns can be used to increase expected frequencies and improve the chi-square approximation.

While the chi-square test allows inference about the existence of association from sample to population, it cannot be used to assess the strength of relationship. For this purpose, we employ the phi coefficient, Cramer's V, or the contingency coefficient. Desirable bounds for a measure of strength of relationship among categorical variables are 0, for no relationship, and 1, for perfect association.

When order is relevant for both variables, a measure of relationship is called a correlation coefficient. For two variables with ordered categories, gamma is used to estimate population correlation. Gamma is based on counting the number of pairs of observations that are concordant (in the same order) and the number that are discordant (in the opposite order). Gamma is tested with an approximate normal variable.

Pearson's r is used to measure correlation between two normally distributed numeric variables. It is computed by dividing the sample covariance by the standard deviation of each variable, producing a result in the range from −1 to +1. Spearman's r_s is obtained by converting each variable to ranks before computing the correlation, and thus can be used for data that is not normally distributed. Both forms of correlation can be tested with a t test.

Correlation assesses only the degree of linear relationship between two variables. Since correlation is unaffected by linear transformation of the variables, it measures how close the points lie to a straight line, not how steep the line is. Outliers, points lying outside the pattern of most of the data, can seriously affect correlation. If an outlier is noticed in a plot of the data, we should attempt to ascertain the cause of the unusual observation.

Using only a subgroup from the population in our sample can produce misleading results about the nature of population relationships. The true correlation can be underestimated or overestimated, depending on the patterns of association for the subgroup and population. Correlation between two variables does not imply that one of them causes the other.

In a broader sense, t tests, one-way analysis of variance, and their nonparametric equivalents can also be viewed as tests for relationships between variables. These procedures are used to examine association when one variable is categorical and the other is numeric. For two variables measured at different levels of measurement, the choice of a test for relationship is somewhat subjective, but we should attempt to utilize all information available in the observed measurements.

SAS, SPSS, and MINITAB programs were used to demonstrate the chi-square, gamma, Pearson, and Spearman approaches to testing for relationship between variables.

8

Simple Linear Regression

In statistics, the term **regression** is equivalent to the word prediction in English. This strange expression does not imply that anything is moving backwards or getting younger. But the term does have a logical explanation, as we will see in Section 8.4. **Simple linear regression** deals with prediction from a single predictor variable. Methods for handling more than one predictor will be discussed in the next chapter, under the heading of multiple linear regression.

The predictor variable is called the **independent variable**, and is labeled with the letter X. The target variable, the one we wish to predict, is called the **dependent variable**, and is labeled Y. In some settings, it is possible for an experimenter to set values of X independently, upon which we expect Y to depend. We must start with a set of data containing measurements on both X and Y. These data are used to obtain an equation relating Y to X, and the equation may allow us to predict Y when only X is available.

As the expression suggests, simple linear regression employs a straight-line equation for prediction. While some pairs of variables may be related in a linear fashion, others probably are not. Typically, we do not know the form of the relationship until we look at a random sample of data. However, in many circumstances, a straight line serves as a good first approximation. One of the topics for later discussion in this chapter is the examination of the linearity assumption, with an eye toward improving the prediction model if a straight line seems inappropriate.

8.1 EQUATION OF A LINE

To study the relationship between X and Y, we must start with a set of observations for which both variables are available. Each observation may be plotted as a point in a two-dimensional graph, according to its X and Y coordinates. When the points fall exactly in a line, an equation can be written that precisely relates the X and Y coordinates. In its most familiar form, the equation of a straight line is $Y = mX + b$, where m is the **slope** of the line and b is the **intercept**. Slope refers to both direction of slant and to steepness. The intercept is the point where the line crosses the y axis.

An example will be helpful in understanding these concepts. Consider the equation $Y = 2X + 1$, in which the slope is 2 and the intercept is 1. We will substitute various X values into this equation, determine Y for each, and then plot the resulting (X,Y) pairs. Substituting the value $X = 1$, we obtain $Y = 2(1) + 1 = 3$. One of the points we wish to plot is $(X,Y) = (1,3)$. Similarly, $X = 2$ yields $Y = 2(2) + 1 = 5$, and we plot the point (2,5). The list below contains these points and several additional (X,Y) pairs satisfying our equation.

(X,Y)
(1,3)
(2,5)
(0.5,2)
(0,1)
(−1,−1)
(−2.5,−4)

As plotted in Figure 8.1(a), these points all lie exactly on a straight line. Every pair of X and Y values satisfying our equation corresponds to a point on this line. Also, every point on this line has X and Y coordinates related by $Y = 2X + 1$. The equation describes the line precisely.

A line with a positive slope, such as our example with $m = 2$, is slanted upward to the right. The slope can be interpreted as "rise over run." Because $m = 2$, the line rises two units for every unit of horizontal distance.

The intercept is the point where the line crosses the Y axis, at $Y = 1$ in our example. When the line crosses the Y axis, its X coordinate is 0. Substituting $X = 0$ in the equation of a line yields $Y = mX + b = m(0) + b = b$ (i.e., the intercept of the line). Any line with slope of 2 is parallel to our example line. Figure 8.1(b) displays several lines with slope of 2, but differing in their intercepts. Changing the intercept slides the line up or down, but does not affect its steepness.

Changing the slope alters steepness, but does not affect the point where the line crosses the Y axis. Figure 8.1(c) shows several lines with the same intercept, but different positive values for slope. As the slope increases, the line becomes steeper, approaching a vertical line as slope approaches infinity. As the slope decreases, the line becomes less steep. A slope of one corresponds to a line intersecting the axes at a 45 degree angle. Decreasing the slope still further, the line approaches a horizontal line as slope approaches 0. A horizontal line has slope exactly equal to 0. If the slope is close to 0 but negative, the line slants slightly up to the left, as

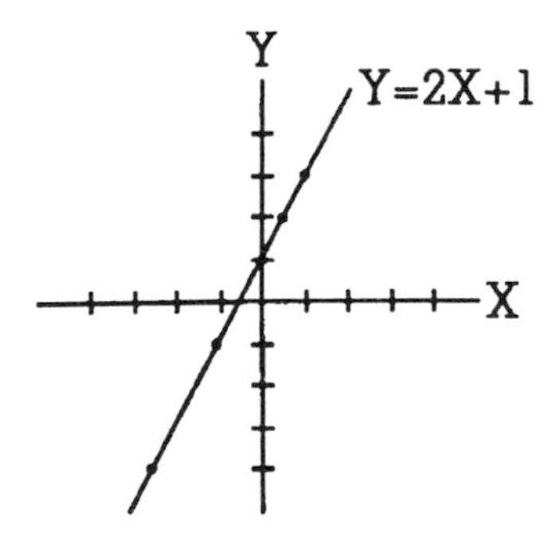

Figure 8.1(a)

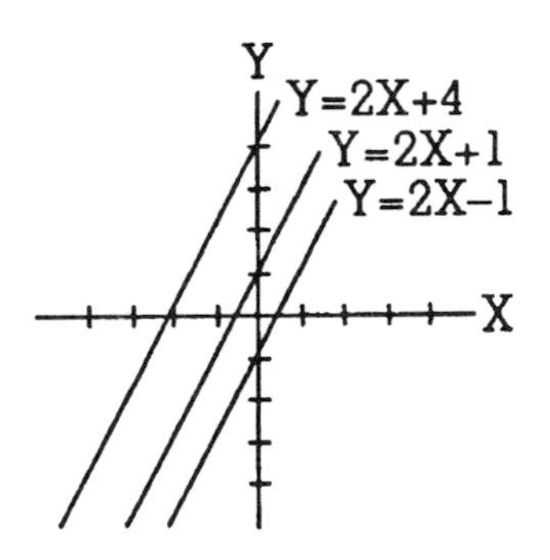

Figure 8.1(b)

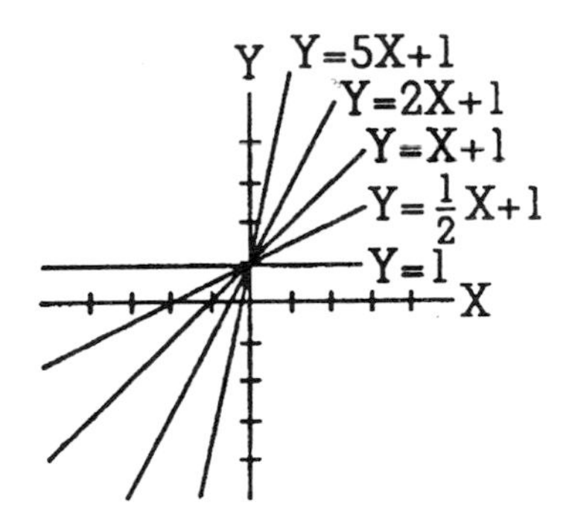

Figure 8.1(c)

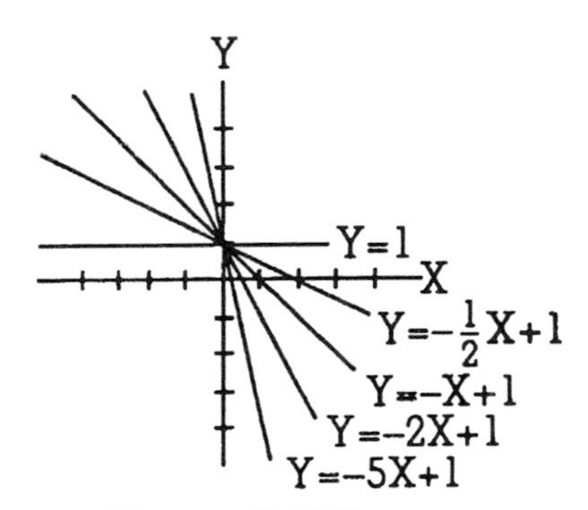

Figure 8.1(d)

FIGURE 8.1. Lines and their equations.

shown in Figure 8.1(d). As with positive slopes, the line becomes steeper as the magnitude of the slope increases, approaching a vertical line as slope approaches negative infinity.

A vertical line represents a point of discontinuity in slope. It is the only line that cannot be written in form $Y = mX + b$. The equation of a vertical line is $X =$ (some number). Since this form of equation does not involve Y, it is not useful as a prediction equation. In regression we are only concerned with lines that can be written in the form $Y = mX + b$.

In statistics we usually do not employ the symbols m for slope and b for intercept; Greek letters are used instead. But, whatever notation is selected, the term that is multiplied times X is the slope, and the additive quantity is the intercept.

8.2 STATISTICAL MODEL

In fitting a straight line to data, the assumption of linearity is technically made for the population, not the sample. That is, we assume that the population mean value of Y is related to X by

$$\mu_{Y|X} = \beta_0 + \beta_1 X. \tag{8.2.1}$$

The notation $\mu_{Y|X}$, instead of simply μ_Y, emphasizes that the mean value of Y depends on X. In this equation, β_0 is the population intercept, since it is the additive component, and β_1 is the population slope, since it is multiplied times X. The population intercept and slope are unknown constants, to be estimated from the sample. (The symbols β_0 and β_1 used in regression bear no relationship to the symbol β, used without a subscript to denote the probability of type two error.) In theory, X is a design variable whose values are known exactly, not a random variable. But in practice, both X and Y are frequently a function of the sample, and both contain random variability.

As we learned in Chapter 7, correlation measures the degree of linear relationship between two variables. Clearly, if there is little linear relationship, it is fruitless to apply a linear model for prediction. Correlation and regression are intimately connected. Theoretically, the two methods apply to different circumstances. Correlation is used when both X and Y are random variables, while regression requires X values to be known exactly. But for most practical purposes, this distinction may be ignored with little consequence. We will discuss the connection between correlation and regression in later sections of this chapter.

In addition to assuming a linear model in the population, we also assume normality and equality of variance. These distributional assumptions apply to the dependent variable, Y. For any given value of X, we require Y to have a normal distribution. The mean of this normal distribution depends on the X value, through the relationship given in (8.2.1). But the variance of Y is assumed to be the same σ^2, regardless of X.

The full set of assumptions is displayed graphically in Figure 8.2. If several observations of Y are obtained with the same X value, we would expect these observations to be normally distributed. In the figure, the points for each X value resemble a sample from a normal distribution. Most of the observations are close to the center, with fewer points trailing off symmetrically toward the extremes. For each value of X, the observations of Y show this characteristic normal pattern.

The mean value of Y, the average height of these normal distributions, changes for different X values. But it changes with X in a linear fashion, and $\mu_{Y|X}$ always falls along the population regression line. Although the mean value of Y changes with X, the variance does not. Each normal distribution is drawn with the same spread. The assumption of equal variances is comparable to the same assumption

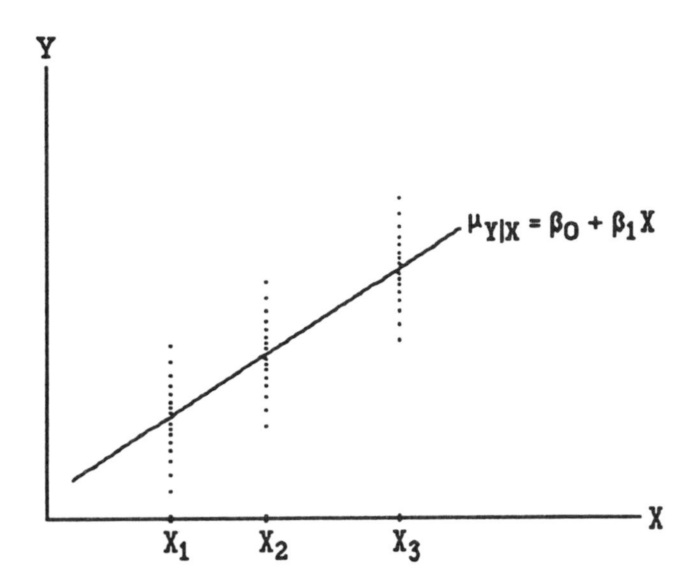

FIGURE 8.2. Graphical display of assumptions.

made in analysis of variance. In Chapter 5, when we used analysis of variance to compare several population means, we assumed that the population variances were equal. Here we assume equal variances for each value of X.

One should always be concerned with long lists of assumptions. How can we detect serious departures from the assumed conditions? What effect do such violations have on the methods we would like to use? And what recourse do we have to redress these problems? We will return to these issues in later sections. For now, we caution that it is often impossible to detect violations of assumptions by simply plotting the raw data. In many experiments, we do not have multiple observations of Y at each X value. Instead, each observation differs in both its X and Y values. In essence, we observe one value of Y for each value of X, drawn at random from the corresponding normal distribution.

In the fact that each X value is unique, Example 8.1 is typical of many regression applications. Replications of Y are not available for each X value, and a plot of the data does not reveal normal distributions. As shown in Figure 8.3, however, a strong linear tendency is evident in the data.

Our first objective is the estimation of the population regression line from the sample data. This is tantamount to estimating the slope and intercept, as these parameters determine the line. Our estimated regression line enables predictions of demand (Y) to be made for various temperatures (X). But, before using the estimated line to make predictions, we should critically examine our estimated

Example 8.1

In an attempt to predict summertime electricity demand, 12 communities widely scattered across the country were selected. In each community, measurements were taken of average daily high temperatures in July and August (X), and electricity demand per residential customer (Y) during the same period.

X = temperature	Y = demand
77	89
86	171
71	58
75	76
85	139
83	130
72	78
79	115
89	182
80	130
84	154
76	105

---

slope and intercept. We may wish to form confidence interval estimates of slope and intercept, or to test hypotheses concerning the values of these parameters. Then we should look at the quality of predictions produced by our estimated line. Finally, if we are satisfied with the precision of our estimated slope and intercept, and with the accuracy of predictions from the model, we may substitute temperature values for additional communities into the equation to predict electricity demand.

8.3 LEAST SQUARES ESTIMATION

To estimate the population regression line, we must estimate its slope and intercept. The estimate of population slope, β_1, will be denoted by b_1, and the estimate of population intercept, β_0, will be denoted by b_0. Employing this notation, the estimated regression line is written as

$$\hat{Y} = b_0 + b_1X, \tag{8.3.1}$$

where Y represents an **estimated,** or **predicted value** of Y. The estimated slope and estimated intercept are known as **regression coefficients** (even though b_0 is not a true coefficient because it is not multiplied times anything).

In Figure 8.3, we could easily draw a line through the center of the data points, and then measure its intercept and slope. But there are two problems with such a subjective approach. First, different people would probably draw slightly different lines. Second, there is no guarantee that our line is the "best possible" in any sense, when we draw it by eye. To alleviate both these problems, we employ an objective approach to finding the best possible line.

Since we will use our line for making predictions, we would like the predicted value, $\hat{Y}$, to be as close as possible to the actual Y value for each observation. Equivalently, we would like $Y-\hat{Y}$ to be as small as possible for each observation. The difference $Y-\hat{Y}$ represents the error in our prediction, frequently called the **residual**. In the sense that Y is the target for prediction, and $\hat{Y}$ is the part of Y we can predict, $Y-\hat{Y}$ is the remaining, or residual, part of Y that we are unable to predict. As shown in Figure 8.4, the residuals are the *vertical* distances from the data points to the line. We are concerned with vertical distances, instead of horizontal, because we are trying to predict Y, not X.

We would like the residual to be small for each observation. Summing these residuals, we might try to find the line that minimizes $\Sigma(Y-\hat{Y})$. Unfortunately, this is not an adequate criterion for choosing a "best fitting" line. Any line that passes through the center of the data (that is, the point with coordinates $\overline{X}$ and $\overline{Y}$) has a sum of residuals equal to 0. For such a line, about half the predictions are too small ($Y-\hat{Y}$ is positive), and half are too large ($Y-\hat{Y}$ is negative), resulting in a

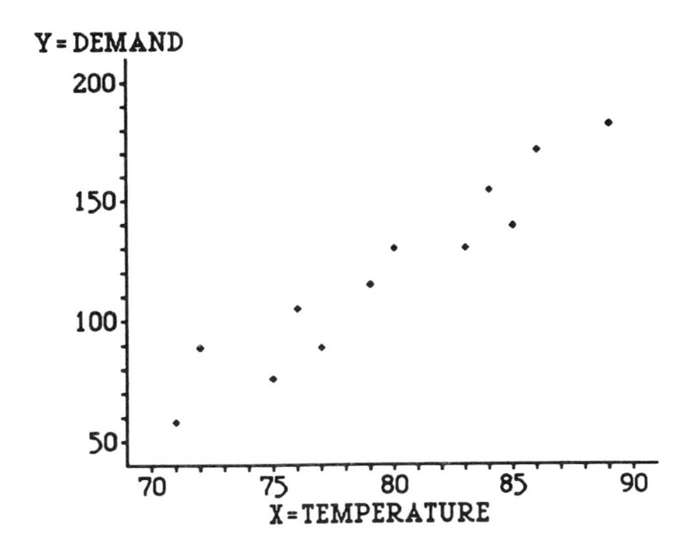

FIGURE 8.3. Data for electricity demand.

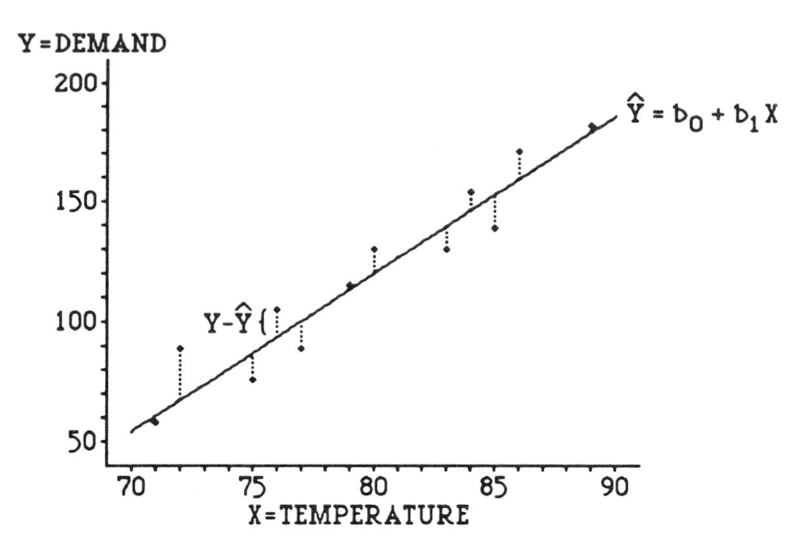

FIGURE 8.4. Least squares approach.

sum of residuals of 0. To avoid cancellation of positives and negatives, we square each residual before summing. Following this logic, we seek the line that minimizes $\Sigma(Y-\hat{Y})^2$. This is called the **least squares** procedure for selecting the best fitting line.

A minimization problem is solved with calculus, which we omit. The solution, however, results in simple formulas for the regression coefficients:

$$b_1 = \frac{S_{XY}}{S_X^2} \tag{8.3.2}$$

and

$$b_0 = \overline{Y} - b_1\overline{X}. \tag{8.3.3}$$

In the formula for slope, S_{XY} is the sample covariance, introduced in Section 7.6, and S_X^2 is the sample variance of the X values.

For the data of Example 8.1, the sample covariance is $S_{XY} = 217.61$, and the sample variance of the X values is $S_X^2 = 32.93$. From (8.3.2) we find the estimated slope to be

$$b_1 = \frac{S_{XY}}{S_X^2} = \frac{217.61}{32.93} = 6.61.$$

Using this estimated slope and the values of $\overline{X}$ (79.75) and $\overline{Y}$ (118.92), we find the estimated intercept from (8.3.3) to be

$$b_0 = \overline{Y} - b_1\overline{X} = 118.92 - (6.61)(79.75) = -408.07.$$

Our estimated regression line, obtained by substituting these b_1 and b_0 values into (8.3.1), is

$$\hat{Y} = -408.07 + 6.61X.$$

In practice, we need not do these computations by hand. We input the raw data to the computer and read the estimated slope and intercept directly from the printout. We still must examine the equation to assess the quality of our estimates of slope and intercept, and the accuracy of predictions. But if the equation passes this scrutiny, it is easily used to obtain predictions. By substituting the temperature (X), we immediately find the predicted electricity demand ($\hat{Y}$).

8.4 REGRESSION TOWARD THE MEAN

As we learned in Section 7.6, the Pearson correlation (r) is computed from the quantities S_{XY}, S_X and S_Y. Since the estimated slope (b_1) is based on the first two of these terms, correlation and slope are related. The precise relationship is

$$b_1 = r\left(\frac{S_Y}{S_X}\right).$$

Because the sample standard deviations S_X and S_Y are always positive, slope and correlation must have the same sign. The slope of the estimated regression line is positive when X and Y are positively correlated, and negative when they are negatively correlated.

Replacing b_1 in our estimated regression line with this function of r, and doing some additional manipulations, we arrive at the equation

$$\left(\frac{\hat{Y}-\overline{Y}}{S_Y}\right) = r\left(\frac{X-\overline{X}}{S_X}\right). \qquad (8.4.1)$$

We usually standardize a variable, say X, by subtracting its mean and dividing by its standard deviation to obtain a Z score:

$$Z_X = \frac{X - \mu_X}{\sigma_X}.$$

But since μ and σ are unknown, we often replace them with their estimates to approximate the Z score:

$$Z_X = \frac{X - \overline{X}}{S_X}.$$

Similarly, the approximate Z score for the predicted value ($\hat{Y}$) can be written

$$Z_{\hat{Y}} = \frac{\hat{Y} - \overline{Y}}{S_Y}.$$

Substituting these expressions into (8.4.1), we obtain

$$Z_{\hat{Y}} = rZ_X. \tag{8.4.2}$$

This is an interesting and revealing equation. Consider a situation in which X and Y are positively related, but the correlation is not perfect—that is, the correlation is in the range $0 < r < 1$. Suppose we look at an observation that is higher than average on the predictor variable X. Since X is greater than $\overline{X}$, the approximate Z score for X is positive. Because r and Z_X are both positive, equation (8.4.2) guarantees that $Z_{\hat{Y}}$, their product, is also positive. Therefore, the predicted value $\hat{Y}$ is above the average of the Y scores. Logically, an observation that is high on X is predicted to be high on Y.

But since r is less than 1, and Z_X is multiplied by r to obtain $Z_{\hat{Y}}$ in equation (8.4.2), it follows that $Z_{\hat{Y}}$ is less than Z_X. In terms of Z scores, the predicted value $\hat{Y}$ is closer to the Y mean than X is to the X mean. For an observation with a high X value, we predict Y to be high, but, relative to the standard deviations, not as high as X. In similar fashion, if X were low, we would predict Y to be low, but not as low as X. Our predictions all appear to *regress toward the mean.*

This peculiar behavior of predicted values is a result of using the least squares criterion to determine our regression line. In least squares, we minimize the sum of *squared* errors. Thus, we must be very careful not to make any extreme errors—the square of an extreme error is huge. The way to avoid extreme errors is to avoid extreme predictions—that is, we simply predict Y to fall a little closer to the mean than was X. In this respect, least squares is a somewhat conservative procedure. When applied to reasonably well-behaved data, it is unlikely that extreme prediction errors will be obtained.

8.5 DISTRIBUTIONS OF REGRESSION COEFFICIENTS

We use the least squares regression coefficients b_1 and b_0 to estimate the population slope (β_1) and population intercept (β_0), respectively. Before discussing the distributions of these estimators, it will be helpful to review a simpler situation, namely the estimation of the population mean (μ) for a single random variable. Typically, we use the sample average, $\overline{X}$, as our estimate. In earlier chapters, we learned that $\overline{X}$ has a normal distribution with mean $\mu_{\overline{X}} = \mu$, and variance $\sigma^2_{\overline{X}} = \sigma^2/n$. Increasing the sample size decreases the variance of $\overline{X}$ and makes it a more precise estimator of μ.

In regression, the estimates of slope and intercept have properties similar to those of $\overline{X}$. Each has a normal distribution, centered at the corresponding population parameter. From a single random sample, we find one value for b_1 and one value for b_0. Drawing another random sample from the population would probably result in slightly different values for b_1 and b_0. Repeating this process for many samples would yield many different values for b_1 and b_0. A plot of the b_1 values would approximate the probability distribution for b_1. We would find most of the values near the center, with fewer values, symmetrically distributed, toward the tails. The center of this distribution is the population slope, β_1. In similar fashion, the b_0 distribution is normal with center at the population intercept, β_0.

We will not write out the expressions for the variances of b_1 and b_0, as they are somewhat complicated, and in practical applications the computer does all the actual computations for us. Nonetheless, it is important to discuss some of the elements involved in these variances. Each variance involves sample size in the denominator, just as the variance of $\overline{X}$ has n in its denominator. Increasing the sample size therefore decreases these variances, and makes b_1 and b_0 more precise estimators of population slope and intercept. But these variances also involve S^2_X in their denominators. To obtain small variances for our estimators, we would want to *increase* the variance of the X values.

In many regression applications, the values of X are outside our control, and there is little we can do to increase their spread. But in some situations, we are in complete control of the X values, and can choose them to maximize their variance. Suppose, for example, we are trying to predict the height of a potted plant (Y) from the amount of water it is given (X). We might start with a group of identical seedlings, each in its own pot, and use a different amount of water in each pot. In most experiments, there is a range of X values of interest. Here, we clearly cannot put a negative amount of water in a pot, so the lower limit for X is 0. Depending on the type of plant and size of pot, an upper limit is also appropriate. (A plant given too much water does not grow to giant height—it dies.)

To minimize the variances of b_1 and b_0, we must maximize the spread of the X values. This is accomplished by giving half the plants 0 water, and the other half the upper limit. However, this strategy essentially yields only two data points, through which a straight line automatically fits. If we *know* that plant height relates to water in a linear fashion, this is the optimal strategy. But, in most applications, we do not know that a linear model is appropriate. The best practical compromise is to spread the X values more or less evenly over the range of interest. This produces a relatively large value for S_X^2 and, therefore, relatively small variances for our regression coefficients, while at the same time allowing us to examine the assumption of a linear model.

If enough plants are available, it is advantageous to use two or more separate plants for each amount of water. Such replications serve a dual purpose. First, we can compute a sample variance among the Y values separately for each value of X. Comparison of these variances provides information about the validity of the equal variance assumption. Second, if this assumption appears to be satisfied, we can compute a pooled estimate of variance that is independent of the model. Some computer programs include an option allowing this **pure error** term to be used in a test of model adequacy. That is, we can test the null hypothesis that a linear model is appropriate. Selection of X values is not always within our control, but when we have options in the design of an experiment, we should exercise them wisely.

When we are concerned with an unknown population mean, the normal distribution of $\overline{X}$ enables us to draw inferences from the sample to the population. We standardize $\overline{X}$ and then replace the unknown variance σ^2 with an estimate, thereby obtaining a t distribution. This t distribution forms the basis of confidence intervals and hypothesis tests for μ. Similarly, in regression, the normal distributions of b_1 and b_0 allow us to perform inference about the population slope and intercept. Each regression coefficient is standardized and then converted to a t distribution by inserting an estimate for the unknown variance σ^2.

The variance estimate used in regression can be logically explained by relating it to a regular sample variance. In Section 2.2, we introduced the usual sample variance, given by

$$S^2 = \frac{\Sigma(X-\overline{X})^2}{n-1}.$$

A brief review of the rationale for this estimate is instructive. The population variance is defined to be the *long run* average squared distance from the mean, and a reasonable estimate is the *sample* average squared distance to the mean—that is, $\Sigma(X-\mu)^2$ divided by n. But since the true mean μ is generally unknown, we use $\overline{X}$ as its estimate. In doing this, we convert the denominator from n to n−1, obtaining the usual sample variance. As we learned in previous chapters, the denominator of a variance estimate is its degrees of freedom. In general, *the degrees of freedom for a variance estimate is the number of observations minus the number of parameters that must be estimated.* For a regular sample variance, we have n observations, and must estimate the single parameter μ with $\overline{X}$, yielding n−1 degrees of freedom.

Now let us consider the regression case. The normal variable in regression is Y, with mean $\mu_{Y|X} = \beta_0 + \beta_1 X$. We estimate $\mu_{Y|X}$ with $\hat{Y}$, so the numerator comparable to $\Sigma(X-\overline{X})^2$ is $\Sigma(Y-\hat{Y})^2$. But $\hat{Y}$ is based on replacing the two parameters β_0 and β_1 with our estimates b_0 and b_1. Since we have estimated two parameters, the degrees of freedom for our variance estimate are n−2. Putting degrees of freedom in the denominator, the estimate of σ^2 used in regression is

$$\frac{\Sigma(Y-\hat{Y})^2}{n-2}. \tag{8.5.1}$$

The numerator of this ratio is the sum of squared residuals, abbreviated $SS_{RESIDUAL}$, and the denominator is the corresponding degrees of freedom. Following conventions established in analysis of variance, we call the ratio of SS to df a mean square. The estimate of variance shown in (8.5.1) is denoted by $MS_{RESIDUAL}$.

In the standardized distributions of b_0 and b_1, we replace the unknown variance σ^2 with the estimate $MS_{RESIDUAL}$, thereby obtaining t distributions. Each of these t distributions has n-2 degrees of freedom, since each uses $MS_{RESIDUAL}$ as the variance estimate. Recall that the least squares criterion minimized the sum of squared residuals. Therefore, our regression coefficients can be interpreted as the optimal values in terms of minimizing variance.

Before leaving this section, we note that the rule italicized above can be used to determine degrees of freedom for any variance estimate. For example, consider the one-way analysis of variance design covered in Chapter 5, in which we tested for equality of ***a*** means. In the denominator of our F ratio, we used the variance estimate MS_{WITHIN}. As discussed in Section 5.2, MS_{WITHIN} is the weighted average of separate sample variances in the ***a*** groups. Each variance requires estimation of the group mean, so a total of ***a*** parameters must be estimated in computing MS_{WITHIN}. We stated that $df_{WITHIN} = N-a$, which we now see is the number of observations minus the number of parameters estimated.

8.6 CONFIDENCE INTERVALS AND HYPOTHESIS TESTS ON REGRESSION COEFFICIENTS

Confidence intervals and hypothesis tests for β_0 and β_1 are analogous to those constructed in earlier chapters for μ. The confidence interval for μ is given by $\overline{X} \pm t[SE(\overline{X})]$, where t denotes a value from the t distribution chosen for the desired level of confidence and n−1 degrees of freedom, and $SE(\overline{X})$ is the standard error of $\overline{X}$. We can characterize the form of the confidence interval as

$$(\text{point estimate}) \pm t[SE(\text{point estimate})].$$

To test a hypothesized value of μ, we use the ratio

$$\frac{\overline{X} - \mu}{SE(\overline{X})},$$

obtaining our rejection region from the t distribution. The form of the test statistic is

$$\frac{(\text{point estimate}) - (\text{hypothesized value})}{SE\ (\text{point estimate})}.$$

Following this logic, the confidence interval for β_0 is given by

$$b_0 \pm t[SE(b_0)]. \tag{8.6.1}$$

For the data of Example 8.1, the estimated intercept is $b_0 = -408.07$. The standard error of b_0 is difficult to calculate by hand, but computer outputs (presented in Section 8.10) print the value 43.75 for $SE(b_0)$. Our example has 12 observations and, therefore, $n-2 = 10$ degrees of freedom. For a 95% confidence interval, we put 0.025 in each tail, finding the t value 2.228 from Table 2 in the Appendix. Substituting these values into (8.6.1), we obtain the 95% confidence interval for population intercept:

$$-408.07 \pm 2.228(43.75) = -408.07 \pm 97.48.$$

We are 95% certain that the population intercept β_0 lies between -505.55 and -310.59.

Most computer programs automatically test the hypothesis that each parameter is equal to 0. Subtracting the hypothesized intercept of 0 from the estimate b_0, and dividing by the standard error of b_0, we obtain the t statistic

$$\frac{b_0 - \beta_0}{SE(b_0)} = \frac{-408.07 - 0}{43.75} = -9.33.$$

This is far beyond the critical values of ±2.228 for a two-sided test at the 5% level of significance, leading to rejection of the hypothesis that the population intercept is 0.

We have presented this test because it is produced by all standard computer programs whenever regression is performed. But in the present situation, this is an absurd hypothesis to test. The intercept is the mean value of Y (electricity demand) when X (temperature) is 0. When looking at summertime electricity demand, as in this example, a temperature of 0 is nonsense. And if we were interested in predicting demand during the winter (when temperatures near 0 might be reasonable), we would not use an equation derived solely from temperatures in July and August.

In some circumstances, an intercept of 0 might be a reasonable hypothesis to test. In our example of predicting plant height from water, an intercept of 0 makes some sense—no water yields no growth. Perhaps we could interpret the result as a test to see if any ambient moisture already existed in the soil. Whether the test of zero intercept makes sense or not, all standard computer programs automatically print it. Part of using computer programs intelligently involves distinguishing between meaningful and irrelevant output.

Following the same logic, we can also form a confidence interval and test hypotheses about the population slope. In Example 8.1, the estimated slope is $b_1 = 6.61$, with standard error of $SE(b_1) = 0.547$. The 95% confidence interval for population slope is given by

$$b_1 \pm t[SE(b_1)] = 6.61 \pm 2.228(0.547) = 6.61 \pm 1.22,$$

resulting in the interval $5.39 < \beta_1 < 7.83$.

The test statistic for the null hypothesis of zero slope is

$$\frac{b_1 - \beta_1}{SE(b_1)} = \frac{6.61 - 0}{0.547} = 12.07.$$

This obtained t value far exceeds the critical values of ±2.228 for a two-sided test, so we reject the hypothesis of zero slope. Actually, in this example, we expect higher electricity demand to accompany higher temperatures during the summer, and would probably employ a one-sided test with alternative hypothesis $H_1\colon \beta_1 > 0$. Since the two-sided test was significant, the one-sided test would also lead to rejection of the null hypothesis.

Again, computer programs automatically test the hypothesis that the slope is 0. However, unlike the test on intercept, *the test of zero slope is always meaningful.* If the null hypothesis is accepted, we conclude that population slope (β_1) is 0. Our statistical model, $\mu_{Y|X} = \beta_0 + \beta_1 X$, reduces to $\mu_{Y|X} = \beta_0$. Now the mean value of

Y does *not* depend on X, so we should not be using X to predict Y! A slope of 0 corresponds to a horizontal line, which results in using the same predicted value of Y for any X value. Common sense tells us that we are not really using X to predict Y. In short, if the hypothesis of zero slope is accepted, prediction of Y from X is not warranted. Before using the estimated regression equation to obtain predictions, we should require this test to be significant.

Estimates of slope and intercept are **scale-dependent**. If the scale of X and/or Y is modified by linear transformation, different values of b_1 and b_0 result. Fortunately, the standard error of b_1 is also scale-dependent, and changes induced by linear transformation cancel out in the ratio of b_1 to its standard error. As a result, the t test for zero slope is **scale-independent**. For example, if temperature in Example 8.1 is measured on the Celsius scale, instead of Fahrenheit, the test for zero slope is unaffected. We saw in Section 7.8 that scale changes can dramatically alter the appearance of a graph. However, these changes in appearance do not affect the conclusion about whether prediction is warranted. We should not attempt to determine if population slope is 0 from looking at a graph, nor examine the value of b_1 without considering its standard error.

If either X or Y is multiplied by a constant, the test for population intercept is unaffected. But if a constant is added to either variable, the t value changes. Common sense tells us that the test for zero intercept *should* be sensitive to additive transformations. Adding a constant is equivalent to sliding the regression line up or down, which logically alters its intercept.

8.7 ANALYSIS OF VARIANCE APPROACH

In this section we discuss using analysis of variance logic to produce an F test for the null hypothesis of zero slope. The F test obtained is equivalent to the t test developed in the previous section. The observed F value is the square of the t value, and the critical value for F is the square of the t critical value. Thus, the two tests must reach the same conclusion with regard to acceptance or rejection of the null hypothesis. A similar equivalence between F and t tests was discussed in Section 5.4 in connection with comparing means of two populations. In that situation, the F test from one-way analysis of variance was equivalent to the pooled t test. Just as the F test in one-way analysis of variance enabled generalization of mean comparisons to more than two groups, the F test in regression can be generalized to the case of more than one predictor, as we will see in Chapter 9. In addition, the use of analysis of variance logic in regression produces a natural measure of practical importance, which we discuss in the next section.

In the one-way classification model discussed in Chapter 5, total variability is partitioned into components that represent variability between groups and variability within groups. Instead of partitioning overall variability, however, we perform separate breakdowns of its numerator, or sum of squares, and its denominator, or degrees of freedom. Similarly, in regression we perform two separate partitions—one for sums of squares and one for degrees of freedom.

In regression the overall variability among Y values is the sample variance S_Y^2. The numerator of this variance, $\Sigma(Y-\overline{Y})^2$, is the total sum of squares, denoted by SS_{TOTAL}. The denominator of overall variance, n - 1, is df_{TOTAL}. Each of these terms is partitioned into components associated with regression and residual. We write

$$SS_{TOTAL} = SS_{REGRESSION} + SS_{RESIDUAL}$$

and

$$df_{TOTAL} = df_{REGRESSION} + df_{RESIDUAL}.$$

As in other uses of analysis of variance, mean squares are obtained by dividing each sum of squares by its degrees of freedom. A ratio of mean squares follows an F distribution. To test the null hypothesis of zero slope, we use the test statistic

$$F = \frac{MS_{REGRESSION}}{MS_{RESIDUAL}}. \qquad (8.7.1]$$

The residual term we have seen already. In Section 8.5 we defined $SS_{RESIDUAL}$ to be $\Sigma(Y-\hat{Y})^2$, and $df_{RESIDUAL}$ to be n - 2. As discussed earlier, the residual variance is the part of total variability we *cannot* predict with our regression model.

The formula for $SS_{REGRESSION}$ is also enlightening. It is

$$SS_{REGRESSION} = \Sigma(\hat{Y}-\overline{Y})^2.$$

The sum of squares for regression is the part of total variability we *can* predict with our linear model. To see this, suppose the estimated slope, b_1, is 0. Substituting $b_1 = 0$ into (8.3.3), we find that the estimated intercept is

$$b_0 = \overline{Y}-b_1\overline{X} = \overline{Y}-(0)\overline{X} = \overline{Y}.$$

Now insert the values $b_0 = \overline{Y}$ and $b_1 = 0$ into the estimated regression equation of (8.3.1) to obtain

$$\hat{Y} = b_0+b_1X = \overline{Y}+(0)X = \overline{Y}.$$

When the slope is 0, our best prediction for Y is obtained by ignoring X and simply guessing the average Y value. And when $\hat{Y} = \overline{Y}$ for each observation, $SS_{REGRESSION}$ is 0—we cannot predict *any* of the variability in Y with a horizontal line. If the regression line is nearly horizontal, $SS_{REGRESSION}$ is small. As the line becomes more steeply slanted, $SS_{REGRESSION}$ increases and predictions improve.

Figure 8.5 depicts the partition of sums of squares graphically. The total sum of squares measures spread in Y values, without regard to X. To ignore X, imagine projecting all the data points to the Y axis, as if looking at their shadows cast by a light shining from the right side of the figure. Spread in the Y direction is measured by SS_{TOTAL}. The residual sum of squares is the spread of data points *away from* the regression line. $SS_{RESIDUAL}$ is the part of Y variability not related to X in a linear fashion. If the line were horizontal, spread away from the regression line would be identical to spread projected onto the Y axis. To the extent that the line slopes, residual variability is less than total. Thus, $SS_{REGRESSION}$, the other part of total variability, depends on the slope of the regression line. In essence, part of the spread in Y values is related to the fact that we have used a variety of X values. When X is small, Y tends to be small. When X is large, Y is also large. The relationship between Y and X explains some of the variability observed among Y values.

The regression sum of squares has only one degree of freedom. While this could be determined by subtracting $df_{RESIDUAL} = n - 2$ from $df_{TOTAL} = n - 1$, a more

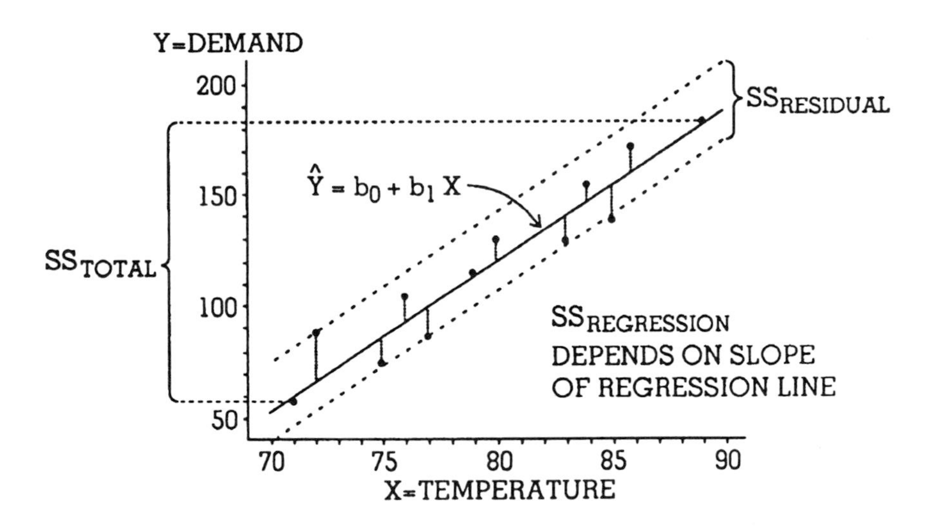

FIGURE 8.5. Analysis of variance in regression.

TABLE 8.1 Anova Summary Table for Predicting Electricity Demand from Temperature

Source	df	SS	MS	F
Regression	1	15,817.91	15,817.91	145.79
Residual	10	1,085.00	108.50	
Total	11	16,902.92		

intuitive explanation exists. As shown in (8.7.1), the regression term appears in the numerator of our test statistic. In general, *the numerator degrees of freedom for an F ratio is equal to the number of questions asked in the null hypothesis.* Our null hypothesis is H_0: $\beta_1 = 0$, where the equal sign is effectively a question. We ask, "Is the slope equal to 0?" Since we are only asking a single question, the regression term has only one degree of freedom.

This logic for numerator degrees of freedom applies to any F test. For example, in one-way analysis of variance we tested the null hypothesis H_0: $\mu_1 = \mu_2 = \mu_3 = \ldots = \mu_a$, which contains $a - 1$ equal signs (one between each pair of means). Each equal sign asks the question, "Are the two sides equal?" and thus the null hypothesis asks $a - 1$ questions. Recall that the F test in the one-way model had $a - 1$ degrees of freedom in the numerator.

In this chapter we have introduced a general framework for both numerator and denominator degrees of freedom in an F ratio. The numerator degrees of freedom is the number of questions asked in the null hypothesis, and the denominator degrees of freedom is the number of observations minus the number of parameters estimated.

In regression, as in earlier uses of analysis of variance, a summary table is used to display the results. For Example 8.1, the anova summary table is shown in Table 8.1. The F ratio of 145.79 has 1 degree of freedom in the numerator and 10 in the denominator. From Table 4a in the Appendix, we find the upper 5% critical value of 4.97. The null hypothesis of zero slope is soundly rejected. We are very sure that some prediction of electricity demand from temperature is possible.

As stated earlier, the analysis of variance approach is equivalent to the t test for zero slope given in Section 8.6. Our obtained F ratio (145.79) is the square of the observed t statistic (12.07), and the critical F value (4.97) is the square of the critical t value (2.228). Thus, the two tests must reach the same conclusion concerning population slope.

8.8 QUALITY OF FIT

The test for zero slope discussed in the previous section pertains to the issue of statistical significance. The smaller the p-value, the more sure we are that some prediction of Y from X is possible. Practical importance, it will be recalled, deals with how wrong the null hypothesis is. The better the quality of predictions, the more wrong the null hypothesis of zero slope is. A good overall measure of quality of predictions is given by

$$R^2 = \frac{SS_{REGRESSION}}{SS_{TOTAL}}. \qquad (8.8.1)$$

The same symbol, R^2, was used in Section 5.3 in connection with one-way analysis of variance. In that context, R^2 was defined as the ratio of $SS_{BETWEEN}$ to SS_{TOTAL}, and interpreted as the proportion of total variability related to differences between group means. In the same sense, R^2 in regression is the proportion of total variability related to different X values. In fact, the same phrase, **coefficient of determination,** is used to reference R^2 in both contexts.

For the data of Example 8.1, we find

$$R^2 = \frac{SS_{REGRESSION}}{SS_{TOTAL}} = \frac{15{,}817.91}{16{,}902.92} = 0.94.$$

Approximately 94 percent of the variability in electricity demand is related to differences in temperature across communities.

The coefficient of determination measures the fit of our regression line to the sample data. We should recognize, however, that we are choosing the line that is optimal for this sample. Values for the regression coefficients (b_0 and b_1) are selected to minimize residuals in the original data. They are probably not identical to the true population slope and intercept (β_0 and β_1). If the fitted regression line is used to predict electricity demand for *other* cities in the population, predictions probably will not be as accurate as in the original sample. A good estimate of the proportion of variability accounted for in the population is given by

$$\text{adjusted } R^2 = \frac{df_{TOTAL}(R^2) - df_{REGRESSION}}{df_{RESIDUAL}}. \qquad (8.8.2)$$

For the data of Example 8.1, we obtain

$$\text{adjusted } R^2 = \frac{11(0.94) - 1}{10} = 0.93.$$

If our fitted regression line is used to predict electricity demand for other cities in the population, approximately 93 percent of the variability in electricity demand would be predictable from temperature. Clearly, predictions in the population would be almost as good as in the sample. A very large value of R^2, as in the present case, also produces a large value of adjusted R^2.

Because we optimize for the sample, adjusted R^2 is always smaller than R^2. Predictions in the population are never as good as in the sample. But the difference between adjusted R^2 and R^2 is negligible if we use a large sample. While this makes good common sense, it can also be seen readily in the formula for adjusted R^2. Substituting $df_{TOTAL} = n - 1$, $df_{REGRESSION} = 1$ and $df_{RESIDUAL} = n - 2$ into (8.8.2), we find

$$\text{adjusted } R^2 = \frac{(n-1)R^2 - 1}{n-2} = \left(\frac{n-1}{n-2}\right)R^2 - \left(\frac{1}{n-2}\right).$$

As n increases, the ratio in the first pair of parentheses approaches one, and the ratio in the second pair approaches 0, so that adjusted R^2 approaches R^2. A prediction equation derived from a very large sample is likely to produce predictions in the population that are nearly as good as those in the sample.

R^2 and adjusted R^2 are useful as measures of overall adequacy of the fitted regression line. But we should also look at predictions for individual observations. For example, the third observation has the values $X = 71$ and $Y = 58$. Using our regression line (from Section 8.3), the predicted electricity demand for this community is

$$\hat{Y} = -408.07 + 6.61X = -408.07 + 6.61(71) = 61.24.$$

The residual, or error of prediction, is

$$Y - \hat{Y} = 58 - 61.24 = -3.24.$$

Relative to the actual demand of 58, the error of prediction is small. This procedure would be tedious to do by hand for each observation, but computers perform such calculations routinely. All standard regression programs print residual values for each observation.

Figure 8.6 shows a plot of the residuals for all observations in Example 8.1. As discussed above in Section 8.3, the sum of residuals and, therefore, the average residual, is 0. A horizontal line has been drawn at the zero point on the vertical axis to reinforce this fact. The data in this example are particularly well-behaved. The residual plot reveals no obvious pattern, and no observations have extremely large residuals.

Residual plots are frequently used to detect unusual or outlying observations and violations of assumptions. Diagnosis of outliers will be discussed in detail in

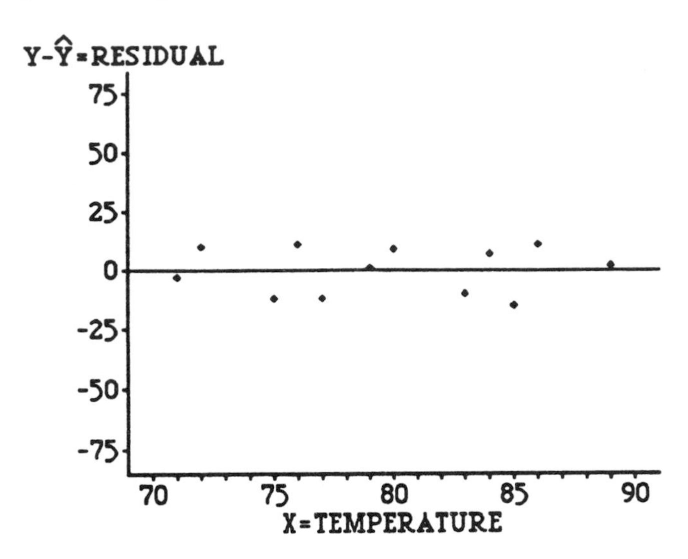

FIGURE 8.6. Residuals for predicting electricity demand.

Section 8.12; here, we focus on the use of residual plots to reveal violations of assumptions. Our coverage of this topic is relatively brief. Users whose data appear to seriously violate the assumptions should read about alternative analyses in more specialized regression books. The basic assumptions for regression are a straight-line relationship, independence of observations, equality of variances, and normality. Figure 8.7 illustrates patterns that suggest violations of assumptions.

The curvature of residuals in Figures 8.7(a) and 8.7(b) indicates that the relationship between X and Y is nonlinear for these two sets of data. While it might be possible to detect curvature in a plot of the raw data, it is generally much easier in a residual plot. In subtracting $\hat{Y}$ from Y, a residual plot *removes the linear tendency*, making other trends more obvious. Curvature can be modeled by adding X^2 as a second predictor. Multiple predictor models are discussed in the next chapter.

Figure 8.7(c) suggests time dependence, also known as **autocorrelation,** since adjacent observations tend to have residuals of the same sign. The **Durbin-Watson test** is specifically designed to detect this sort of pattern. If the test is significant, the model can frequently be improved by adding a lagged Y value as a second predictor.

The pattern of residuals in Figure 8.7(d) strongly suggests heterogeneous variances. The spread among Y values is far greater for large values of X than for small ones. This condition can be handled by using **weighted regression** or by transforming the data. In weighted regression, the sum of weighted squared residuals is minimized. To account for inequality of variance, each observation should be weighted by a term inversely proportional to its variance. Although the formulas change slightly, all the

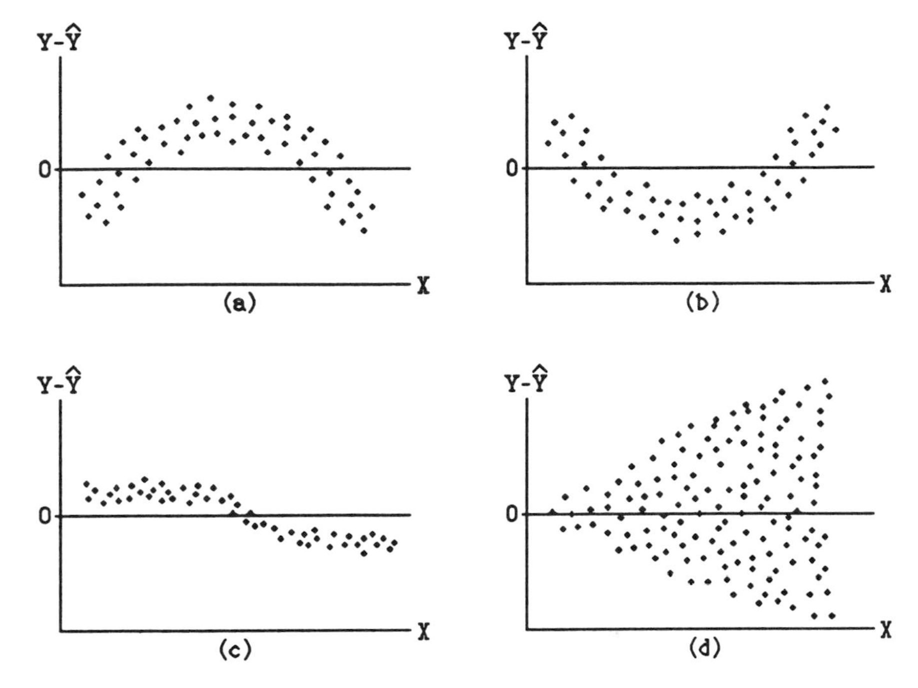

FIGURE 8.7. Patterns in residual plots.

usual regression logic holds for weighted regression. Most standard regression programs have facilities for including weights. However, the user must supply the weights from knowledge of the experimental situation.

When variance generally increases with the mean, as in Figure 8.7(d), a **natural log transformation** of the dependent variable can be used to stabilize variances. Instead of predicting Y from X, we use the dependent variable Y^*, given by $Y^* = \ln(Y)$. Additional discussion of transformations is given in Section 8.14.

To detect departures from normality, a different plot, known as a normal probability plot, can be constructed. Unfortunately, the appearance of non-normality in such plots is often caused by violations of other assumptions, such as curvature or unequal variances. As a result, these plots are not frequently employed.

8.9 CONFIDENCE AND PREDICTION INTERVALS

The point estimate of $\mu_{Y|X}$ is $\hat{Y}$, obtained from our regression line. But we can also construct a confidence interval for $\mu_{Y|X}$. Following the usual procedure, we wish to compute

$$\text{(point estimate)} \pm t[\text{SE(point estimate)}].$$

The standard error of the predicted value, $SE(\hat{Y})$, is called the **standard error of prediction**. Its formula is somewhat complex, so we leave its calculation to computer programs. But we note that since $\hat{Y}$ depends on X, so does its standard error.

To demonstrate this confidence interval, consider the tenth observation in Example 8.1, with temperature (X) of 80. From our estimated regression equation, we find the predicted value

$$\hat{Y} = -408.07 + 6.61X = -408.07 + 6.61(80) = 120.73.$$

The standard error of prediction for this temperature is SE(Y) = 3.01. $M_{RESIDUAL}$ is again used as the estimate of σ^2, so we still use n - 2 = 10 degrees of freedom. Putting 0.025 in each tail of the t distribution, we find the t value of 2.228. Employing the usual format for a confidence interval, we obtain

$$\hat{Y} \pm t[SE(\hat{Y})] = 120.73 \pm 2.228(3.01) = 120.73 \pm 6.71,$$

and the resulting interval $114.02 < \mu_{Y|X} < 127.44$. When temperature is 80 degrees, we are 95% certain that these bounds include the true mean electricity demand. This interval is centered at $\hat{Y}$ = 120.73, the point on the regression line corresponding to X = 80. Figure 8.8 displays the interval graphically.

The standard error of prediction and, therefore, the width of the confidence interval for $\mu_{Y|X}$, depends on the X value at which prediction is made. More

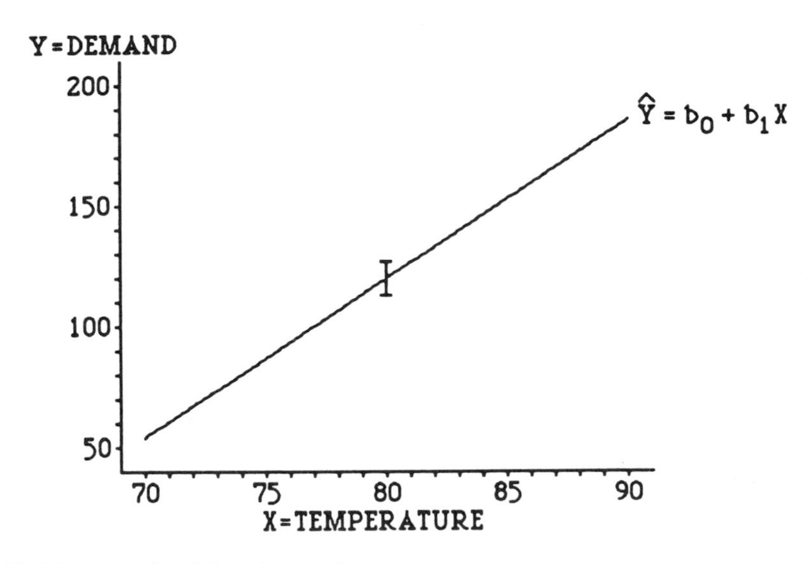

FIGURE 8.8. 95% confidence interval for $\mu_{Y|X}$ when X = 80.

precisely, the standard error of prediction depends on the quantity $(X-\overline{X})$. The further X is from the average of the X values, the wider the interval is. Recall that the estimated regression line passes through the point $(\overline{X},\overline{Y})$. A slight change in slope would not greatly alter predictions for X values near $\overline{X}$, but would dramatically affect predictions far from $\overline{X}$. Since population slope is only estimated, the confidence interval for $\mu_{Y|X}$ must be wider when X is further from $\overline{X}$. On a logical level, when we predict $\mu_{Y|X}$ near the center of the data, we get "help" from both ends. But for extreme X values, we get help from only one side.

Most standard computer programs print the confidence interval for $\mu_{Y|X}$ for each of the original X values. Suppose these intervals are plotted on a common graph, and separate curves are drawn through all the upper limits and through all the lower limits. These curves, called **confidence bands**, are shown in Figure 8.9 for the data of Example 8.1. As discussed above, the confidence bands are narrowest at $X = \overline{X}$, and widen toward extreme X values.

We should exercise great caution when using our regression equation to predict $\mu_{Y|X}$ for values of X outside the range of the original data. There are two reasons why such extrapolation may be hazardous. First, the straight-line relationship between X and Y may not hold for more extreme values of X. Second, even if the linear model is still appropriate, the great width of the confidence interval may render predictions worthless.

A confidence interval is appropriate for predicting $\mu_{Y|X}$ when we are interested in the *average* Y for a certain value of X. In Example 8.1, the power company may want to predict average electricity demand for all communities in a certain climatic region. But

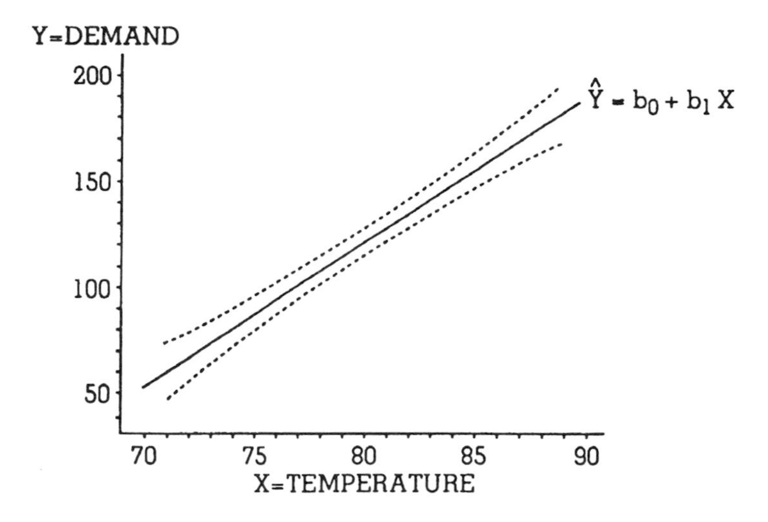

FIGURE 8.9. 95% confidence bands.

the confidence interval is not adequate for predicting a *particular* observation of Y. Demand for any single community is subject to many local factors, and is more difficult to predict than average demand. In essence, we must also consider the random variability associated with the particular observation we are trying to predict.

To put probability bounds on a single prediction, a **prediction interval** is constructed. Again, the formula is complex, so we leave its calculation to computer programs. But prediction intervals share many of the properties of confidence intervals. They are always centered at the point estimate, namely $\hat{Y}$. They are narrowest for X values near $\overline{X}$, and widen toward the extremes. However, for any specific value of X, the prediction interval is wider than the confidence interval, since the prediction interval must allow for the random variability associated with the individual observation.

By constructing prediction intervals for many different values of X and then drawing separate curves through all the upper bounds and through all the lower bounds, we obtain **prediction bands** for a single observation. Roughly 95 percent of all data points should fall within the 95% prediction bands. For the data of Example 8.1, the prediction bands are shown in Figure 8.10, along with the confidence bands for comparison.

8.10 SIMPLE LINEAR REGRESSION IN COMPUTER PROGRAMS

Simple linear regression can be performed with each of the programs, SAS, SPSS, and MINITAB. Each program automatically outputs the regression coefficients,

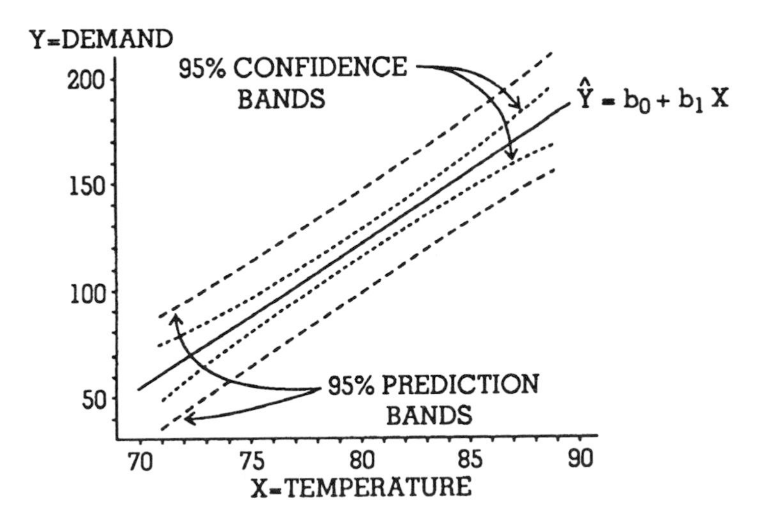

FIGURE 8.10. Confidence and prediction bands.

their standard errors, t statistics for testing that the population intercept and slope are 0, analysis of variance summary table, R^2, and adjusted R^2. Options are available to print predicted values, residuals, and 95% confidence and prediction intervals. The data of Example 8.1 are used to demonstrate these features.

SAS Input

```
OPTIONS LS = 80;
TITLE 'EXAMPLE 8.1 - ELECTRICITY DEMAND';
DATA;
INPUT TEMP DEMAND;
LINES;
77   89
86  171
71   58
75   76
85  139
83  130
72   78
79  115
89  182
80  130
84  154
76  105
PROC REG;
MODEL DEMAND = TEMP / P CLM CLI;
OUTPUT OUT=RESULT RESIDUAL=RESIDUAL;
PROC PLOT;
PLOT RESIDUAL*TEMP = '*';
```

Discussion of SAS Input

For each observation, values of TEMP and DEMAND are entered. Simple linear regression is performed by the SAS procedure REG, which also performs a variety of other regression analyses. In the accompanying MODEL statement, the dependent variable (DEMAND) is on the left side of the equation, and the independent variable (TEMP) is on the right. Options requested are P, to print the predicted and residual values for each observation, CLM, to output 95% confidence intervals, and CLI, to produce 95% prediction intervals.

The OUTPUT statement creates a new dataset, named RESULT, containing both the original variables and the new variable called RESIDUAL. This dataset is used by PROC PLOT to print the residual plot. In the PLOT command, the first variable named is used for the vertical axis, and the second for the horizontal axis. Each observation is indicated in the plot as an asterisk.

SAS Output

```
                    EXAMPLE 8.1 - ELECTRICITY DEMAND                    1
Model: MODEL1
Dependent Variable: DEMAND
                             Analysis of Variance

                        Sum of            Mean
Source       DF        Squares          Square       F Value      Prob>F
Model         1    15817.91322     15817.91322       145.787      0.0001
Error        10     1085.00345       108.50035
C Total      11    16902.91667

          Root MSE       10.41635       R-square       0.9358
          Dep Mean      118.91667       Adj R-sq       0.9294
          C.V.            8.75937

                             Parameter Estimates

                     Parameter        Standard         T for H0:
Variable   DF         Estimate           Error       Parameter=0   Prob > |T|
INTERCEP    1      -408.071774     43.74921725           -9.328        0.0001
TEMP        1         6.608006      0.54728225           12.074        0.0001

    Dep Var   Predict   Std Err  Lower95%   Upper95%  Lower95%   Upper95%
Obs  DEMAND     Value   Predict      Mean       Mean   Predict    Predict
 1  89.0000     100.7     3.363   93.2524      108.2   76.3561      125.1
 2    171.0     160.2     4.554     150.1      170.4     134.9      185.5
 3  58.0000   61.0966     5.655   48.4975    73.6957   34.6882    87.5051
 4  76.0000   87.5286     3.975   78.6720    96.3853   62.6870      112.4
 5    139.0     153.6     4.159     144.3      162.9     128.6      178.6
 6    130.0     140.4     3.494     132.6      148.2     115.9      164.9
 7  78.0000   67.7046     5.199   56.1201    79.2892   41.7649    93.6444
 8    115.0     114.0     3.035     107.2      120.7   89.7864      138.1
 9    182.0     180.0     5.888     166.9      193.2     153.4      206.7
10    130.0     120.6     3.010     113.9      127.3   96.4098      144.7
11    154.0     147.0     3.802     138.5      155.5     122.3      171.7
12    105.0   94.1366     3.641   86.0249      102.2   69.5507      118.7

    Obs   Residual
     1    -11.7447
     2     10.7833
     3     -3.0966
     4    -11.5286
     5    -14.6087
     6    -10.3927
     7     10.2954
     8      1.0393
```

```
  9      1.9593
 10      9.4313
 11      6.9993
 12     10.8634

Sum of Residuals                        0
Sum of Squared Residuals        1085.0035
Predicted Resid SS (Press)      1511.9755
```

```
                 EXAMPLE 8.1 - ELECTRICITY DEMAND                 2
             Plot of RESIDUAL*TEMP. Symbol used is '*'.
```

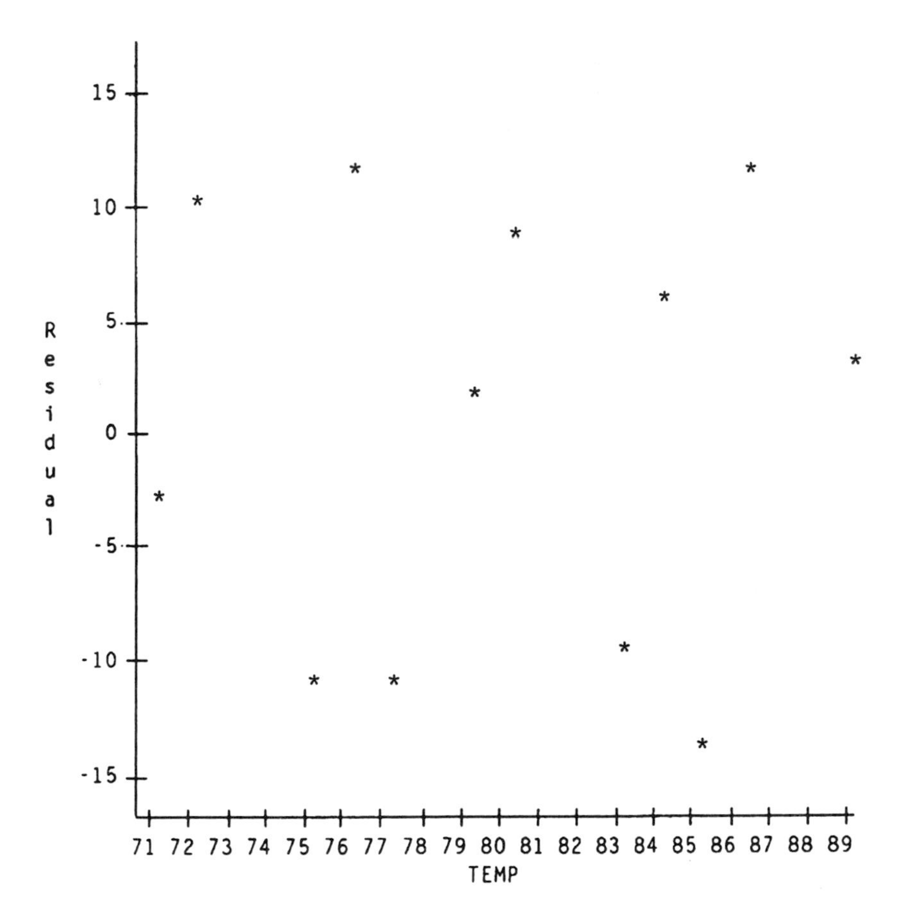

Discussion of SAS Output

At the top of page 1 of the output is the analysis of variance summary table. SAS uses the word Model instead of REGRESSION, and Error instead of

Residual. C Total stands for corrected total, which is the same as total sum of squares. The F test is highly significant (p = 0.0001), leading to rejection of the null hypothesis of zero slope. We are very sure that some prediction of DEMAND is possible from TEMP. Below the anova table are the values of R^2 (0.9358) and adjusted R^2 (0.9294). The numbers shown to the left of these values for Root Mse, Dep Mean, and C.V. have the same interpretations as in one-way analysis of variance (refer to Section 5.8 for further discussion).

Under the heading Parameter Estimates are the regression coefficients, their standard errors, t tests for the corresponding population parameters, and associated two-tailed p-values. The row labeled INTERCEP(T) is for b_0, and the row labeled TEMP refers to the *coefficient* of TEMP—that is, the slope b_1.

In the next block of information, SAS prints, for each observation, the original Y value (Dep Var), the predicted value $\hat{Y}$, its standard error of prediction, the lower and upper bounds of the 95% confidence interval for the mean, and the lower and upper bounds of the 95% prediction interval. The bounds of the confidence interval shown for the tenth observation differ slightly from those computed in Section 8.9 due to the rounding error we incurred by carrying only two decimal places of accuracy.

Below this display are the residual values for each observation. The residual shown for the third observation (-3.0966) differs from the value computed in Section 8.8 (-3.24), due to our rounding error. Note that the sum of residuals is 0, as discussed earlier. The least squares criterion guarantees that no other line can have a smaller sum of squared residuals (1085.0035). (Predicted Resid SS is the sum of squared Press residuals, which are discussed in Section 8.12.)

The plot of residuals on the second page of the output appears somewhat different from our plot in Figure 8.6, due to the elongated scale used by SAS for the vertical axis. But, again, the residuals appear to be randomly scattered about a horizontal line drawn at a residual of 0, indicating that the linear model provides good fit to these data.

SPSS Input

```
SET WIDTH = 80
TITLE 'EXAMPLE 8.1 - ELECTRICITY DEMAND'
DATA LIST FREE / TEMP DEMAND
REGRESSION VARIABLES = TEMP DEMAND
      / DEPENDENT = DEMAND
      / METHOD = ENTER TEMP
      / CASEWISE = ALL DEPENDENT PRED RESID
      / SAVE RESID(RESIDUAL) MCIN ICIN
```

```
BEGIN DATA
77   89
86  171
71   58
75   76
85  139
83  130
72   78
79  115
89  182
80  130
84  154
76  105
END DATA
LIST
PLOT PLOT = RESIDUAL WITH TEMP
FINISH
```

Discussion of SPSS Input

For each observation, values of TEMP and DEMAND are entered in free format. Simple linear regression is performed by the SPSS procedure REGRESSION, which also performs a variety of other regression analyses. The DEPENDENT statement identifies the dependent variable (DEMAND), and the METHOD command specifies that the variable TEMP is to be entered into the model. (Methods other than ENTER are demonstrated in connection with model selection in the next chapter.) The CASEWISE command requests printing of the dependent variable, predicted value (PRED) and residual (RESID) for each observation. Without the keyword ALL, only unusual values would be printed. Residual values are stored internally by the SAVE command for later use in plotting.

Also stored are bounds of the 95% confidence interval for each observation (keyword MCIN on the SAVE command) and bounds of the 95% prediction intervals (keyword ICIN). Output of these values could have been requested on the CASEWISE command, but the SPSS manual warns that printing is restricted by the width of the paper, and values that do not fit on a single line will be truncated. To avoid losing some of these bounds, they are saved in a SAVE command for later printing. After data entry, the LIST procedure requests a printout of all variables for each observation, which now includes the new variables added to the dataset by the SAVE command. The PLOT command is used to obtain the residual plot. The first variable named is used for the vertical axis, and the second for the horizontal axis.

SPSS Output

```
              EXAMPLE 8.1 - ELECTRICITY DEMAND        Page 1

              * * * MULTIPLE REGRESSION * * *
Listwise Deletion of Missing Data
Equation Number 1  Dependent Variable..  DEMAND

Block Number 1.  Method:  Enter    TEMP

Variable(s) Entered on Step Number
   1..    TEMP

Multiple R            .96737
R Square              .93581
Adjusted R Square     .92939
Standard Error      10.41635

Analysis of Variance
                  DF      Sum of Squares        Mean Square
Regression         1         15817.91322        15817.91322
Residual          10          1085.00345          108.50035

F =  145.78675   Signif F =  .0000
-----------------Variables in the Equation------------------
Variable              B          SE B       Beta        T  Sig T

TEMP           6.608006       .547282   .967373  12.074  .0000
(Constant)  -408.071774     43.749217            -9.328  .0000
End Block Number  1  All requested variables entered.
```

```
             EXAMPLE 8.1 - ELECTRICITY DEMAND         Page 2

             * * * MULTIPLE REGRESSION * * *
Equation Number 1     Dependent Variable..     DEMAND

Casewise Plot of Standardized Residual

*: Selected     M: Missing

       -3.0     0.0     3.0
Case #   O:       :       :O   DEMAND        *PRED        *RESID
 1       .     *          .     89.00     100.7447      -11.7447
 2       .           *    .    171.00     160.2167       10.7833
 3       .       *        .     58.00      61.0966       -3.0966
```

```
4          .    *          .     76.00        87.5286      -11.5286
5          .   *           .    139.00       153.6087      -14.6087
6          .    *          .    130.00       140.3927      -10.3927
7          .          *    .     78.00        67.7046       10.2954
8          .        *      .    115.00       113.9607        1.0393
9          .         *     .    182.00       180.0407        1.9593
10         .          *    .    130.00       120.5687        9.4313
11         .         *     .    154.00       147.0007        6.9993
12         .           *   .    105.00        94.1366       10.8634
Case #    O:        :     :O    DEMAND          *PRED        *RESID
        -3.0      0.0    3.0
```

```
              EXAMPLE 8.1 - ELECTRICITY DEMAND            Page 3
              * * * MULTIPLE REGRESSION * * *
Equation Number 1  Dependent Variable..  DEMAND

Residuals Statistics:

                  Min.          Max          Mean     Std Dev     N
*PRED          61.0966     180.0407      118.9167     37.9209    12
*ZPRED         -1.5248       1.6119         .0000      1.0000    12
*SEPRED         3.0101       5.8881        4.1478       .9795    12
*ADJPRED       62.3904     179.1207      118.7738     37.9274    12
*RESID        -14.6087      10.8634         .0000      9.9316    12
*ZRESID        -1.4025       1.0429         .0000       .9535    12
*SRESID        -1.5297       1.1511         .0061      1.0350    12
*DRESID       -17.3793      13.7114         .1429     11.7231    12
*SDRESID       -1.6581       1.1724        -.0075      1.0628    12
*MAHAL           .0019       2.5982         .9167       .8850    12
*COOK D          .0005        .2219         .0897       .0752    12
*LEVER           .0002        .2362         .0833       .0805    12
Total Cases =   12

               * * * * * * * * * * * * * * *

From Equation  1:  5 new variables have been created.
Name         Contents
RESIDUAL     Residual
LMCI_1       95% L CI for DEMAND mean
UMCI_1       95% U CI for DEMAND mean
LICI_1       95% L CI for DEMAND individual
UICI_1       95% U CI for DEMAND individual
```

EXAMPLE 8.1 - ELECTRICITY DEMAND Page 4

TEMP	DEMAND	RESIDUAL	LMCI_1	UMCI_1	LICI_1	UICI_1
77.00	89.00	-11.74465	93.25041	108.23889	76.34974	125.13956
86.00	171.00	10.78330	150.06640	170.36701	134.87944	185.55396
71.00	58.00	-3.09662	48.49422	73.69902	34.68131	87.51192
75.00	76.00	-11.52864	78.66973	96.38755	62.68054	112.37675
85.00	139.00	-14.60870	144.33943	162.87796	128.61135	178.60604
83.00	130.00	-10.39268	132.60635	148.17902	115.90646	164.87891
72.00	78.00	10.29538	56.11704	79.29221	41.75814	93.65111
79.00	115.00	1.03934	107.19685	120.72448	89.78015	138.14118
89.00	182.00	1.95928	166.91782	193.16362	153.37316	206.70827
80.00	130.00	9.43133	113.86007	127.27727	96.40354	144.73379
84.00	154.00	6.99931	138.52807	155.47331	122.28767	171.71371
76.00	105.00	10.86335	86.02282	102.25048	69.54433	118.72897

Number of cases read: 12 Number of cases listed: 12

Discussion of SPSS Output

On page 1 of the output, SPSS prints the values of Multiple R (square root of R^2), R^2, adjusted R^2, and the Standard Error (square root of the residual mean square). Beneath these values is the analysis of variance summary table. The F test is highly significant, leading to rejection of the null hypothesis of zero slope. The printed p-value of 0.0000 implies that the tail area beyond F is closer to 0.0000 than to 0.0001. We are very sure that some prediction of DEMAND is possible from TEMP.

At the bottom of this page are the regression coefficients (in the column labeled B), their standard errors (labeled SE B), t tests for the corresponding population parameters (T), and associated two-tailed p-values (Sig T). The row labeled TEMP refers to the *coefficient* of TEMP (that is, the slope b_1), and the row labeled (Constant) is for b_0. If X and Y were both standardized before performing least squares, the slope would be .967373 (BETA), and no intercept term would be needed, since the means of standardized variables are 0.

The second page shows, for each observation, a plot of the standardized residual (to be discussed in Section 8.12), the value of Y (DEMAND), the predicted value (*PRED), and the residual (*RESID). The asterisks indicate that the last two of these are temporary variables, not available for use by later procedures unless saved with a SAVE command. On page 3, the smallest, largest, average, and standard deviation of predicted values and residual values are printed. Also included in this display are similar statistics for a variety of other diagnostic variables, some of which will be discussed in Section 8.12. At the bottom of this page we see the variables stored by the SAVE command.

The fourth page results from the LIST command, and shows the values of all variables stored for each observation. Included in this display are the lower and

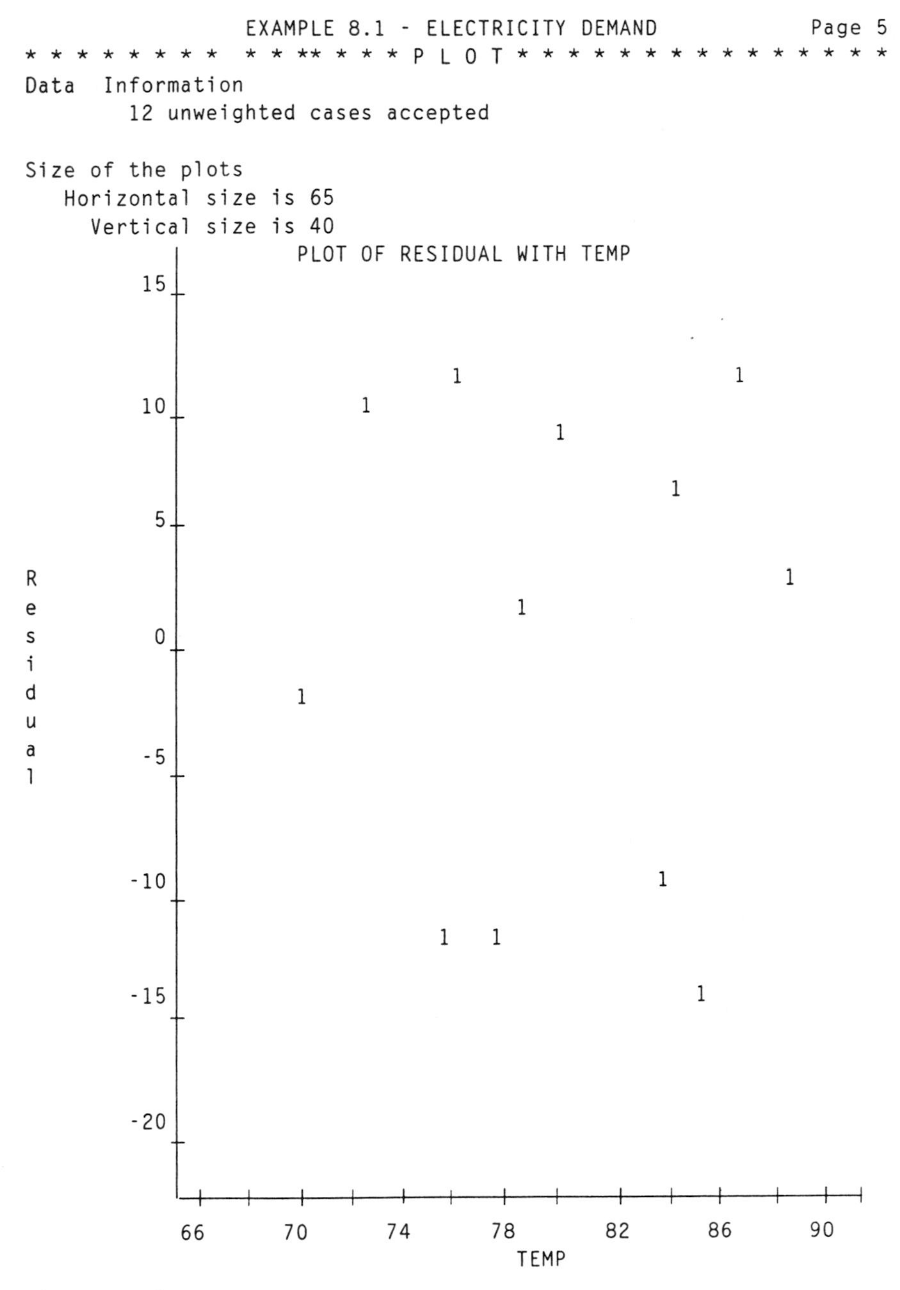

EXAMPLE 8.1 - ELECTRICITY DEMAND Page 5

* * * * * * * * * * * ** * * * P L O T * * * * * * * * * * * * * * *

Data Information
12 unweighted cases accepted

Size of the plots
Horizontal size is 65
Vertical size is 40

12 cases plotted.

upper bounds of the 95% confidence interval for the mean (LMCI_1 and UMCI_1, respectively), and bounds of the 95% prediction interval (LICI_1 and UICI_1). The bounds of the confidence interval shown for the tenth observation differ slightly from those computed in Section 8.9 due to rounding error we incurred by carrying only two decimal places of accuracy.

The plot of residuals on the fifth page of the output appears somewhat different from our plot in Figure 8.6, due to the elongated scale used by SPSS for the vertical axis. But, again, the residuals appear to be randomly scattered about a horizontal line drawn at a residual of 0, indicating that the linear model provides good fit to these data.

MINITAB Transcript

```
MTB > NAME C1 = 'TEMP'
MTB > NAME C2 = 'DEMAND'
MTB > READ C1-C2
DATA > 77  89
DATA > 86 171
DATA > 71  58
DATA > 75  76
DATA > 85 139
DATA > 83 130
DATA > 72  78
DATA > 79 115
DATA > 89 182
DATA > 80 130
DATA > 84 154
DATA > 76 105
DATA > END
     12 ROWS READ

MTB > NAME C3 = 'RESIDUAL'
MTB > BRIEF 3
MTB > REGRESS 'DEMAND' 1 'TEMP';
SUBC > RESIDUALS 'RESIDUAL';
SUBC > PREDICT 'TEMP'.

The regression equation is
DEMAND = -408 + 6.61 TEMP
```

Predictor	Coef	Stdev	t-ratio	p
Constant	-408.07	43.75	-9.33	0.000
TEMP	6.6080	0.5473	12.07	0.000

s = 10.42 R-sq = 93.6% R-sq(adj) = 92.9%

Analysis of Variance

SOURCE	DF	SS	MS	F	p
Regression	1	15818	15818	145.79	0.000
Error	10	1085	109		
Total	11	16903			

Obs.	TEMP	DEMAND	Fit	Stdev.Fit	Residual	St.Resid
1	77.0	89.00	100.74	3.36	-11.74	-1.19
2	86.0	171.00	160.22	4.55	10.78	1.15
3	71.0	58.00	61.10	5.65	-3.10	-0.35
4	75.0	76.00	87.53	3.97	-11.53	-1.20
5	85.0	139.00	153.61	4.16	-14.61	-1.53
6	83.0	130.00	140.39	3.49	-10.39	-1.06
7	72.0	78.00	67.70	5.20	10.30	1.14
8	79.0	115.00	113.96	3.03	1.04	0.10
9	89.0	182.00	180.04	5.89	1.96	0.23
10	80.0	130.00	120.57	3.01	9.43	0.95
11	84.0	154.00	147.00	3.80	7.00	0.72
12	76.0	105.00	94.14	3.64	10.86	1.11

Fit	Stdev.Fit	95% C.I.	95% P.I.
100.74	3.36	(93.25, 108.24)	(76.35, 125.14)
160.22	4.55	(150.07, 170.37)	(134.88, 185.55)
61.10	5.65	(48.49, 73.70)	(34.68, 87.51)
87.53	3.97	(78.67, 96.39)	(62.68, 112.38)
153.61	4.16	(144.34, 162.88)	(128.61, 178.61)
140.39	3.49	(132.61, 148.18)	(115.91, 164.88)
67.70	5.20	(56.12, 79.29)	(41.76, 93.65)
113.96	3.03	(107.20, 120.72)	(89.78, 138.14)
180.04	5.89	(166.92, 193.16)	(153.37, 206.71)
120.57	3.01	(113.86, 127.28)	(96.40, 144.73)
147.00	3.80	(138.53, 155.47)	(122.29, 171.71)
94.14	3.64	(86.02, 102.25)	(69.54, 118.73)

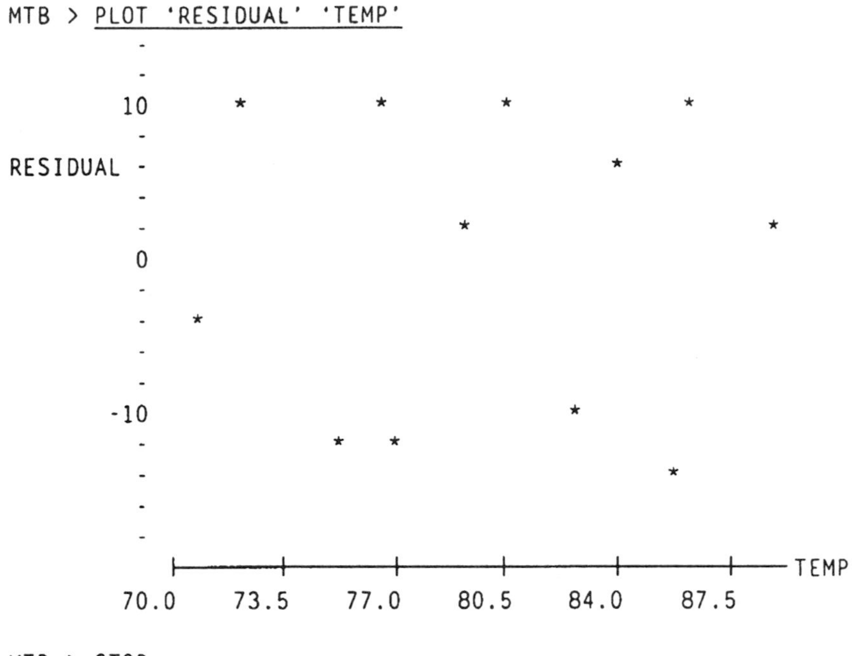

MTB > STOP

Discussion of MINITAB Transcript

The values of TEMP are read into column C1 on the worksheet, and the values of DEMAND into column C2. Column C3 is named RESIDUAL and is used during regression for storing residual values. The MINITAB command BRIEF 3 requests that results be printed for all observations, rather than only unusual values. The REGRESS command lists the dependent variable, the number of predictors, and the names of predictors. In our case, DEMAND is to be predicted from the single predictor TEMP. The first subcommand stores residual values, and the second requests printing of predicted values, residuals, 95% confidence intervals, and 95% prediction intervals.

MINITAB first shows the estimated regression equation, then prints a table containing the regression coefficients, their standard errors (labeled Stdev), t tests for the corresponding population parameters, and associated two-tailed p-values. The row labeled Constant is for b_0, and the row labeled TEMP refers to the *coefficient* of TEMP—that is, the slope b_1. Below this are the values of s (square root of the residual mean square), R^2, and adjusted R^2.

In the analysis of variance summary table, MINITAB uses the word Error instead of residual. The F test is highly significant, leading to rejection of the null hypothesis of zero slope. The printed p-value of 0.000 implies that the tail area beyond F is closer to 0.000 than to 0.001. We are very sure that some prediction of DEMAND is possible from TEMP.

The subsequent table lists for each observation the values of X (TEMP), Y (DEMAND), the predicted value $\hat{Y}$ (Fit), its standard error of prediction (Stdev. Fit), the residual and standardized residual (to be discussed in Section 8.12). The predicted value and standard error of prediction are repeated below, along with the lower and upper bounds of the 95-percent confidence interval for the mean, and the lower and upper bounds of the 95-percent prediction interval. The bounds of the confidence interval shown for the tenth observation differ slightly from those computed in Section 8.9, due to the rounding error we incurred by carrying only two decimal places of accuracy.

In the PLOT command, the first variable named is used for the vertical axis and the second for the horizontal axis. The plot of residuals appears somewhat different from our plot in Figure 8.6, due to the elongated scale used by MINITAB for the vertical axis. But, again, the residuals appear to be randomly scattered about a horizontal line drawn at a residual of 0, indicating that the linear model provides good fit to these data.

8.11 REGRESSION AND CORRELATION

As stated in Section 8.2, theoretically, correlation and regression apply to different situations. Correlation is employed when both X and Y are random variables, while regression assumes that X values are known exactly. However, for most practical purposes, this distinction is of little importance.

We learned in Section 8.4 that slope (b_1) and correlation (r) have the same sign. But the relationship between correlation and regression is much deeper than mere sign agreement. In the correlation setting, we test for zero population correlation using a t test based on r (see Section 7.6). In regression, we test for zero population slope using a t test based on b_1 (Section 8.6). In fact, these two t tests are algebraically equivalent. If the same set of data is used for both tests, identical t values are obtained and comparable conclusions are reached. This result should not be surprising. The hypotheses of zero correlation and zero slope both logically imply no linear relationship between X and Y.

As further evidence of the intimate relationship between correlation and regression, we note that the squared sample correlation between two variables is identical to the coefficient of determination—that is, $r^2 = R^2$. (There is some logic in statistical notation.)

On the other hand, if X values can be chosen by the researcher, regression, unlike correlation, offers the opportunity to draw a causal conclusion. Example 8.2 helps demonstrate this point. In this example, it is likely that values of X (overtime) and Y (productivity) were merely recorded for each of the ten employees. In this case, both X and Y are random variables, and correlation is appropriate.

Analysis of these data reveals a negative relationship between overtime and productivity. As expected, productivity per hour appears to decrease as the amount of overtime increases. While a test for correlation supports the negative relationship between overtime and productivity, it does not allow us to draw a *causal* conclusion. Specifically, we cannot assert that increased overtime *leads* to decreased productivity. In fact, the direction of causation may be exactly the reverse. It is possible that slower workers had to put in more overtime to finish assigned jobs. Reduced productivity may have caused greater overtime.

Now suppose that values of X (overtime) are within the researcher's control. Each employee may be told to work a certain number of overtime hours. Suppose further that assignments of the ten values of X are made *randomly* to the ten employees. Under these conditions, we would expect worker speed to roughly balance out across the ten values of X. Some slow workers would be assigned a small number of overtime hours, and some would be assigned a large number. Now if we find a negative relationship between overtime and productivity, we can conclude that increasing overtime *causes* reduced productivity.

Random assignment of X values in regression is similar to random assignment to groups when comparing two means. As discussed in Section 4.6, random assignment to groups allows us to determine that our treatments caused the

---

Example 8.2

A large corporation was interested in the merit of overtime work and suspected that productivity would suffer as the amount of overtime increased. Ten employees each worked different numbers of overtime hours per week (X), and observations were made of their productivity per hour (Y) during the overtime period.

X=overtime	Y=productivity
16.0	54
7.0	61
14.5	12
4.0	52
19.0	28
12.0	39
2.5	63
28.0	19
10.0	73
13.5	47

difference in means. In any experiment, random assignment is the key to drawing a causal conclusion. In correlation, unlike regression, X is a random variable whose values are simply observed, and thus cannot be randomly assigned. Not every regression application lends itself to control of X values. But when this option is available, it should be exercised intelligently, to produce the most meaningful conclusions.

8.12 DIAGNOSTICS FOR INDIVIDUAL OBSERVATIONS

In any application of regression, plots of both the raw data and residuals should be examined for evidence of unusual observations. For Example 8.1, these plots show no reason for concern about any of the data points. Example 8.2, on the other hand, has some worrisome observations. The data plot, shown in Figure 8.11, indicates that observation #3 departs substantially from the general linear trend of the other points. The same may be true, to a lesser extent, for observation #9 and, possibly, observation #1. In addition, observation #8, while in line with most of the other points, is rather far from the bulk of the data.

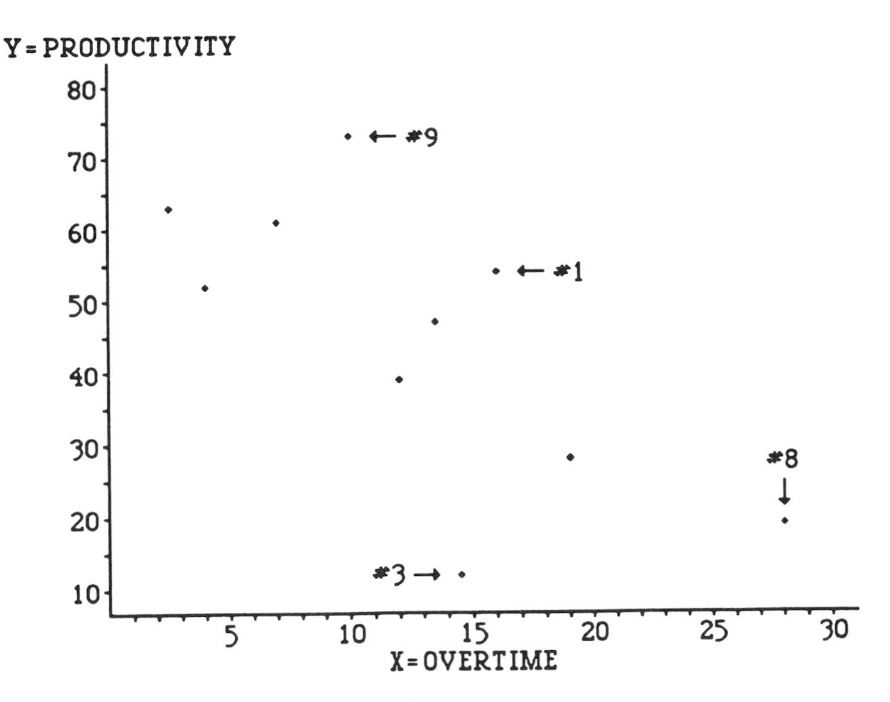

FIGURE 8.11. Productivity of overtime work.

The residual plot, displayed in Figure 8.12, raises similar concerns. Large (in magnitude) residuals for observations #1, #3, and #9 stand out clearly. Observation #8, on the other hand, has a residual near 0. Each of these four data points merits closer scrutiny.

Detection of unusual observations, or **outliers,** is facilitated by special diagnostic tools available in most regression programs. Unfortunately, a bewildering variety of these statistics exists, and the terminology is far from standardized. Two programs may compute the same quantity, but label it differently. Users of computer packages should be careful to understand the diagnostics produced by their programs.

An observation may be an outlier in either location, or pattern, or both. Location refers to the separate X and Y coordinates. If an observation has an extreme value on X and/or Y, its location qualifies it for special examination. Pattern refers to the tendency established by the bulk of the data. If an observation seems to violate this trend, it is inconsistent with the pattern. Any point that is remote from the rest of the data is cause for concern.

Figure 8.13 displays some of the possibilities for outliers in terms of location

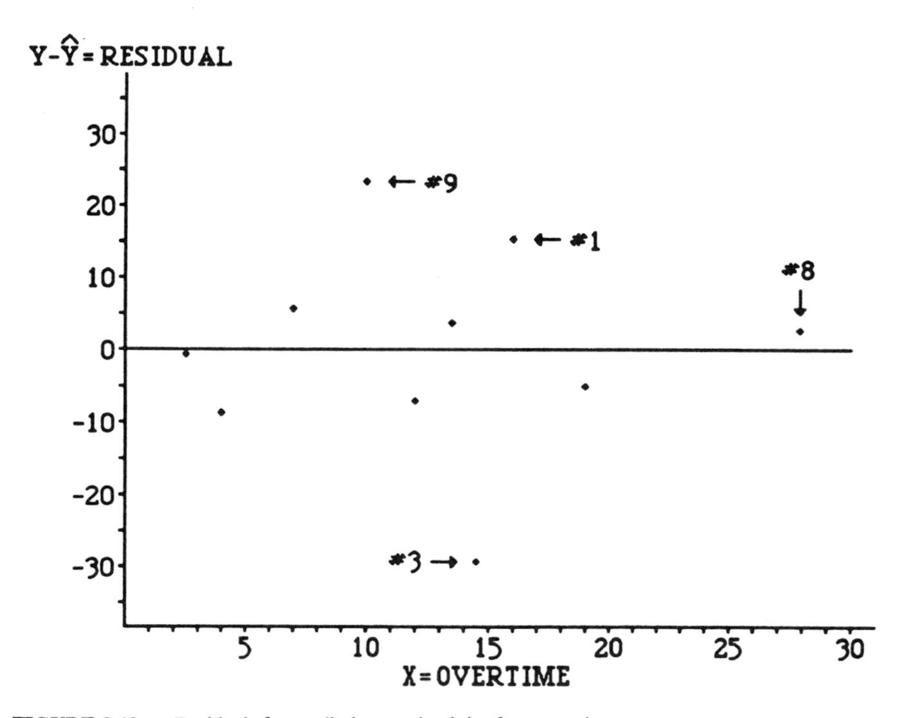

FIGURE 8.12. Residuals for predicting productivity form overtime.

and pattern. In each drawing, the circled observation has unusual characteristics. In Figure 8.13(a), the outlier is extreme in both its X and Y coordinates, but not in pattern. While this observation is remote, its removal probably would not greatly alter the regression line. The circled observation in Figure 8.13(b) is extreme on both its coordinates and is also an outlier in terms of pattern. Inclusion of this point may strongly influence the fitted line. If the point were even further right and higher, it might dramatically change the regression line, single-handedly causing the line to have positive slope. While the outlier in Figure 8.13(c) is extreme in its X coordinate, its Y value is near the average. Similarly, the circled observation in Figure 8.13(d) is extreme in Y, but not in X. Generally, when an observation is extreme in only one coordinate, it is also inconsistent in pattern. Finally, in Figure 8.13(e), we see an outlier in pattern that is not extreme in either variable.

We discuss in this section four diagnostic statistics that are useful in outlier detection. While our terminology is chosen for maximal consistency with computer packages, be wary—some programs even reverse the labels used here. The next section clarifies terminology for users of SAS, SPSS, and MINITAB. In our discussion we use n as the number of observations, and p as the number of parameters. For simple linear regression, p is equal to 2, the two parameters being slope and intercept. These same diagnostics can be used in multiple regression where p is greater than 2.

Leverage measures how unusual the X value is for an observation. It is similar to a standardized X score, and is not influenced at all by the Y value of the observation. An observation is considered unusual if its leverage exceeds 2p/n. For Example 8.2, this threshold is $2(2)/10 = 0.4$.

The **standardized residual** indicates if an observation is unusual in its Y value. Unlike leverage, however, the Y value is evaluated relative to its X coordinate. A large standardized residual suggests that the Y value of an observation is extreme, given the X value of the observation. An observation extreme in Y may not have a large standardized residual if its X value is also extreme. Computed by dividing the residual by its standard error, a standardized residual whose absolute value is greater than 2.0 suggests closer scrutiny of the data point.

Cook's D combines the information from leverage and the standardized residual. It can be interpreted as a measure of how different the regression coefficients (β_0 and β_1) would be if this observation were deleted. A data point should be considered unusual if its Cook's D value exceeds the 50th percentile of an F distribution with p degrees of freedom in the numerator and n−p in the denominator. For Example 8.2, the degrees of freedom are $p = 2$ in the numerator and $n-p = 10 - 2 = 8$ in the denominator, and the critical F value is 0.76.

An entirely different approach is incorporated in computation of the **studentized residual**. For each observation separately, least squares is employed on the rest of the data with the observation deleted. The obtained regression

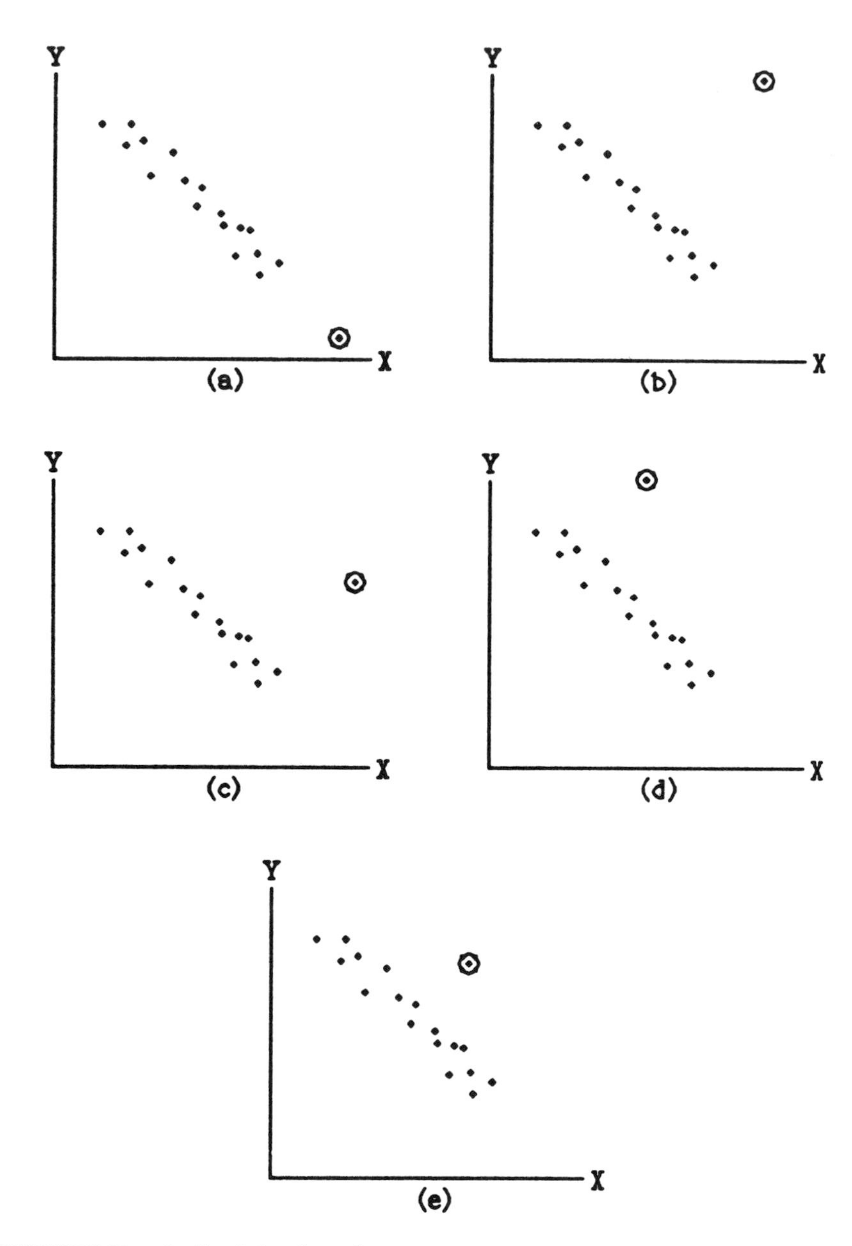

FIGURE 8.13. Outliers in location and pattern.

equation is used to predict the deleted observation, and the residual is determined. This residual is known as a **PRESS residual**. (PRESS stands for prediction sum of squares.) The residual is studentized by dividing it by its standard error, also computed with the observation deleted. When a substantial outlier is deleted, the regression coefficients change greatly. The resulting equation poorly predicts the deleted observation, resulting in a large studentized residual.

The studentized residual has another interesting property that makes it especially useful—it has a t distribution with n-p-1 degrees of freedom. Thus, it can be used to test the null hypothesis that an observation is not an outlier. A significant result implies that the tested observation *is* an outlier. While this may seem to be the ideal solution to outlier detection, we caution against indiscriminate use of this test. If the test is employed for all observations in a data set, there is a very large chance of making at least one type one error—that is, we are quite likely to find at least one "outlier" even when the data are well-behaved. If this observation is deleted and the regression is computed anew, the next most extreme observation may then have a significant t value. Two options exist to resolve this dilemma. The first is to employ this test only when an observation is suspected to be an outlier *a priori* on experimental grounds. The second is to use the test on all observations, but with Bonferroni critical limits (see Section 6.13 for discussion of the Bonferroni procedure).

For the data of Example 8.2, these four diagnostics are shown in Table 8.2. Observation #8 has leverage higher than 0.4, reflecting its unusual X value. Observation 3 has standardized residual greater than 2.0 in magnitude, suggesting an unusual Y value for its location (i.e., for its X value). The standardized residual is close to 2.0 for observation #9. Cook's D does not single out any data points as unusual, as no values approach the threshold of 0.76. Thus, the fitted regression equation would probably not change greatly if any single observation were deleted.

TABLE 8.2 Diagnostics for Individual Observations When Predicting Productivity from Overtime

Obs.	X	Y	Leverage	Standardized Residual	Cook's D	Studentized Residual
1	16.0	54	0.1221	1.0806	0.081	1.0938
2	7.0	61	0.1628	0.4136	0.017	0.3911
3	14.5	12	0.1067	-2.0445	0.250	-2.7677
4	4.0	52	0.2471	-0.6665	0.073	-0.6415
5	19.0	28	0.1793	-0.3673	0.015	-0.3465
6	12.0	39	0.1008	-0.4857	0.013	-0.4611
7	2.5	63	0.3026	-0.0446	0.000	-0.0417
8	28.0	19	0.5633	0.2568	0.043	0.2412
9	10.0	73	0.1138	1.6280	0.170	1.8622
10	13.5	47	0.1014	0.2617	0.004	0.2458

For testing the studentized residuals, degrees of freedom are n−p−1 = 10 − 2 − 1 = 7. Critical values for a two-sided test at the 5% level of significance are ±2.365, so only observation #3 is confirmed as an outlier by its studentized residual.

Diagnostic tools for detection of unusual observations and outliers may uncover data points worthy of further investigation. If this investigation reveals that an outlier is actually erroneous (for instance, because experimental procedure was not followed), the offending data point may be rightfully deleted. But statistical analysis alone should not result in the removal of an observation. Unusual data points may be harbingers of an inadequate model. If there is a pattern to the outliers, transformation of the data may be in order. Transformations are discussed in Section 8.14.

8.13 DIAGNOSTICS IN COMPUTER PROGRAMS

Each of the programs, SAS, SPSS, and MINITAB, has options for computing leverage, the standardized residual, Cook's D, and the studentized residual. The data of Example 8.2 are used to illustrate these features.

SAS Input

```
OPTIONS LS = 80;
TITLE 'EXAMPLE 8.2 - PRODUCTIVITY OF OVERTIME WORK';
DATA;
INPUT OVERTIME PRODUCT;
LINES;
16.0   54
 7.0   61
14.5   12
 4.0   52
19.0   28
12.0   39
 2.5   63
28.0   19
10.0   73
13.5   47
PROC REG;
MODEL PRODUCT = OVERTIME / R INFLUENCE;
```

Discussion of SAS Input

On the MODEL statement, the R option requests predicted values, standard errors of prediction, residuals, standardized residuals, and Cook's D for each observation.

Beware: the SAS manual refers to standardized residuals as "studentized residuals." The INFLUENCE option produces leverage, the studentized residual (which, to add to the confusion, the SAS manual also calls a "studentized residual"), and several other measures.

SAS Output

```
        EXAMPLE 8.2 - PRODUCTIVITY OF OVERTIME WORK                1
Model: MODEL1
Dependent Variable: PRODUCT
                       Analysis of Variance

                    Sum of          Mean
Source   DF        Squares         Square        F Value      Prob>F
Model     1     1738.31383     1738.31383          7.520      0.0254
Error     8     1849.28617      231.16077
C Total   9     3587.60000

        Root MSE       15.20397       R-square        0.4845
        Dep Mean       44.80000       Adj R-sq        0.4201
        C.V.           33.93744
                       Parameter Estimates
                 Parameter    Standard      T for H0:
Variable   DF     Estimate       Error    Parameter=0  Prob > |T|
INTERCEP    1    68.188290  9.79069363          6.965      0.0001
OVERTIME    1    -1.848877  0.67421889         -2.742      0.0254
```

```
        EXAMPLE 8.2 - PRODUCTIVITY OF OVERTIME WORK      Standardized
     Dep Var   Predict  Std Err                Std Err     Student
Obs  PRODUCT     Value  Predict   Residual    Residual    Residual
  1  54.0000   38.6063    5.312    15.3937      14.246       1.081
  2  61.0000   55.2462    6.134     5.7538      13.912       0.414
  3  12.0000   41.3796    4.967   -29.3796      14.370      -2.045
  4  52.0000   60.7928    7.558    -8.7928      13.192      -0.667
  5  28.0000   33.0596    6.438    -5.0596      13.774      -0.367
  6  39.0000   46.0018    4.828    -7.0018      14.417      -0.486
  7  63.0000   63.5661    8.363    -0.5661      12.697      -0.045
  8  19.0000   16.4197   11.412     2.5803      10.047       0.257
  9  73.0000   49.6995    5.129    23.3005      14.313       1.628
 10  47.0000   43.2285    4.842     3.7715      14.412       0.262
```

```
                           Cook's                Hat Diag      Cov
Obs   -2-1 0 1 2              D     Rstudent       H          Ratio     Dffits

  1  |      |**    |       0.081     1.0938     0.1221      1.0851     0.4078
  2  |      |      |       0.017     0.3911     0.1628      1.4941     0.1724
  3  |  ****|      |       0.250    -2.7677     0.1067      0.3334    -0.9567
  4  |     *|      |       0.073    -0.6415     0.2471      1.5475    -0.3676
  5  |      |      |       0.015    -0.3465     0.1793      1.5382    -0.1620
  6  |      |      |       0.013    -0.4611     0.1008      1.3682    -0.1544
  7  |      |      |       0.000    -0.0417     0.3026      1.8719    -0.0275
  8  |      |      |       0.043     0.2412     0.5633      2.9421     0.2740
  9  |      |***   |       0.170     1.8622     0.1138      0.6591     0.6673
 10  |      |      |       0.004     0.2458     0.1014      1.4288     0.0826

       INTERCEP    OVERTIME
Obs    Dfbetas     Dfbetas
  1     0.0302      0.1734
  2     0.1597     -0.1071
  3    -0.2455     -0.2402
  4    -0.3619      0.2836
  5     0.0344     -0.1077
  6    -0.0877      0.0140
  7    -0.0273      0.0225
  8    -0.1598      0.2485
  9     0.5097     -0.2325
 10     0.0318      0.0098

Sum of Residuals                         0
Sum of Squared Residuals         1849.2862
Predicted Resid SS (Press)       2415.9764
```

Discussion of SAS Output

From the anova table on the first page of the output, we see that the regression is significant at the 0.0254 level, with $R^2 = 0.4845$.

The second page of the output results from the R and INFLUENCE options. Shown in the first chart for each observation are the original Y value (Dep Var), the predicted value and its standard error, the residual and its standard error, and the standardized residual (labeled Student Residual). Below this, SAS plots the standardized residual and prints the values of Cook's D, the studentized residual (labeled Rstudent), and leverage (Hat Diag), along with several other diagnostic variables.

SPSS Input

```
SET WIDTH = 80
TITLE 'EXAMPLE 8.2 - PRODUCTIVITY OF OVERTIME WORK'
DATA LIST FREE / OVERTIME PRODUCT
REGRESSION VARIABLES = OVERTIME PRODUCT
     / DEPENDENT = PRODUCT
     / METHOD = ENTER OVERTIME
     / CASEWISE = ALL PLOT (SDRESID) DEPENDENT PRED RESID SDRESID
     / SAVE PRED (PREDICT) RESID (RESIDUAL) SRESID (STANDRES)
                 SDRESID (STUDRES) LEVER (LEVERAGE) COOK (COOKSD)
BEGIN DATA
16.0  54
 7.0  61
14.5  12
 4.0  52
19.0  28
12.0  39
 2.5  63
28.0  19
10.0  73
13.5  47
END DATA
LIST
FINISH
```

Discussion of SPSS Input

SPSS provides a confusing array of possible diagnostics, including Z scores for several variables. These are computed by dividing by the standard deviation of all observations on the variable, rather than by the standard error for each observation separately. None of these Z scores have been discussed in this chapter.

As noted in Section 8.10, specification of numerous diagnostics on the CASEWISE option is pointless, as printing is restricted by the width of the paper and values that do not fit on a single line will not be printed. To avoid losing some of our diagnostics, they are saved in a SAVE command for later printing. On the CASEWISE and SAVE commands: DEPENDENT is the Y variable, PRED refers to predicted scores, RESID to residual scores, SDRESID is the studentized residual (which the SPSS manual refers to as a "studentized deleted residual"), SRESID is the standardized residual, LEVER is leverage, and COOK requests Cook's D. In our example, the CASEWISE command specifies that SDRESID is the variable to be plotted for each observation. On the CASEWISE command, the keyword ALL requests printing of listed diagnostics for each observation; the default is to print them only for unusual observations. The labels in parentheses on the SAVE command are user-specified names employed in printouts.

After data entry, the LIST procedure requests a printout of all variables for each observation, which now included the new variables added to the dataset by the SAVE command.

SPSS Output

```
        EXAMPLE 8.2 - PRODUCTIVITY OF OVERTIME WORK    Page 1
               * * * MULTIPLE REGRESSION * * *
Listwise Deletion of Missing Data
Equation Number 1   Dependent Variable..   PRODUCT

Block Number 1.   Method:   Enter    OVERTIME

Variable(s) Entered on Step Number
  1..     OVERTIME
Multiple R              .69608
R Square                .48453
Adjusted R Square       .42010
Standard Error        15.20397

Analysis of Variance
               DF     Sum of Squares      Mean Square
Regression      1         1738.31383       1738.31383
Residual        8         1849.28617        231.16077

F =     7.51993     Signif F =     .0254

------------------Variables in the Equation----------------
Variable              B        SE B        Beta         T   Sig T

OVERTIME    -1.848877    .674219   -.696085    -2.742   .0254
(Constant)  68.188290   9.790694                6.965   .0001
End Block Number   1   All requested variables entered.
```

```
        EXAMPLE 8.2 - PRODUCTIVITY OF OVERTIME WORK    Page 2
                 * * * MULTIPLE REGRESSION * * *
Equation Number 1    Dependent Variable..    PRODUCT

Casewise Plot of Studentized Deleted (Press) Residual

*: Selected    M: Missing

     -3.0    0.0   3.0
Case # O:.....:.....:O  PRODUCT     *PRED     *RESID  *SDRESID
     1 .     .  *  .      54.00   38.6063   15.3937    1.0938
     2 .     .*    .      61.00   55.2462    5.7538     .3911
     3 . *   .     .      12.00   41.3796  -29.3796   -2.7677
     4 .   * .     .      52.00   60.7928   -8.7928    -.6415
     5 .    *.     .      28.00   33.0596   -5.0596    -.3465
     6 .    *.     .      39.00   46.0018   -7.0018    -.4611
     7 .     *     .      63.00   63.5661    -.5661    -.0417
     8 .     .*    .      19.00   16.4197    2.5803     .2412
     9 .     .   * .      73.00   49.6995   23.3005    1.8622
    10 .     .*    .      47.00   43.2285    3.7715     .2458
Case # O:.....:.....:O  PRODUCT     *PRED     *RESID  *SDRESID
     -3.0    0.0   3.0
```

```
        EXAMPLE 8.2 - PRODUCTIVITY OF OVERTIME WORK    Page 3
                 * * * MULTIPLE REGRESSION * * *
Equation Number 1  Dependent Variable..  PRODUCT

Residuals Statistics:
```

	Min	Max	Mean	Std Dev	N
*PRED	16.4197	63.5661	44.8000	13.8977	10
*ZPRED	-2.0421	1.3503	.0000	1.0000	10
*SEPRED	4.8278	11.4115	6.4983	2.1092	10
*ADJPRED	13.0909	63.8117	44.6527	14.9230	10
*RESID	-29.3796	23.3005	.0000	14.3344	10
*ZRESID	-1.9324	1.5325	.0000	.9428	10
*SRESID	-2.0445	1.6280	.0032	1.0067	10
*DRESID	-32.8899	26.2928	.1473	16.3835	10
*SDRESID	-2.7677	1.8622	-.0425	1.2167	10
*MAHAL	.0075	4.1701	.9000	1.3019	10

```
*COOK D          .0004      .2497      .0665      .0826    10
*LEVER           .0008      .4633      .1000      .1447    10
Total Cases =     10

                         ***************

From Equation 1:   6 new variables have been created.
NAME         Contents
----         --------
PREDICT      Predicted Value
RESIDUAL     Residual
STANDRES     Studentized Residual
STUDRES      Studentized Deleted (Press) Residual
COOKSD       Cook's Distance
LEVERAGE     Leverage
```

EXAMPLE 8.2. - PRODUCTIVITY OF OVERTIME WORK   Page 4

OBS	OVERTIME	PRODUCT	PREDICT	RESIDUAL	STANDRES	STUDRES	COOKSD	LEVERAGE
1	16.00	54.00	38.6063	15.3937	1.0806	1.0938	.0812	.0221
2	7.00	61.00	55.2462	5.7539	.4136	.3911	.0166	.0628
3	14.50	12.00	41.3796	-29.3796	-2.0446	-2.7677	.2497	.0067
4	4.00	52.00	60.7928	-8.7928	-.6665	-.6415	.0729	. 1471
5	19.00	28.00	33.0596	-5.0596	-.3673	-.3466	.0147	.0793
6	12.00	39.00	46.0018	-7.0018	-.4857	-.4611	.0132	.0008
7	2.50	63.00	63.5661	-.5661	-.0446	-.0417	.0004	.2026
8	28.00	19.00	16.4197	2.5803	.2568	.2412	.0426	.4633
9	10.00	73.00	49.6995	23.3005	1.6280	1.8622	.1702	.0138
10	13.50	47.00	43.2285	3.7716	.2617	.2458	.0039	.0014

Number of cases read: 10   Number of cases listed: 10

Discussion of SPSS Output

On the first page of the output, we see that the regression is significant at the 0.0254 level, with $R^2 = 0.48453$. On page 2, SPSS plots the studentized residuals and prints the dependent variable (PRODUCT), predicted values (PRED), residuals (RESID), and studentized residuals (SDRESID). As mentioned in Section 8.10, asterisks indicate temporary variables, not available for use by later procedures unless saved with a SAVE command. Page 3 lists the full set of diagnostic variables available in SPSS, and prints the smallest observation, largest, average, and standard deviation for each diagnostic variable. At the bottom of this page, we see the variables stored by the SAVE command, along with the user-specified names supplied in parentheses on the SAVE command.

The fourth page of output results from the LIST command, and shows all variables stored for each observation. The last four of these are the standardized residual, studentized residual, Cook's D, and leverage, all labeled according to their user-specified names. SPSS uses a slightly different definition of leverage and 1/n should be added to the printed values to correspond to our usage.

MINITAB Transcript

```
MTB > NAME C1 = 'OVERTIME'
MTB > NAME C2 = 'PRODUCT'
MTB > READ C1-C2
DATA > 16.0  54
DATA >  7.0  61
DATA > 14.5  12
DATA >  4.0  52
DATA > 19.0  28
DATA > 12.0  39
DATA >  2.5  63
DATA > 28.0  19
DATA > 10.0  73
DATA > 13.5  47
DATA > END
     10 ROWS READ
MTB > NAME C3 = 'LEVERAGE'
MTB > NAME C4 = 'STUDRES'
MTB > NAME C5 = 'COOKSD'
MTB > BRIEF 3
MTB > REGRESS 'PRODUCT' 1 'OVERTIME';
SUBC > HI C3;
SUBC > TRESIDUALS C4;
SUBC > COOKD C5.
The regression equation is
PRODUCT = 68.2 - 1.85 OVERTIME

Predictor       Coef       Stdev     t-ratio        p
Constant      68.188       9.791        6.96    0.000
OVERTIME     -1.8489      0.6742       -2.74    0.025

s = 15.20     R-sq = 48.5%     R-sq(adj) = 42.0%

Analysis of Variance
```

SOURCE	DF	SS	MS	F	p
Regression	1	1738.3	1738.3	7.52	0.025
Error	8	1849.3	231.2		
Total	9	3587.6			

Obs.	OVERTIME	PRODUCT	Fit	Stdev.Fit	Residual	St.Resid
1	16.0	54.00	38.61	5.31	15.39	1.08
2	7.0	61.00	55.25	6.13	5.75	0.41
3	14.5	12.00	41.38	4.97	-29.38	-2.04R
4	4.0	52.00	60.79	7.56	-8.79	-0.67
5	19.0	28.00	33.06	6.44	-5.06	-0.37
6	12.0	39.00	46.00	4.83	-7.00	-0.49
7	2.5	63.00	63.57	8.36	-0.57	-0.04
8	28.0	19.00	16.42	11.41	2.58	0.26
9	10.0	73.00	49.70	5.13	23.30	1.63
10	13.5	47.00	43.23	4.84	3.77	0.26

R denotes an obs. with a large st. resid.

MTB > PRINT C1-C5

ROW	OVERTIME	PRODUCT	LEVERAGE	STUDRES	COOKSD
1	16.0	54	0.122069	1.09376	0.081176
2	7.0	61	0.162775	0.39109	0.016629
3	14.5	12	0.106730	-2.76773	0.249729
4	4.0	52	0.247136	-0.64154	0.072914
5	19.0	28	0.179293	-0.34655	0.014739
6	12.0	39	0.100831	-0.46114	0.013225
7	2.5	63	0.302591	-0.04171	0.000431
8	28.0	19	0.563345	0.24123	0.042548
9	10.0	73	0.113810	1.86220	0.170181
10	13.5	47	0.101421	0.24584	0.003865

MTB > STOP

Discussion of MINITAB Transcript

After data entry, columns C3, C4, and C5 are named according to the values to be stored in them. As noted in Section 8.10, the MINITAB command BRIEF 3 requests that results be printed for all observations, rather than only unusual values.

The HI subcommand to REGRESS stores leverage in column C3, the TRESIDUALS subcommand stores studentized residuals in C4 (TRESIDUALS stands for residuals having a t distribution), and the COOKD subcommand stores Cook's D in C5. These values are stored, but not automatically printed, by MINITAB.

From the analysis of variance table, we see that the regression is significant at the 0.025 level, with $R^2 = 0.485$. For each observation, MINITAB prints the values of X (OVERTIME), Y (PRODUCT), the predicted value (Fit), standard error of prediction (Stdev. Fit), the residual and standardized residual (St. Resid). The standardized residual for the third observation is marked with an "R," since it exceeds 2.0 in magnitude. The PRINT command is used to print all variables stored for each observation, including X and Y, leverage, the studentized residual, and Cook's D.

8.14 TRANSFORMATIONS

Transformations of the data are primarily used to make statistical assumptions more reasonable. In Section 8.8 we discussed one such application, the natural log transformation, to remove heterogeneity of variance. In other cases, a plot of the data may reveal a distinctly nonlinear pattern that can be linearized by transforming X and/or Y. Common transformations include the natural logarithm, reciprocating a variable, and taking the square or square root. Depending on the form of relationship, either X or Y or both may be transformed, using the same or different functions. Researchers interested in such transformations should refer to specialized texts on regression for guidance in selecting an appropriate transformation.

The **Box-Cox transformation** includes a whole class of functional forms, indexed by an additional parameter, which are applied to the dependent variable. This parameter and, therefore, the form of transformation can be estimated from the data. However, the additional parameter is involved in a nonlinear fashion, and the least squares technique cannot be used for estimation. Other procedures, such as maximum likelihood, are employed for estimating the new parameter and the regression coefficients.

A few cautionary comments are in order with regard to transformations in general. Occasionally, a transformation eliminates one violation of the regression assumptions, while introducing another. All assumptions should be rechecked after the data have been transformed. When only the independent variable (X) is transformed, this is typically not a problem. But transformation of Y also changes its distribution. If Y appears to be normally distributed, a transformed variable such as $Y^* = \sqrt{Y}$ *cannot* have a normal distribution.

When Y is transformed, the prediction equation is appropriate for predicting the transformed Y value, not the original Y. To obtain predictions of the original Y variable, it is necessary to apply the inverse transformation to predicted values. For example, if Y is transformed by the natural log function to $Y^* = \ln(Y)$, we obtain a prediction equation for Y^*, not for Y—that is, $\hat{Y}^*$ predicts $Y^* = \ln(Y)$. To predict Y, we must take the inverse of the natural logarithm, the exponential function. The predicted Y value is $\hat{Y} = e^{\hat{Y}^*}$. If we wish to compare two models for the same data, the comparison should be made in terms of predicted values for the original dependent variable. We should compare values of $\hat{Y}$, not $\hat{Y}^*$.

In addition, when least squares is applied to transformed Y values, it minimizes the sum of squared residuals in the transformed values, not in Y itself. For example, suppose Y is transformed to $Y^* = 1/Y$ before regression, and then the inverse transformation is applied to the results to obtain predicted Y scores. The $\hat{Y}$ values obtained in this manner do not minimize the sum of squared errors in predicting Y. Transforming the data may yield a more nearly linear model, but worse predictions in terms of the original dependent variable.Transformations, especially to Y, should be handled with extreme care. Extra precautions should be employed in checking assumptions and evaluating adequacy of predictions.

8.15 SUMMARY

Simple linear regression is used to predict the dependent variable from the independent variable with a linear model. Technically, values of the independent variable are known exactly, while the dependent variable is assumed to be normally distributed. The least squares procedure, used to estimate the slope and intercept of the regression line, minimizes the sum of squared errors of prediction, or residuals. Least squares produces somewhat conservative predictions in that predictions lie closer to the mean (in standardized form) than do values on the predictor variable.

The estimated slope and intercept of the regression line have normal distributions centered at the population slope and intercept, respectively. Variances of these estimates can be reduced by increasing sample size or by increasing variability among values of the predictor variable. To estimate the unknown variance of the dependent variable, the sum of squared residuals is divided by its degrees of freedom. Standardizing the estimated slope and intercept, and inserting the variance estimate, leads to t statistics used in constructing confidence intervals and testing hypotheses on the population slope and intercept. Rejection of the hypothesis of zero slope implies that predictions from the linear model are warranted. Although the estimated slope and intercept are scale-dependent, the test for zero slope is not affected by linear transformation.

Alternatively, the hypothesis of zero population slope can be tested with an F test, based on analyzing variance. Total variability in the dependent variable is partitioned into two portions—one related to the independent variable (regression), and the other representing spread away from the regression line (residual). A general framework for degrees of freedom in analysis of variance was presented. Numerator degrees of freedom are the same as the number of questions asked in the null hypothesis, and denominator degrees of freedom are equal to the number of observations minus the number of parameters estimated.

Quality of predictions within the sample is assessed in a global manner by the coefficient of determination (R^2). Adjusted R^2, the proportion of variability predictable in the population, is less than R^2, since least squares chooses the regression coefficients to provide optimal fit to the sample. The residual plot is used to examine predictions for individual observations. This plot highlights outliers and also serves to check model assumptions. Patterns in the residual plot may suggest transformation of the data or addition of predictors, to improve accuracy of predictions.

A confidence interval can be constructed for the mean value of the dependent variable at a given value of the predictor. To obtain probability bounds on predictions of individual observations, a prediction interval is employed. The prediction interval is wider than the confidence interval due to random variability associated with the individual observation. Both types of intervals are narrowest when the predictor variable takes a value near the center of the sample data.

Regression and correlation are intimately connected. The square of the correlation between the dependent and independent variables is the coefficient of determination. The correlation and the slope of the regression line have the same sign, and t tests based on these estimates are equivalent. Logically, it is impossible to predict one variable from another with a linear model unless they share a linear relationship. On the other hand, regression and correlation differ, in that regression allows for a causal conclusion when values of the independent variable are randomly assigned.

An outlier is an observation that differs from the bulk of the data in either location or pattern, or both. Regression diagnostics used for outlier detection include leverage, the standardized residual, Cook's D, and the studentized residual. The latter has a t distribution and provides a test to determine if an observation is an outlier. Outliers detected by these diagnostic tools merit further investigation, but should be discarded only if they are found to be flawed.

Transformations can be employed to meet model assumptions or to linearize data. But they have drawbacks if the dependent variable is transformed. Least squares is applied to the transformed scores, and predictions of the original variable must be made using the inverse transformation. In addition, a transformation of the data that solves one problem may introduce another. SAS, SPSS, and MINITAB programs were used to demonstrate simple linear regression and various diagnostic statistics.

9

Multiple Linear Regression

Simple linear regression employs a single predictor, or independent variable X, to predict the dependent variable Y. In **multiple linear regression,** several independent variables are used simultaneously to predict the dependent variable. We denote the independent variables by X_1, X_2, ..., X_k, where k is the number of predictors. Our objective is to find a single equation to predict Y from all of the X variables. Each observation used in obtaining this equation must include values for Y and for each of the X variables.

If the **full model**, including all of the X variables, produces good prediction, we might logically ask whether all the predictors are really needed. Perhaps a subset of predictors can be selected that is essentially as good as the full model. The task of variable selection is complicated by correlations among the predictors. These relationships also impede interpretations concerning specific predictors in the full model.

This chapter begins by discussing the fitting and examination of the full model. We then present a test procedure for comparing models, and consider the problem of variable selection, using this test.

9.1 STATISTICAL MODEL

In simple linear regression, the population mean value of Y is assumed to be related to the predictor X through the equation:

$$\mu_{Y|X} = \beta_0 + \beta_1 X,$$

where β_0 is the intercept and β_1 is the slope. When several predictors are used in multiple linear regression, we extend this model to

$$\mu_Y = \beta_0 + \beta_1 X_1 + \beta_2 X_2 + \ldots + \beta_k X_k. \tag{9.1.1}$$

It is still correct to call β_0 the intercept, but the other β's are no longer appropriately labeled "slopes." Instead, we refer to the set of β's, including the intercept, as **regression coefficients**. These are population values that we wish to estimate from sample data.

As in simple regression, the values of the predictors $X_1, X_2, \ldots, X_k$ are assumed to be known, measured without error. The dependent variable Y is a random variable, assumed to be normally distributed, with mean depending on the values of the X variables, but constant variance.

The multiple linear regression model of (9.1.1) is enormously general. For example, it includes polynomial regression as a special case. Suppose we are trying to predict Y from a single predictor, X_1, but we do not believe that a linear model is appropriate. A quadratic equation can be fit to the data by defining a second predictor, X_2, whose values are obtained by squaring the X_1 values. Computer programs treat such problems the same as if two separate predictors were employed. All the program needs are values for each of the X variables. It does not care if each X_2 value is the square of the corresponding X_1 value. We could fit a cubic model by defining a third predictor, X_3, whose values are the cubes of the X_1 values. Continuing in this fashion, we can fit a polynomial model of any desired order. By the same logic, we could fit a trigonometric model, for example with $X_2 = \sin(X_1)$, or a logarithmic model using $X_2 = \log(X_1)$.

If such general forms of equations can be employed, why is the technique called multiple *linear* regression? The answer is that Y must be a linear function of the β's—that is, each β must be multiplied by an X variable, and the results summed. An example of a nonlinear model is

$$\mu_Y = \frac{\beta_0 + \beta_1 X_1}{\beta_2 X_2}.$$

Researchers who need to fit nonlinear models should consult more advanced books on this subject. The least squares approach, discussed in this chapter, is adequate for any model that is linear in the β's. Although the linear model encompasses a wide variety of functional forms, most applications involve k separate predictor variables, as shown originally in (9.1.1).

In Example 9.1, six separate predictor variables are used to obtain a regression equation for predicting gas mileage of automobiles. While most of these variables are straightforward, the fifth predictor, TRANS, deserves special mention. Transmission type is not a numeric variable, but categorical, with two categories: manual and automatic. We cannot insert a categorical variable into an equation; the computer must have numeric values. To satisfy this requirement, the **dummy**

Example 9.1

In an attempt to predict gas mileage, data were obtained for 50 cars on the variables listed below.

Y = gas mileage (MPG)
X_1 = number of cylinders (CYLIN)
X_2 = weight in pounds (WEIGHT)
X_3 = engine displacement in cubic inches (DISP)
X_4 = advertised net horsepower (HP)
X_5 = transmission type: 1=manual, 2=automatic (TRANS)
X_6 = number of forward gears (GEARS)

The six independent variables—CYLIN, WEIGHT, DISP, HP, TRANS, and GEARS—are employed in multiple linear regression to predict the dependent variable, MPG. (The raw data for this example are shown in computer inputs in Section 9.5.)

---

variable X_5 is assigned the value 1 for each car with manual transmission, and 2 for automatic transmission.

But why use the values 1 and 2? Why not 4 and 7? There are no "correct" numeric values for manual and automatic transmission types. The only way we can assign values is if it does not matter what values are selected. Suppose, instead of X_5, we use the variable X_5^*, with values 4 for manual, and 7 for automatic transmission. This substitution does not affect statistical tests of the regression coefficients.

To see this, consider Figure 9.1, which shows a plot with horizontal axis labeled X_5 and vertical axis X_5^*. Only two data points must be plotted—one for manual transmission with $(X_5, X_5^*) = (1, 4)$, and one for automatic transmission with $(X_5, X_5^*) = (2, 7)$. Since there are only two points, a straight line can be found connecting them. Thus, there must exist a slope, m, and an intercept, b, such that $X_5^* = mX_5 + b$. In short, X_5^* is a linear transformation of X_5. Recall from Section 7.8 that correlation is unaffected by linear transformation. The correlation between X_5^* and Y is identical to the correlation between X_5 and Y. Correlations between transmission type and other X variables are also unchanged.

A categorical variable with only two categories can be employed in multiple regression by assigning any convenient values to its categories. But if the variable has more than two categories, a single dummy variable is not sufficient. In general, the number of dummy variables needed to represent a categorical variable is one fewer than the number of categories. For two categories, a single dummy variable is adequate, but for three categories, two dummy variables are needed. This is

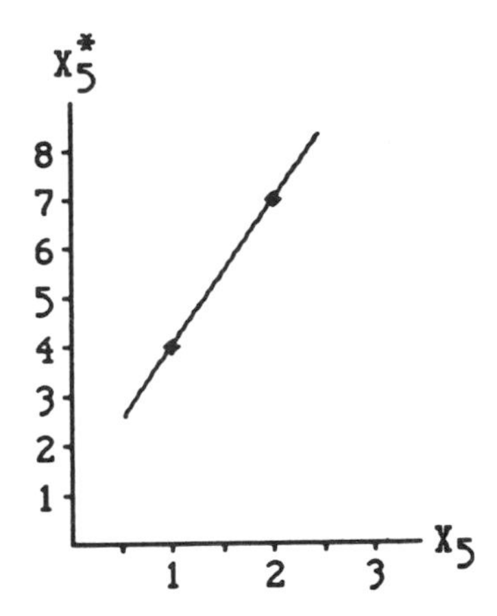

FIGURE 9.1. Dummy variable equivalence.

because three points do not necessarily fall on the same line, but they do lie in a two-dimensional plane.

When including a categorical variable with more than two categories in multiple regression, selection of values for dummy variables requires some planning. While this is generally not a problem in fitting the full model, it can pose difficulties in model selection where each dummy variable is treated separately. Depending on how the dummy variables are defined, it may not make sense to discard some of the dummy variables corresponding to a categorical variable, while keeping others in the model. Researchers who wish to include categorical variables having more than two categories in multiple regression models should refer to more advanced regression books on how to treat dummy variables.

Before leaving this topic, however, we wish to again emphasize the generality of the multiple linear regression model. It includes all of analysis of variance! Each factor in analysis of variance is a categorical variable and can be included in multiple regression by defining appropriate dummy variables. In fact, most computer programs perform computations for analysis of variance by internally defining dummy variables and then treating the problem as if it were multiple regression.

9.2 LEAST SQUARES ESTIMATION

The population regression coefficients $\beta_0, \beta_1, \beta_2, \ldots, \beta_k$ are estimated with values $b_0, b_1, b_2, \ldots, b_k$ computed from the sample. These values enable us to write the estimated regression equation

$$\hat{Y} = b_0 + b_1X_1 + b_2X_2 + \ldots + b_kX_k.$$

As in simple linear regression, we use least squares to determine the estimated regression coefficients by minimizing the sum of squared residuals, $\Sigma(Y-\hat{Y})^2$. Unlike simple regression, however, it is difficult to write formulas for the estimated regression coefficients in multiple regression without using matrix notation. In practice, this poses no problems, since computers perform the calculations for us.

For the data in Example 9.1, the estimated regression equation is

$$\hat{Y} = 43.36 + (0.761)X_1 + (-0.00488)X_2 + (-0.0176)X_3 + (-0.0313)X_4 + (-2.61)X_5 + (0.812)X_6.$$

Replacing the symbols for the variables with their names, we write

$$\widehat{MPG} = 43.36 + (0.761)CYLIN + (-0.00488)WEIGHT + (-0.0176)DISP + (-0.0313)HP + (-2.61)TRANS + (0.812)GEARS. \quad (9.2.1)$$

We caution that *the relative importance of predictors cannot be determined from the size of their coefficients*. For example, it is incorrect to conclude that TRANS is a more important predictor of gas mileage than WEIGHT because the magnitude of its coefficient is larger. There are three reasons why relative importance cannot be judged from the coefficients. The most obvious reason is that the coefficients are scale-dependent. For example, WEIGHT is recorded in pounds and has values ranging from 1845 to 3855 for these 50 cars. If, instead, WEIGHT had been listed in tons, its values would all have been smaller by a factor of 2000. But it would still have been just as important in predicting MPG, so its coefficient would have increased by a factor of 2000 to compensate—that is, the product $b_2 X_2$ in the model is unchanged by altering the scale of X_2.

A second and less obvious reason for this caution is that the regression coefficients depend on the correlations among the predictors. For example, DISP may have a small coefficient partly because it is related to CYLIN, which has a large coefficient.

Thirdly, the regression coefficients are affected by the sample data used to obtain them. A single unusual observation may have a large influence on one or more regression coefficients. For instance, observation 48 in Example 9.1 is a Ford Mustang, with horsepower of 225. No other car in this sample has horsepower over 200. As a result, the coefficient of HP is strongly influenced by this single observation.

Even the signs of the regression coefficients may be affected by these factors. In general, mileage decreases as engine size increases. For these 50 cars, MPG is negatively correlated with CYLIN (see Table 9.2). Yet CYLIN has a positive coefficient in the regression equation. Thus, it is impossible to determine even the *direction* of association between the dependent variable and the predictors based on their coefficients.

Least squares guarantees that the set of regression coefficients shown in (9.2.1) provides the best prediction of gas mileage from these independent variables. But any attempt to draw conclusions about individual predictors based on their coefficients is fraught with danger. In fact, the notion of "relative importance" is difficult to define. Because of correlations among predictors, the importance of any particular variable often depends on which other predictors are included in the model. Nonetheless, we may still ask whether prediction accuracy is seriously affected by deletion of some predictors. Methods for answering this question are addressed in later sections of this chapter. First, however, we should examine the full model to be certain that prediction is adequate from the complete set of independent variables.

9.3 EXAMINING THE FULL MODEL

Least squares produces the best set of regression coefficients in terms of minimizing squared errors of prediction. But even the best set of coefficients may not provide adequate prediction. In simple linear regression, we use the model $\mu_{Y|X} = \beta_0 + \beta_1 X$, and test the null hypothesis H_0: $\beta_1 = 0$. If this hypothesis is accepted, X drops out of the model, and we should not attempt to predict Y from X. In short, no prediction is possible with a linear model. In multiple linear regression, the null hypothesis that no prediction is possible requires that *all* of the X variables drop out of the model. Accordingly, we wish to test

$$H_0\colon \beta_1 = \beta_2 = \ldots = \beta_k = 0. \tag{9.3.1}$$

If this null hypothesis is accepted, all of the X variables drop out of the equation, and no prediction of Y is possible from this entire set of independent variables. Rejection of this hypothesis implies that the regression coefficients are not *all* zero; at least one of the coefficients is non-zero, and the corresponding X variable contributes to the prediction of Y. But we will not know from this overall test *which* predictors are worth including in the model.

An analysis of variance approach is used to test this null hypothesis. The partition of sums of squares is identical to that in simple linear regression. Total sum of squares is broken down into sums of squares for regression and residual. As in simple regression, the regression sum of squares is the portion of Y variability that can be predicted from the model, while residual is the portion that cannot be predicted. A comparable partition of degrees of freedom is also employed. Mean squares for regression and residual are again defined as the ratios of sums of squares to degrees of freedom. The F ratio of $MS_{REGRESSION}$ to $MS_{RESIDUAL}$ is our test statistic.

Degrees of freedom for this F test can be found from the rules stated in Section 8.7. Numerator degrees of freedom is the number of questions asked in the null hypothesis, and denominator degrees of freedom is the number of observations minus the number of parameters estimated. The hypothesis in (9.3.1) has k equal signs and thus asks k questions. Accordingly, $df_{REGRESSION} = k$. We have estimated k+1 parameters ($\beta_0, \beta_1, \beta_2, \ldots, \beta_k$), so $df_{RESIDUAL} = n-(k+1)$. Example 9.1 has k=6 predictors and n=50 observations, resulting in 6 degrees of freedom for regression and 43 for residual. Table 4a in the Appendix does not list 43 denominator degrees of freedom, but using the line for 40 degrees of freedom we find the value 2.34, which suffices as an approximate 5% cutoff for our test.

The analysis of variance summary table for Example 9.1 is shown in Table 9.1. The obtained F value of 38.81 far exceeds the critical value of 2.34, so we are very confident that the null hypothesis is wrong. Some prediction of gas mileage from these six predictors is possible.

Unlike the F test in simple linear regression, this F test cannot be interpreted as the square of a t test. Most computer programs print a t test for each predictor, but these are not directly related to the overall F test just demonstrated. These t tests, which are often misunderstood, are discussed in Section 9.6.

A small p-value for this overall F test means we are very sure that the null hypothesis of no predictability is wrong. But it does not tell us how good the predictions are. To measure the quality of prediction in the sample, we use the coefficient of determination (R^2), introduced in Section 8.8. For our example, we find $R^2 = 0.84$. Approximately 84 percent of the variability in gas mileage is predictable from the six independent variables employed in our equation.

As discussed in Section 8.8, R^2 measures the fit of our regression equation to the sample data. But since values for the regression coefficients are selected to minimize residuals in the sample, they are probably not identical to the true population coefficients. To estimate the proportion of variability accounted for in the population, we use adjusted R^2. In Example 9.1, adjusted $R^2 = 0.82$. If the fitted regression equation is used to predict gas mileage for other cars (of the same model year), we would expect to account for approximately 82 percent of the variability in gas mileage.

In simple linear regression, prediction accuracy for individual observations is examined by means of a residual plot. The residual, $Y-\hat{Y}$ is plotted on the vertical axis, and the predictor, X, on the horizontal axis. Comparable plots of residuals against each X variable are also used in multiple regression. These residual plots are useful for diagnosing nonlinearity in each predictor. But they do not assess the impact of each predictor in the full model due to correlations among predictors.

TABLE 9.1 Anova Summary Table for Predicting Gas Mileage

Source	df	SS	MS	F
Regression	6	842.50	140.42	38.81
Residual	43	155.58	3.62	
Total	49	998.08		

A different form of plot, known as a **partial regression plot**, is used to examine the role of each predictor in the full model. For each independent variable X, the dependent variable Y is predicted from all *other* X variables, and residuals of predictions from this equation are obtained. These values are used for the vertical axis. In addition, the independent variable X is also predicted from all other X variables. Residuals of predictions from this equation are employed for the horizontal axis. Each data point is plotted according to these two coordinates.

If a line is fitted to the points in a partial regression plot, it will pass through the origin and have slope equal to the coefficient of the predictor in the multiple regression equation. Appreciable spread away from this line indicates that this predictor is not providing much unique information—that is, information unavailable in other predictors. In addition, outliers in this plot represent observations that are particularly influential in terms of this predictor.

Figure 9.2 displays the partial regression plot for the first predictor in Example

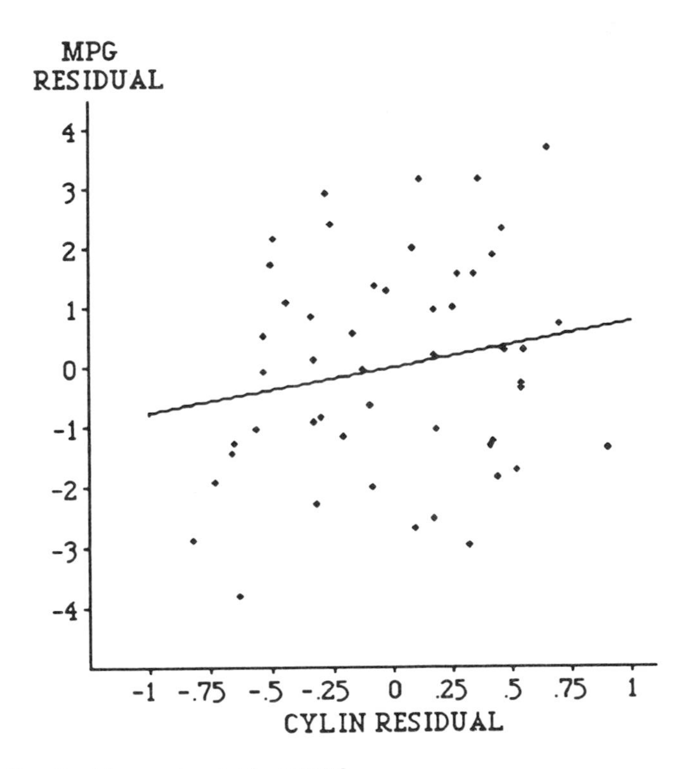

FIGURE 9.2. Partial regression plot for CYLIN.

9.1, CYLIN. For each observation, the vertical coordinate is its residual when MPG is predicted from the other five X variables (WEIGHT, DISP, HP, TRANS, GEARS). The horizontal coordinate is the residual when CYLIN is predicted from these same five variables. The line that best fits these points passes through the origin and has slope 0.761, the same as the coefficient of CYLIN in the multiple regression equation (9.2.1). Since there is substantial spread away from this line, we conclude that CYLIN is not providing a great deal of unique information in the prediction of MPG, due to its correlations with other predictors. The lack of obvious outliers suggests that the coefficient of CYLIN is not heavily influenced by a few data points.

9.4 MULTICOLLINEARITY

Multicollinearity refers to correlations among the predictors. If a pair of X variables is highly correlated, they are virtually interchangeable. A large coefficient for one and a small coefficient for the other may be reversed without greatly altering predictions of Y. But multicollinearity involves more than just pairwise correlations. Multicollinearity exists if any of the X variables is strongly related to a linear combination of the other X variables. In this case, many different sets of regression coefficients produce nearly identical predictions of Y.

Multicollinearity seriously hampers interpretations of individual regression coefficients. Substantial multicollinearity tends to result in overestimating the population regression coefficients. In addition to being too large, estimated regression coefficients have inflated variances. If we compute a confidence interval for a regression coefficient affected by multicollinearity, the interval is extremely wide. The estimated coefficients are also unstable, in that minor perturbations in the data produce large changes in estimated regression coefficients.

In general, multicollinearity does not reduce the overall quality of predictions within the range of the sample data. R^2, for example, is not affected by multicollinearity. But individual predictions can be seriously affected, and prediction intervals (see Section 8.9) may be extremely wide. The risks of extrapolation are compounded by multicollinearity.

In some experiments, multicollinearity can be completely avoided. When values of the X variables are assigned by the researcher, it is possible to guarantee that each of the predictors used in multiple regression is uncorrelated with all others. To obtain this pattern of zero correlations, each value of every X variable must be used in conjunction with each value of every other X variable equally often. Example 9.2 describes an experiment that contains three uncorrelated predictors. Each combination of values for these three predictors is used exactly twice. Clearly, this approach can only be used when the X variables are entirely within the researcher's control. Many regression situations do not allow this degree of

Example 9.2

A laboratory experiment was conducted to study the height (Y) of a certain species of plant in relationship to the volume of water given weekly (X_1), the number of daily hours of sunlight it receives (X_2), and the amount of fertilizer in the soil (X_3). A total of 36 seedlings was employed, two for each combination of X values shown below. The height of each plant was recorded after six weeks of treatment.

Water (X_1)	Sunlight (X_2)	Fertilizer(X_3)
2	6.0	1.4
2	6.0	2.1
2	6.0	3.3
2	8.5	1.4
2	8.5	2.1
2	8.5	3.3
2	12.0	1.4
2	12.0	2.1
2	12.0	3.3
4	6.0	1.4
4	6.0	2.1
4	6.0	3.3
4	8.5	1.4
4	8.5	2.1
4	8.5	3.3
4	12.0	1.4
4	12.0	2.1
4	12.0	3.3

control. Once again, our point is simply to emphasize that when you have control over the design of an experiment, use it wisely.

In most practical applications of regression, some correlations exist among the predictor variables. Modest correlations are typically not a problem, but strong interrelationships produce the undesirable effects of multicollinearity that were noted above. Several diagnostic statistics can be used to detect serious multicollinearities. All the diagnostics are functions of the correlations among the predictors. Table 9.2 displays these correlations for Example 9.1, as well as the correlation of each predictor with the dependent variable. All pairs of the variables CYLIN, WEIGHT, DISP, and HP are highly correlated. TRANS and GEARS are strongly (negatively) related. These correlations suggest potential multicollinearity problems.

In addition to examining pairwise correlations among the predictors, we should

TABLE 9.2 Correlations Among Variables in Predicting Gas Mileage

	MPG	CYLIN	WEIGHT	DISP	HP	TRANS	GEARS
MPG	1.00	-0.65	-0.84	-0.74	-0.75	-0.71	0.43
CYLIN	-0.65	1.00	0.78	0.93	0.75	0.37	-0.04
WEIGHT	-0.84	0.78	1.00	0.86	0.79	0.50	-0.13
DISP	-0.74	0.93	0.86	1.00	0.78	0.41	-0.10
HP	-0.75	0.75	0.79	0.78	1.00	0.41	-0.07
TRANS	-0.71	0.37	0.50	0.41	0.41	1.00	-0.82
GEARS	0.43	-0.04	-0.13	-0.10	-0.07	-0.82	1.00

be concerned with more complex relationships. If any of the predictors is related to a linear combination of the others, multicollinearity may be a problem. To uncover such patterns, we can try to predict each X variable from the other X variables by using multiple regression. If the R^2 for any of these multiple regressions is large, the predictor playing the role of dependent variable is related to a linear combination of the other predictors. To some extent, this variable is redundant on the other predictors.

For each predictor, its R^2 serves as a measure of the degree of its redundancy on the other predictors. **Tolerance,** defined as $1 - R^2$, reflects the degree of uniqueness of the predictor. Most computer programs automatically delete variables whose tolerance is less than 0.0001, since computations are unreliable in the presence of such strong linear dependencies. Instead of the tolerance, other programs print the **variance inflation factor** for each variable, obtained as the reciprocal of tolerance. A tolerance of less than 0.0001 corresponds to a variance inflation factor greater than 10,000. While this limit implies a severe multicollinearity problem, some experts suggest that there is cause for concern if any predictor has a variance inflation factor greater than 10. Table 9.3 shows the tolerance and variance inflation factor for each predictor in Example 9.1. Only DISP has a variance inflation factor above 10 for these data.

TABLE 9.3 Tolerance and Variance Inflation Factors for Predictors of Gas Mileage

	Tolerance	Variance Inflation Factor
CYLIN	0.120	8.33
WEIGHT	0.178	5.61
DISP	0.081	12.27
HP	0.322	3.10
TRANS	0.150	6.65
GEARS	0.193	5.18

Multicollinearity exists if any of the predictors is related to a linear combination of other predictors. But several different such linear relationships may exist within a single set of variables. For each linear relationship, the **condition number** is a value that summarizes the degree of correlation among the predictors. A condition number greater than 30 is cause for concern about multicollinearity. For Example 9.1, the largest condition number is 62.87, corresponding to a strong linear dependence among the predictors.

The most obvious remedy for multicollinearity is to delete some of the predictors. In Example 9.1, the large variance inflation factor for DISP suggests that this variable might be removed without substantial loss of predictive power. But other variables are also involved in the pattern of high correlations. It is difficult to determine how many, and which variables to delete by mere inspection. To assist us in these decisions, it would be nice to have a statistical test to determine if the loss in predictability is significant when one or more variables is removed. Section 9.6 presents this test.

Some researchers object to the deletion of any predictors, even in the face of strong multicollinearity. A few alternative strategies have been proposed in recent years that reduce multicollinearity without eliminating any variables. The most common of these are **ridge regression** and **principal components regression**. Each of these techniques, however, has substantial drawbacks. Ridge regression programs require an additional parameter, generally selected subjectively by the user. Principal components regression employs a smaller set of predictors, each of which is a function of all the original predictor variables.

In addition, both ridge regression and principal components regression produce **biased estimates** of the coefficients—that is, each regression coefficient estimated from the sample has a probability distribution that is *not* centered at the corresponding population coefficient. In Section 8.5 we learned that least squares produces unbiased estimates of slope and intercept in simple linear regression. The same is true in multiple regression. When using least squares, the probability distribution for each estimated regression coefficient is centered at the value of the corresponding population coefficient. Bias in the estimated regression coefficients is clearly not advantageous.

In light of these drawbacks, ridge regression and principal components regression are highly controversial. Instead of these techniques, most statisticians recommend variable deletion to reduce multicollinearity. Strategies for selecting variables to be removed from the model are discussed in Section 9.7.

9.5 MULTIPLE LINEAR REGRESSION IN COMPUTER PROGRAMS

Multiple linear regression can be performed in each of the programs, SAS, SPSS, and MINITAB. Each program automatically outputs the regression coefficients, analysis of variance summary table, R^2, and adjusted R^2. Options are available to print predicted values, residuals, and various diagnostic statistics. The data of Example 9.1 are used to demonstrate these features.

SAS Input

```
OPTIONS LS = 80;
TITLE 'EXAMPLE 9.1 - GAS MILEAGE';
DATA;
INPUT OBS MPG CYLIN WEIGHT DISP HP TRANS GEARS;
LINES;
 1  27  4  2670  121  108  2  3
 2  23  6  3610  232  140  2  4
 3  25  4  2910  153  100  2  3
 4  21  6  3480  180  158  2  4
 5  23  6  3480  180  190  2  4
 6  26  4  2285   97   82  2  3
 7  20  8  3850  302  150  2  4
 8  28  4  2350   98   74  2  3
 9  32  4  2440  113  103  1  5
10  24  4  2575  116  120  2  4
11  33  4  2275   97   90  1  5
12  23  4  2935  135  130  2  4
13  27  4  2885  153  100  1  5
14  23  4  2985  141  114  2  4
15  30  4  2695  133  110  1  5
16  21  6  2950  204  160  2  4
17  27  4  2920  122  115  2  4
18  18  8  3855  305  170  2  4
19  23  4  2770  138  160  2  3
20  23  6  3325  231  165  2  4
21  21  4  2920  146  138  2  4
22  28  4  2485  109   97  1  5
23  25  4  2745  122  102  2  4
24  26  4  2655  133   95  2  3
25  20  6  3265  163  160  2  4
26  24  4  2750  141   98  2  3
27  22  6  3200  180  160  2  4
28  29  4  2390   97  102  2  3
29  23  4  2885  143  110  2  4
30  33  4  2345  114   90  1  5
```

```
31   22   6   3190   182   140   2   4
32   32   4   2260    91    92   2   4
33   25   4   2295   109    90   2   3
34   26   4   2330   109   100   2   3
35   35   4   2075    89    78   1   5
36   21   6   3220   189   135   2   4
37   37   4   1845    81    63   1   5
38   22   4   2975   153   150   2   3
39   33   4   2560    97   113   1   5
40   21   6   3110   181   142   2   4
41   27   4   2710   125   140   1   5
42   24   4   2960   133   145   2   4
43   26   4   2840   107    92   1   5
44   24   4   2775   146   140   1   5
45   26   4   2920   132   125   2   4
46   33   4   2170    97   108   1   5
47   21   6   3295   191   125   2   4
48   19   8   3310   302   225   2   4
49   22   6   3145   180   150   2   4
50   20   8   3320   305   170   2   4
PROC REG;
MODEL MPG = CYLIN WEIGHT DISP HP TRANS GEARS /
     P TOL VIF COLLIN PARTIAL;
MODEL MPG = CYLIN WEIGHT;
MODEL MPG = CYLIN WEIGHT DISP HP TRANS GEARS /
     SELECTION = FORWARD SLENTRY = 0.25;
MODEL MPG = CYLIN WEIGHT DISP HP TRANS GEARS /
     SELECTION = BACKWARD SLSTAY = 0.05;
MODEL MPG = CYLIN WEIGHT DISP HP TRANS GEARS /
     SELECTION = STEPWISE SLENTRY = 0.15 SLSTAY = 0.15;
MODEL MPG = CYLIN WEIGHT DISP HP TRANS GEARS /
     SELECTION = RSQUARE BEST = 3 ADJRSQ CP MSE;
```

Discussion of SAS Input

The SAS procedure REG performs multiple linear regression, as well as a variety of other regression analyses. Multiple MODEL statements are allowed in PROC REG, with each MODEL statement resulting in a separate analysis. In this example, the first MODEL statement is used to fit the full model. It has the dependent variable (MPG) on the left side of the equation, and all independent variables (CYLIN, WEIGHT, DISP, HP, TRANS, and GEARS) on the right. Options requested are P, to print the predicted and residual values for each observation, TOL, to obtain the tolerance for each predictor, VIF, for variance inflation factors, COLLIN, for multicollinearity diagnostics, and PARTIAL, to produce partial regression plots.

The second MODEL statement requests prediction of MPG using only CYLIN and WEIGHT. Results from this analysis are used for demonstration purposes in the next section. The remaining four MODEL statements perform various stepwise regression routines and are discussed in Section 9.8, where the output from these statements is presented.

SAS Output

```
                        EXAMPLE 9.1 - GAS MILEAGE                         1
Model: MODEL1
Dependent Variable: MPG

                          Analysis of Variance

                          Sum of           Mean
Source          DF       Squares         Square     F Value      Prob>F
Model            6     842.50491      140.41748      38.811      0.0001
Error           43     155.57509        3.61803
C Total         49     998.08000

       Root MSE          1.90211      R-square        0.8441
       Dep Mean         25.28000      Adj R-sq        0.8224
       C.V.              7.52417

                          Parameter Estimates
                    Parameter       Standard      T for H0:
Variable     DF      Estimate          Error     Parameter=0    Prob > |T|
INTERCEP      1     43.361217     4.90649066          8.838        0.0001
CYLIN         1      0.761102     0.61378047          1.240        0.2217
WEIGHT        1     -0.004885     0.00142435         -3.429        0.0013
DISP          1     -0.017621     0.01647747         -1.069        0.2909
HP            1     -0.031349     0.01452760         -2.158        0.0366
TRANS         1     -2.614422     1.58119797         -1.653        0.1055
GEARS         1      0.811758     0.88522342          0.917        0.3643

                                        Variance
Variable     DF        Tolerance       Inflation
INTERCEP      1                       0.00000000
CYLIN         1       0.12004816      8.32999052
WEIGHT        1       0.17841071      5.60504472
```

DISP	1	0.08149687	12.27040988
HP	1	0.32223420	3.10333290
TRANS	1	0.15042640	6.64776948
GEARS	1	0.19302104	5.18078237

Collinearity Diagnostics

Number	Eigenvalue	Condition Number	Var Prop INTERCEP	Var Prop CYLIN	Var Prop WEIGHT	Var Prop DISP
1	6.78907	1.00000	0.0001	0.0002	0.0001	0.0002
2	0.11357	7.73158	0.0046	0.0042	0.0000	0.0283
3	0.06649	10.10482	0.0011	0.0056	0.0002	0.0166
4	0.01922	18.79491	0.0027	0.0456	0.0000	0.0392
5	0.00691	31.35419	0.0000	0.5629	0.1918	0.2357
6	0.00302	47.37675	0.1661	0.3081	0.7220	0.6269
7	0.00172	62.86623	0.8254	0.0734	0.0859	0.0531

Number	Var Prop HP	Var Prop TRANS	Var Prop GEARS
1	0.0004	0.0002	0.0001
2	0.0059	0.0007	0.0195
3	0.0022	0.0889	0.0124
4	0.8436	0.0057	0.0008
5	0.0796	0.0079	0.0000
6	0.0532	0.0212	0.0063
7	0.0150	0.8755	0.9608

EXAMPLE 9.1 - GAS MILEAGE 2

Obs	Dep Var MPG	Predict Value	Residual
1	27.0000	25.0520	1.9480
2	23.0000	19.8352	3.1648
3	25.0000	23.5666	1.4334
4	21.0000	20.8223	0.1777
5	23.0000	19.8191	3.1809
6	26.0000	28.1706	-2.1706
7	20.0000	18.6381	1.3619
8	28.0000	28.0863	-0.0863
9	32.0000	30.7112	1.2888
10	24.0000	26.0397	-2.0397
11	33.0000	32.2066	0.7934
12	23.0000	23.6330	-0.6330

```
13   27.0000   27.9267   -0.9267
14   23.0000   23.7846   -0.7846
15   30.0000   28.8937    1.1063
16   21.0000   22.9256   -1.9256
17   27.0000   24.4055    2.5945
18   18.0000   17.9339    0.0661
19   23.0000   22.6338    0.3662
20   23.0000   20.4613    2.5387
21   21.0000   23.2616   -2.2616
22   28.0000   30.7499   -2.7499
23   25.0000   25.6679   -0.6679
24   26.0000   25.3214    0.6786
25   20.0000   22.1093   -2.1093
26   24.0000   24.6223   -0.6223
27   22.0000   22.1273   -0.1273
28   29.0000   27.0307    1.9693
29   23.0000   24.3632   -1.3632
30   33.0000   31.5651    1.4349
31   22.0000   22.7679   -0.7679
32   32.0000   28.8967    3.1033
33   25.0000   27.6595   -2.6595
34   26.0000   27.1751   -1.1751
35   35.0000   33.7007    1.2993
36   21.0000   22.6547   -1.6547
37   37.0000   35.4354    1.5646
38   22.0000   21.6816    0.3184
39   33.0000   30.0934    2.9066
40   21.0000   23.1136   -2.1136
41   27.0000   28.0209   -1.0209
42   24.0000   23.0758    0.9242
43   26.0000   29.2078   -3.2078
44   24.0000   27.3334   -3.3334
45   26.0000   23.9158    2.0842
46   33.0000   32.1552    0.8448
47   21.0000   22.5666   -1.5666
48   19.0000   18.9247    0.0753
49   22.0000   22.7094   -0.7094
50   20.0000   20.5472   -0.5472
Sum of Residuals                      0
Sum of Squared Residuals       155.5751
Predicted Resid SS (Press)     211.2438
```

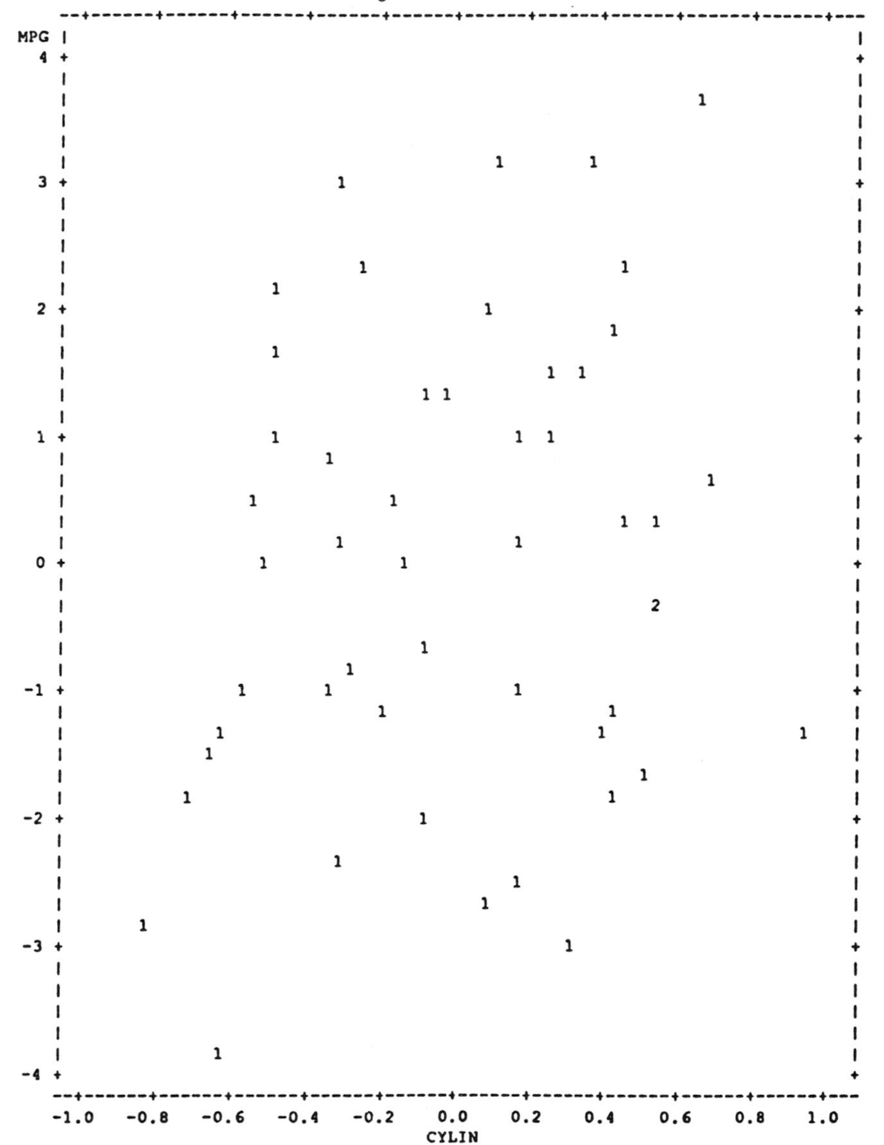
EXAMPLE 9.1 - GAS MILEAGE
3
Partial Regression Residual Plot
MPG
4
3
2
1
0
-1
-2
-3
-4
-1.0
-0.8
-0.6
-0.4
-0.2
0.0
0.2
0.4
0.6
0.8
1.0
CYLIN

```
                    EXAMPLE 9.1 - GAS MILEAGE                          4
Model: MODEL2
Dependent Variable: MPG

                          Analysis of Variance

                   Sum of         Mean
Source     DF     Squares       Square      F Value     Prob>F
Model       2   702.25844    351.12922       55.787     0.0001
Error      47   295.82156      6.29408
C Total    49   998.08000

           Root MSE       2.50880     R-square      0.7036
           Dep Mean      25.28000     Adj R-sq      0.6910
           C.V.           9.92405

                          Parameter Estimates

                    Parameter      Standard       T for H0:
Variable     DF      Estimate         Error     Parameter=0    Prob > |T|

INTERCEP      1     49.222219    2.38160730          20.668        0.0001
CYLIN         1      0.072630    0.45049030           0.161        0.8726
WEIGHT        1     -0.008542    0.00127445          -6.702        0.0001
```

Discussion of SAS Output

Much of the SAS output parallels that for simple linear regression, demonstrated in Section 8.10. At the top of page 1 is the analysis of variance summary table. The F test is highly significant (p = 0.0001), leading to rejection of the null hypothesis of no predictability. We are very sure that some prediction of MPG is possible from the full model containing all six predictors. Below the anova table are the values of R^2 and adjusted R^2. Regression coefficients are shown under the heading Parameter Estimates. The t tests for each coefficient, which are commonly misunderstood, are discussed in Section 9.6. Tolerance values and variance inflation factors are printed below the parameter estimates.

Collinearity diagnostics printed by SAS include additional statistics not discussed above. Each row of the table corresponds to a *different* near linear dependency among the predictors. For each row, the condition number measures the severity of dependence, and the variance proportions indicate which variables are involved. The last row, with condition number 62.86623, reflects the high correlation between TRANS and GEARS. The sixth row corresponds primarily to the relationship between WEIGHT and DISP.

On page 2, SAS prints the original Y value (Dep Var), the predicted value and residual for each observation. The high R^2 is reflected in the fact that most residuals are quite small compared with the actual Y values.

The third page of the output displays the partial regression plot for the first independent variable (CYLIN). Comparable plots were printed for all other predictors, but are not shown here. (We have taken the liberty of renumbering succeeding pages of output after deleting these plots.)

The second MODEL statement requested prediction of MPG by using only the two variables CYLIN and WEIGHT. Results for this regression, displayed on page 4 of the output, are discussed in the next section.

SPSS Input

```
SET WIDTH = 80
TITLE 'EXAMPLE 9.1 - GAS MILEAGE'
DATA LIST FREE / OBS MPG CYLIN WEIGHT DISP HP TRANS GEARS
REGRESSION VARIABLES = MPG TO GEARS
        / STATISTICS = DEFAULTS TOL COLLIN
        / DEPENDENT = MPG
        / METHOD = ENTER CYLIN TO GEARS
        / CASEWISE = ALL DEPENDENT PRED RESID
        / PARTIALPLOT
        / STATISTICS = DEFAULTS
        / DEPENDENT = MPG
        / METHOD = ENTER CYLIN WEIGHT
BEGIN DATA
 1   27   4   2670   121   108   2   3
 2   23   6   3610   232   140   2   4
 3   25   4   2910   153   100   2   3
 4   21   6   3480   180   158   2   4
 5   23   6   3480   180   190   2   4
 6   26   4   2285    97    82   2   3
 7   20   8   3850   302   150   2   4
 8   28   4   2350    98    74   2   3
 9   32   4   2440   113   103   1   5
10   24   4   2575   116   120   2   4
11   33   4   2275    97    90   1   5
12   23   4   2935   135   130   2   4
13   27   4   2885   153   100   1   5
14   23   4   2985   141   114   2   4
15   30   4   2695   133   110   1   5
```

```
16   21   6   2950   204   160   2   4
17   27   4   2920   122   115   2   4
18   18   8   3855   305   170   2   4
19   23   4   2770   138   160   2   3
20   23   6   3325   231   165   2   4
21   21   4   2920   146   138   2   4
22   28   4   2485   109    97   1   5
23   25   4   2745   122   102   2   4
24   26   4   2655   133    95   2   3
25   20   6   3265   163   160   2   4
26   24   4   2750   141    98   2   3
27   22   6   3200   180   160   2   4
28   29   4   2390    97   102   2   3
29   23   4   2885   143   110   2   4
30   33   4   2345   114    90   1   5
31   22   6   3190   182   140   2   4
32   32   4   2260    91    92   2   4
33   25   4   2295   109    90   2   3
34   26   4   2330   109   100   2   3
35   35   4   2075    89    78   1   5
36   21   6   3220   189   135   2   4
37   37   4   1845    81    63   1   5
38   22   4   2975   153   150   2   3
39   33   4   2560    97   113   1   5
40   21   6   3110   181   142   2   4
41   27   4   2710   125   140   1   5
42   24   4   2960   133   145   2   4
43   26   4   2840   107    92   1   5
44   24   4   2775   146   140   1   5
45   26   4   2920   132   125   2   4
46   33   4   2170    97   108   1   5
47   21   6   3295   191   125   2   4
48   19   8   3310   302   225   2   4
49   22   6   3145   180   150   2   4
50   20   8   3320   305   170   2   4
END DATA
REGRESSION VARIABLES = MPG TO GEARS
     / CRITERIA = PIN (0.25) POUT (0.99)
     / DEPENDENT = MPG
     / METHOD = FORWARD CYLIN TO GEARS
     / CRITERIA = PIN (0.01) POUT (0.05)
```

```
      / DEPENDENT = MPG
      / METHOD = BACKWARD CYLIN TO GEARS
      / CRITERIA = PIN (0.15) POUT (0.15001)
      / DEPENDENT = MPG
      / METHOD = STEPWISE CYLIN TO GEARS
FINISH
```

Discussion of SPSS Input

Multiple linear regression is performed by the SPSS procedure REGRESSION, which also performs a variety of other regression analyses. The STATISTICS subcommand requests printing of the tolerance value for each predictor (TOL) and various multicollinearity diagnostics (COLLIN). MPG is identified as the DEPENDENT variable, and the METHOD subcommand specifies that all six predictors (CYLIN TO GEARS) are to be entered in the model. The CASEWISE subcommand requests printing of the dependent variable, predicted value, and residual for each observation. PARTIALPLOT requests partial regression plots.

Prediction of MPG using only the two variables CYLIN and WEIGHT is requested in the last three lines above BEGIN DATA. Tolerance values and multicollinearity diagnostics are not requested. Results from this model are used for demonstration purposes in the next section.

After data entry, REGRESSION is used to perform various stepwise regression routines. These procedures are discussed in Section 9.8, where the output from these commands is presented.

SPSS Output

```
                    EXAMPLE 9.1 - GAS MILEAGE                    Page 1


         * * * * M U L T I P L E   R E G R E S S I O N * * * *
Listwise Deletion of Missing Data
Equation Number 1     Dependent Variable..   MPG


Block Number   1.     Method:   Enter
    CYLIN      WEIGHT      DISP      HP      TRANS      GEARS


Variable(s) Entered on Step Number
1..      GEARS
2..      CYLIN
3..      HP
```

```
4..     WEIGHT
5..     TRANS
6..     DISP

Multiple R           .91876
R Square             .84413
Adjusted R Square    .82238
Standard Error      1.90211

Analysis of Variance
               DF   Sum of Squares    Mean Square
Regression      6        842.50491      140.41748
Residual       43        155.57509        3.61803

F =     38.81053    Signif F =    .0000

- - - - - - - - - - - - Variables in the Equation - - - - - - - - -
Variable          B        SE B     Beta  Tolerance  VIF      T    Sig T

GEARS       .811758   .885223  .125668 .193021  5.181    .917   .3643
CYLIN       .761102   .613780  .215479 .120048  8.330   1.240   .2217
HP         -.031349   .014528 -.228875 .322234  3.103  -2.158   .0366
WEIGHT     -.004885   .001424 -.488836 .178411  5.605  -3.429   .0013
TRANS     -2.614422  1.581198 -.256673 .150426  6.648  -1.653   .1055
DISP       -.017621   .016477 -.225541 .081497 12.270  -1.069   .2909
(Constant) 43.361217 4.906491                            8.838   .0000

Collinearity Diagnostics

Number  Eigenval   Cond                 Variance Proportions
                   Index   Constant  CYLIN    WEIGHT    DISP     HP
  1     6.78907    1.000   .00006   .00017   .00009   .00021  .00044
  2      .11357    7.732   .00462   .00417   .00005   .02833  .00591
  3      .06649   10.105   .00114   .00557   .00016   .01663  .00220
  4      .01922   18.795   .00272   .04559   .00004   .03915  .84364
  5      .00691   31.354   .00001   .56295   .19177   .23570  .07960
  6      .00302   47.377   .16609   .30813   .72200   .62688  .05317
  7      .00172   62.866   .82536   .07343   .08589   .05310  .01504

Number    TRANS     GEARS
  1      .00019    .00011
  2      .00071    .01955
  3      .08892    .01245
  4      .00566    .00079
  5      .00790    .00001
```

```
  6      .02116    .00633
  7      .87546    .96076

End Block Number   1   All requested variables entered.
```

```
                EXAMPLE 9.1 - GAS MILEAGE                 Page 2
    * * * * M U L T I P L E   R E G R E S S I O N * * * *
Casewise Plot of Standardized Residual
*: Selected   M: Missing
       -3.0       0.0       3.0
Case #  0:........:........:0    MPG        *PRED      *RESID
 1       .        .   *    .    27.00     25.0520      1.9480
 2       .        .     *       23.00     19.8352      3.1648
 3       .        .  *     .    25.00     23.5666      1.4334
 4       .        *        .    21.00     20.8223       .1777
 5       .        .     *  .    23.00     19.8191      3.1809
 6       .     *  .        .    26.00     28.1706     -2.1706
 7       .        .  *     .    20.00     18.6381      1.3619
 8       .        *        .    28.00     28.0863      -.0863
 9       .        .  *     .    32.00     30.7112      1.2888
10       .     *  .        .    24.00     26.0397     -2.0397
11       .        .*       .    33.00     32.2066       .7934
12       .      * .        .    23.00     23.6330      -.6330
13       .      * .        .    27.00     27.9267      -.9267
14       .      * .        .    23.00     23.7846      -.7846
15       .        .  *     .    30.00     28.8937      1.1063
16       .     *  .        .    21.00     22.9256     -1.9256
17       .        .    *   .    27.00     24.4055      2.5945
18       .        *        .    18.00     17.9339       .0661
19       .        .  *     .    23.00     22.6338       .3662
20       .        .    *   .    23.00     20.4613      2.5387
21       .    *   .        .    21.00     23.2616     -2.2616
22       .    *   .        .    28.00     30.7499     -2.7499
23       .       *.        .    25.00     25.6679      -.6679
24       .        .*       .    26.00     25.3214       .6786
25       .     *  .        .    20.00     22.1093     -2.1093
26       .       *.        .    24.00     24.6223      -.6223
27       .        *        .    22.00     22.1273      -.1273
28       .        .   *    .    29.00     27.0307      1.9693
29       .      * .        .    23.00     24.3632     -1.3632
```

```
30        .         . *       .   33.00     31.5651      1.4349
31        .        *.         .   22.00     22.7679      -.7679
32        .         .    *    .   32.00     28.8967      3.1033
33        .    *    .         .   25.00     27.6595     -2.6595
34        .      * .          .   26.00     27.1751     -1.1751
35        .         , *       .   35.00     33.7007      1.2993
36        .      * .          .   21.00     22.6547     -1.6547
37        .         . *       .   37.00     35.4354      1.5646
38        .         .*        .   22.00     21.6816       .3184
39        .         .     *   .   33.00     30.0934      2.9066
40        .     *   .         .   21.00     23.1136     -2.1136
41        .      * .          .   27.00     28.0209     -1.0209
42        .         .*        .   24.00     23.0758       .9242
43        .   *     .         .   26.00     29.2078     -3.2078
44        .   *     .         .   24.00     27.3334     -3.3334
45        .         .   *     .   26.00     23.9158      2.0842
46        .         .*        .   33.00     32.1552       .8448
47        .      * .          .   21.00     22.5666     -1.5666
48        .         *         .   19.00     18.9247       .0753
49        .        *.         .   22.00     22.7094      -.7094
50        .        *.         .   20.00     20.5472      -.5472
Case #  O:........:........:O      MPG      *PRED       *RESID
      -3.0       0.0       3.0
```

```
                EXAMPLE 9.1 - GAS MILEAGE              Page 3
    * * * * M U L T I P L E   R E G R E S S I O N * * * *

Residuals Statistics:

               Min         Max        Mean    Std Dev    N

*PRED      17.9339     35.4354     25.2800     4.1466   50
*RESID     -3.3334      3.1809       .0000     1.7819   50
*ZPRED     -1.7716      2.4491       .0000     1.0000   50
*ZRESID    -1.7525      1.6723       .0000      .9368   50

Total Cases =    50
```

EXAMPLE 9.1 - GAS MILEAGE Page 4

Standardized Partial Regression Plot
Across - CYLIN Down - MPG

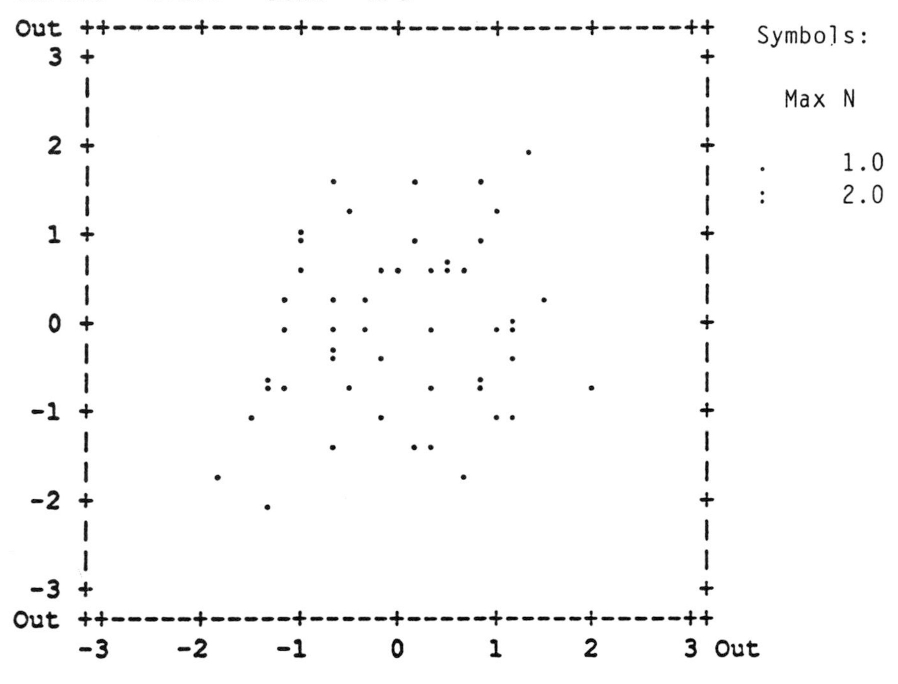

EXAMPLE 9.1 - GAS MILEAGE Page 5

```
   * * * * M U L T I P L E  R E G U L A T I O N * * * *
Equation Number 2   Dependent Variable..   MPG

Block Number   1.     Method:   Enter     CYLIN   WEIGHT

Variable(s) Entered on Step Number
   1..   WEIGHT
   2..   CYLIN

Multiple R            .83881
R Square              .70361
Adjusted R Square     .69100
Standard Error       2.50880
```

```
Analysis of Variance
                 DF      Sum of Squares       Mean Square
Regression        2           702.25844         351.12922
Residual         47           295.82156           6.29408

F =     55.78726        Signif F =    .0000

---------- Variables in the Equation ---------
Variable                B         SE B        Beta          T    Sig T

WEIGHT          -.008542      .001274    -.854807     -6.702    .0000
CYLIN            .072630      .450490     .020563       .161    .8726
(Constant)     49.222219     2.381607                 20.668    .0000

End Block Number   1   All requested variables entered.
```

Discussion of SPSS Output

Much of the SPSS output parallels that for simple linear regression, demonstrated in Section 8.10. On page 1 of the output, SPSS prints the values of Multiple R (square root of R^2), R^2, adjusted R^2, and the Standard Error (square root of the residual mean square). Beneath these values is the analysis of variance summary table. The F test is highly significant, leading to rejection of the null hypothesis of no predictability. We are very sure that some prediction of MPG is possible from the full model containing all six predictors.

The chart below this contains the regression coefficients (B) and their standard errors (SE B). If all variables were standardized before performing least squares, the coefficients would be as shown in the column labeled Beta; no intercept term would be needed, since the means of standardized variables are 0. Although standardized regression coefficients eliminate the problem of scale-dependence, they still cannot be used to judge the relative importance of predictors, due to correlations among predictors. Tolerance values and variance inflation factors for each predictor are printed in the next two columns. The t tests for each coefficient, which are commonly misunderstood, are discussed in Section 9.6.

Collinearity diagnostics printed by SPSS include additional statistics not discussed above. Each row of the table corresponds to a *different* near linear dependency among the predictors. For each row, the condition index measures the severity of dependence, and the variance proportions indicate which variables are involved. The last row, with condition index 62.866, reflects the high correlation between TRANS and GEARS. The sixth row corresponds primarily to the relationship between WEIGHT and DISP.

The second page shows for each observation a plot of the standardized residual (see Section 8.12), the value of Y (MPG), the predicted value, and the residual. The high R^2 is reflected in the fact that most residuals are quite small compared with the actual Y values. On page 3, the smallest, largest, average, and standard deviation of these values are printed. Also included in this display are similar statistics for predicted and residual values converted to Z scores.

The fourth page of the output displays the partial regression plot for the first independent variable (CYLIN). Comparable plots were printed for all other predictors, but are not shown here. (We have taken the liberty of renumbering succeeding pages of output after deleting these plots.)

Prediction of MPG using only the two variables CYLIN and WEIGHT was requested as a separate analysis. Results from this regression, displayed on page 5, are discussed in the next section.

MINITAB Transcript

```
MTB > NAME C1='OBS' C2='MPG' C3='CYLIN' C4='WEIGHT' &
CONT > C5 = 'DISP' C6 = 'HP' C7 = 'TRANS' C8 = 'GEARS'
MTB > READ C1-C8
DATA >  1  27  4  2670  121  108  2  3
DATA >  2  23  6  3610  232  140  2  4
DATA >  3  25  4  2910  153  100  2  3
DATA >  4  21  6  3480  180  158  2  4
DATA >  5  23  6  3480  180  190  2  4
DATA >  6  26  4  2285   97   82  2  3
DATA >  7  20  8  3850  302  150  2  4
DATA >  8  28  4  2350   98   74  2  3
DATA >  9  32  4  2440  113  103  1  5
DATA > 10  24  4  2575  116  120  2  4
DATA > 11  33  4  2275   97   90  1  5
DATA > 12  23  4  2935  135  130  2  4
DATA > 13  27  4  2885  153  100  1  5
DATA > 14  23  4  2985  141  114  2  4
DATA > 15  30  4  2695  133  110  1  5
DATA > 16  21  6  2950  204  160  2  4
DATA > 17  27  4  2920  122  115  2  4
DATA > 18  18  8  3855  305  170  2  4
DATA > 19  23  4  2770  138  160  2  3
DATA > 20  23  6  3325  231  165  2  4
```

```
DATA > 21   21   4   2920   146   138   2   4
DATA > 22   28   4   2485   109    97   1   5
DATA > 23   25   4   2745   122   102   2   4
DATA > 24   26   4   2655   133    95   2   3
DATA > 25   20   6   3265   163   160   2   4
DATA > 26   24   4   2750   141    98   2   3
DATA > 27   22   6   3200   180   160   2   4
DATA > 28   29   4   2390    97   102   2   3
DATA > 29   23   4   2885   143   110   2   4
DATA > 30   33   4   2345   114    90   1   5
DATA > 31   22   6   3190   182   140   2   4
DATA > 32   32   4   2260    91    92   2   4
DATA > 33   25   4   2295   109    90   2   3
DATA > 34   26   4   2330   109   100   2   3
DATA > 35   35   4   2075    89    78   1   5
DATA > 36   21   6   3220   189   135   2   4
DATA > 37   37   4   1845    81    63   1   5
DATA > 38   22   4   2975   153   150   2   3
DATA > 39   33   4   2560    97   113   1   5
DATA > 40   21   6   3110   181   142   2   4
DATA > 41   27   4   2710   125   140   1   5
DATA > 42   24   4   2960   133   145   2   4
DATA > 43   26   4   2840   107    92   1   5
DATA > 44   24   4   2775   146   140   1   5
DATA > 45   26   4   2920   132   125   2   4
DATA > 46   33   4   2170    97   108   1   5
DATA > 47   21   6   3295   191   125   2   4
DATA > 48   19   8   3310   302   225   2   4
DATA > 49   22   6   3145   180   150   2   4
DATA > 50   20   8   3320   305   170   2   4
DATA > END
      50 ROWS READ

MTB > BRIEF 3
MTB > REGRESS 'MPG' 6 C3-C8;
SUBC > VIF.

The regression equation is
MPG = 43.4 + 0.761 CYLIN - 0.00488 WEIGHT - 0.0176 DISP
- 0.0313 HP - 2.61 TRANS + 0.812 GEARS
```

Predictor	Coef	Stdev	t-ratio	p	VIF
Constant	43.361	4.906	8.84	0.000	
CYLIN	0.7611	0.6138	1.24	0.222	8.3
WEIGHT	-0.004885	0.001424	-3.43	0.001	5.6
DISP	-0.01762	0.01648	-1.07	0.291	12.3
HP	-0.03135	0.01453	-2.16	0.037	3.1
TRANS	-2.614	1.581	-1.65	0.106	6.6
GEARS	0.8118	0.8852	0.92	0.364	5.2

s = 1.902 R-sq = 84.4% R-sq(adj) = 82.2%

Analysis of Variance

SOURCE	DF	SS	MS	F	p
Regression	6	842.50	140.42	38.81	0.000
Error	43	155.58	3.62		
Total	49	998.08			

SOURCE	DF	SEQ SS
CYLIN	1	419.53
WEIGHT	1	282.73
DISP	1	7.73
HP	1	18.98
TRANS	1	110.50
GEARS	1	3.04

Obs.	CYLIN	MPG	Fit	Stdev.Fit	Residual	St.Resid
1	4.00	27.000	25.052	0.591	1.948	1.08
2	6.00	23.000	19.835	0.697	3.165	1.79
3	4.00	25.000	23.567	0.767	1.433	0.82
4	6.00	21.000	20.822	0.751	0.178	0.10
5	6.00	23.000	19.819	0.918	3.181	1.91
6	4.00	26.000	28.171	0.699	-2.171	-1.23
7	8.00	20.000	18.638	0.983	1.362	0.84
8	4.00	28.000	28.086	0.728	-0.086	-0.05
9	4.00	32.000	30.711	0.530	1.289	0.71
10	4.00	24.000	26.040	0.647	-2.040	-1.14
11	4.00	33.000	32.207	0.579	0.793	0.44
12	4.00	23.000	23.633	0.590	-0.633	-0.35
13	4.00	27.000	27.927	0.819	-0.927	-0.54
14	4.00	23.000	23.785	0.645	-0.785	-0.44

```
15     4.00   30.000   28.894   0.602    1.106    0.61
16     6.00   21.000   22.926   0.613   -1.926   -1.07
17     4.00   27.000   24.406   0.577    2.594    1.43
18     8.00   18.000   17.934   0.858    0.066    0.04
19     4.00   23.000   22.634   0.919    0.366    0.22
20     6.00   23.000   20.461   0.566    2.539    1.40
21     4.00   21.000   23.262   0.678   -2.262   -1.27
22     4.00   28.000   30.750   0.537   -2.750   -1.51
23     4.00   25.000   25.668   0.616   -0.668   -0.37
24     4.00   26.000   25.321   0.624    0.679    0.38
25     6.00   20.000   22.109   0.815   -2.109   -1.23
26     4.00   24.000   24.622   0.661   -0.622   -0.35
27     6.00   22.000   22.127   0.565   -0.127   -0.07
28     4.00   29.000   27.031   0.669    1.969    1.11
29     4.00   23.000   24.363   0.679   -1.363   -0.77
30     4.00   33.000   31.565   0.574    1.435    0.79
31     6.00   22.000   22.768   0.508   -0.768   -0.42
32     4.00   32.000   28.897   0.844    3.103    1.82
33     4.00   25.000   27.660   0.657   -2.660   -1.49
34     4.00   26.000   27.175   0.638   -1.175   -0.66
35     4.00   35.000   33.701   0.696    1.299    0.73
36     6.00   21.000   22.655   0.486   -1.655   -0.90
37     4.00   37.000   35.435   0.895    1.565    0.93
38     4.00   22.000   21.682   0.848    0.318    0.19
39     4.00   33.000   30.093   0.649    2.907    1.63
40     6.00   21.000   23.114   0.490   -2.114   -1.15
41     4.00   27.000   28.021   0.729   -1.021   -0.58
42     4.00   24.000   23.076   0.646    0.924    0.52
43     4.00   26.000   29.208   0.821   -3.208   -1.87
44     4.00   24.000   27.333   0.776   -3.333   -1.92
45     4.00   26.000   23.916   0.575    2.084    1.15
46     4.00   33.000   32.155   0.667    0.845    0.47
47     6.00   21.000   22.567   0.594   -1.567   -0.87
48     8.00   19.000   18.925   1.236    0.075    0.05 X
49     6.00   22.000   22.709   0.513   -0.709   -0.39
50     8.00   20.000   20.547   1.020   -0.547   -0.34

X denotes an obs. whose X value gives it large influence.

MTB > BRIEF 2
MTB > REGRESS 'MPG' 2 'CYLIN' 'WEIGHT'
```

```
The regression equation is
MPG = 49.2 + 0.073 CYLIN - 0.00854 WEIGHT

Predictor        Coef       Stdev    t-ratio        p
Constant       49.222       2.382      20.67    0.000
CYLIN          0.0726      0.4505       0.16    0.873
WEIGHT      -0.008542    0.001274      -6.70    0.000

s = 2.509      R-sq = 70.4%      R-sq(adj) = 69.1%

Analysis of Variance

SOURCE        DF        SS        MS        F        p
Regression     2    702.26    351.13    55.79    0.000
Error         47    295.82      6.29
Total         49    998.08

SOURCE        DF    SEQ SS
CYLIN          1    419.53
WEIGHT         1    282.73

Obs.  CYLIN     MPG      Fit    Stdev.Fit   Residual   St.Resid
33     4.00  25.000   29.910      0.592     -4.910     -2.01R
37     4.00  37.000   33.753      1.076      3.247      1.43 X
39     4.00  33.000   27.646      0.427      5.354      2.17R
48     8.00  19.000   21.530      1.103     -2.530     -1.12 X
50     8.00  20.000   21.445      1.097     -1.445     -0.64 X

R denotes an obs. with a large st. resid.
X denotes an obs. whose X value gives it large influence.
```

Discussion of MINITAB Transcript

The first line of this transcript and its continuation line name columns to store the observation number (OBS), the dependent variable (MPG), and all six independent variables (CYLIN, WEIGHT, DISP, HP, TRANS, and GEARS). After data entry, the MINITAB command BRIEF 3 requests that results be printed for all observations, rather than only unusual values. The REGRESS command lists the dependent variable (MPG), the number of predictors (6), and the columns in which the predictors are stored (C3–C8). Variance inflation factors are requested in the subcommand.

Much of the output parallels that for simple linear regression, as demonstrated in Section 8.10. MINITAB first shows the estimated regression equation, then prints a table containing the regression coefficients, their standard errors (Stdev), t tests with associated p-values, and variance inflation factors. The t tests for each coefficient, which are commonly misunderstood, are discussed in Section 9.6. Below this are the values of s (square root of the residual mean square), R^2, and adjusted R^2. The F test in the analysis of variance summary table is highly significant, leading to rejection of the null hypothesis of no predictability. We are very sure that some prediction of MPG is possible from the full model containing all six predictors.

The sequential sums of squares shown below the anova table (SEQ SS) are only useful if there is a natural order among the predictors. This ordering may be based on the cost of obtaining the predictors, their logical relationship to the dependent variable, or any other reasonable scheme. In this case, F tests for individual predictors are constructed by dividing each sequential sum of squares by $MS_{RESIDUAL}$. Examination of these tests begins with the *last* predictor and proceeds upwards, in an attempt to eliminate the most expensive (least logical, etc.) predictors first. In our example there is no natural order for the predictors, so these sequential tests are not a reasonable approach to model selection.

The following table lists for each observation the values of the first predictor (CYLIN), the dependent variable (MPG), the predicted value (Fit), its standard error of prediction (Stdev. Fit), the residual, and the standardized residual (see Section 8.12). The high R^2 is reflected in the fact that most residuals are quite small compared with the actual Y values. Observation 48 is singled out as having an unusual X value. As noted in Section 9.2, this car has an unusual value for HP.

A second REGRESS command requests prediction of MPG by using only the two variables CYLIN and WEIGHT. The preceding BRIEF 2 command requests that predicted values and residuals be printed only for unusual observations. Results from this model are used for demonstration purposes in the next section.

9.6 COMPARING MODELS

In examining the full model, we test the null hypothesis that no prediction of the dependent variable is possible from the complete set of independent variables. A significant F test in the analysis of variance tells us that prediction is possible, but does not imply that all the predictors are *necessary*. We might logically ask whether a subset of X variables could provide predictions essentially as good as the full set. The statistical test for comparing the full model to a sub-model is called a **partial F test**, since it corresponds to testing part of the model.

Selection of an appropriate subset is a difficult problem, and we will defer its discussion until the next section. But, to demonstrate the partial F test, suppose we wish to compare the full set of six predictors in Example 9.1 with a subset containing only the first two predictors. These models can be written

$$\text{full: } \mu_Y = \beta_0 + \beta_1 X_1 + \beta_2 X_2 + \beta_3 X_3 + \beta_4 X_4 + \beta_5 X_5 + \beta_6 X_6$$
$$\text{sub: } \mu_Y = \beta_0 + \beta_1 X_1 + \beta_2 X_2.$$

Comparing these models is equivalent to testing in the full model

$$H_0: \beta_3 = \beta_4 = \beta_5 = \beta_6 = 0.$$

If this null hypothesis is accepted, the full model reduces to the sub-model.

To test this hypothesis, we must run both the full model and sub-model on the computer. For each model, the total sum of squares is analyzed into regression and residual portions. The total sum of squares is merely the spread among all Y values, and is the same in the two models. Computer outputs displayed in the previous section list a total sum of squares of 998.08 for both full and sub-models. However, the decomposition of this total into regression and residual components differs for the two models. Recall that regression is the portion of Y variability that can be predicted from the X variables, while residual is the portion that cannot be predicted. Logically, the full model predicts better than any sub-model, and has a larger regression sum of squares than the sub-model. The difference in regression sums of squares between the full and sub-models reflects how much better the full model predicts than the sub-model. This difference, called a **partial sum of squares,** pertains directly to the null hypothesis, and is used in the numerator of our F statistic for comparing the models.

In Example 9.1, the regression sum of squares for the full model is 842.50 (see Table 9.1). For the sub-model containing the first two predictors, the regression sum of squares is 702.26, as shown in printouts of the previous section. The partial sum of squares is the difference, 842.50 - 702.26 = 140.24. This is the extra part of Y variability predictable from the full model, but not from the sub-model. This value is the numerator sum of squares in the partial F test.

An F statistic is the ratio of two mean squares, where each mean square is a sum of squares divided by its degrees of freedom. We have computed the numerator sum of squares, and now must determine its degrees of freedom. As we learned earlier, the degrees of freedom for the numerator of an F ratio is the number of questions asked in the null hypothesis. Our null hypothesis, as shown above, has four equal signs, and therefore asks four questions. The numerator mean square for the partial F test is the partial sum of squares divided by its degrees of freedom, or 140.24/4 = 35.06.

The denominator of the partial F test is the residual mean square for the full model. From Table 9.1, we find this value to be 3.62. Taking the ratio, we obtain the F value of 35.06/3.62 = 9.69. As explained in the last paragraph, this ratio has 4 degrees of freedom in the numerator. It has 43 degrees of freedom in the denominator, which is the degrees of freedom for the residual mean square from the full model that is used in the denominator of the ratio. Table 4a in the Appendix does not list 43 denominator degrees of freedom, but, using the line for 40 degrees of freedom, we find the value 2.61, which suffices as an approximate 5% cutoff for our test. Our obtained F statistic of 9.69 far exceeds this critical value, so we confidently reject the null hypothesis. We are very sure that the full model is better than the sub-model containing only the first two predictors.

The partial F test can be used to compare any two models, as long as one of them is a subset of the other. We simply call the more complete model "full," and the subset model "sub." For example, we could compare the model containing predictors X_1, X_2, X_4, and X_6 against the model with only X_2 and X_6 by treating the former as the full model and the latter as the sub-model. This test is not valid, however, for comparing two models unless one is a subset of the other. We could not use this test, for example, to compare the model with X_1, X_2, X_4, and X_6 against the model with X_2 and X_5.

When comparing two models with a partial F test, R^2 is always larger for the more complete model. Common sense tells us that we obtain better prediction by using a full set of predictors rather than only some of them. The partial F ratio can be interpreted as a test of whether R^2 for the more complete model is *significantly* larger than that for the subset.

In our example we compared the full model with all six predictors against the sub-model containing only the first two. We can view this comparison as if we started with the sub-model and then *added* the other four predictors. The partial F test determines whether it is worth adding predictors X_3, X_4, X_5, and X_6, given that we started with X_1 and X_2 already in the model. We represent the test symbolically as 3,4,5,6|1,2, where the vertical bar is read "given." A significant result implies that something in the variables X_3, X_4, X_5, and X_6 contributes to the prediction of Y above and beyond what can be predicted using only X_1 and X_2. Clearly, this conclusion is not particularly useful. In essence, selecting the sub-model with only the first two predictors corresponds to testing the other four predictors all at once.

To maximize the utility of partial F tests, we should test only one predictor at a time. For example, we might test 6|1,2,3,4,5. This test compares the full model with all six predictors against the sub-model with the first five. The null hypothesis for this test is H_0: $\beta_6 = 0$. Since we are testing only one parameter, this F statistic has one degree of freedom in the numerator. As we have seen in other situations (see Sections 5.4 and 8.7), an F test with one degree of freedom in the numerator can be interpreted as the square of a t test. Taking the square root of the F ratio yields the t value. Since the F ratio is called a partial F test, we refer to its square root as a **partial t test**. This partial t test is automatically printed for each predictor by all standard computer programs, but it is frequently misinterpreted by users.

For each predictor separately, the partial t test asks whether it contributes to the prediction of Y above and beyond the other predictors in the model. For the first predictor, we test 1|2,3,4,5,6; for the second, we test 2|1,3,4,5,6; for the third, we test 3|1,2,4,5,6; and so forth. Great care must be exercised in examining the whole set of partial t tests. These tests are heavily influenced by correlations among predictors.

We can see the effects of correlations among predictors graphically in Figure 9.3, which depicts multiple regression with two predictors. The variables X_1, X_2, and Y are each drawn as a rectangle, with overlaps representing correlation. In this picture, X_1 and X_2 each correlates well with Y, since they overlap substantially with Y, but they also correlate strongly with each other. The partial t test for 1|2 assesses the *additional* part of Y that can be predicted by adding X_1 to a model already containing X_2. As shown in the figure, this is a small region. Likewise, the partial t test for 2|1 assesses the *additional* predictive value of adding X_2 to a model already containing X_1, which is also small. Both these partial t tests would be non-significant in this situation. But this does not imply that both variables should be dropped out of the model! Clearly, either X_1 or X_2 is useful in predicting Y, but not both.

For Example 9.1, partial t tests for each variable are shown in computer outputs of the previous section. The partial t value for TRANS is -1.653, with a two-tailed p-value of 0.1055. For GEARS, the partial t value is 0.917 and the two-tailed p-value is 0.3643. Since both p-values are greater than 5 percent, both tests are nonsignificant. Hasty analysis leads to the conclusion that both variables can be eliminated from the model without much loss in predictability. More careful scrutiny, however, reveals that this example is comparable to the situation represented in Figure 9.3. As shown in Table 9.2, TRANS and GEARS are highly correlated. Thus, we should keep one or the other of these variables in the model,

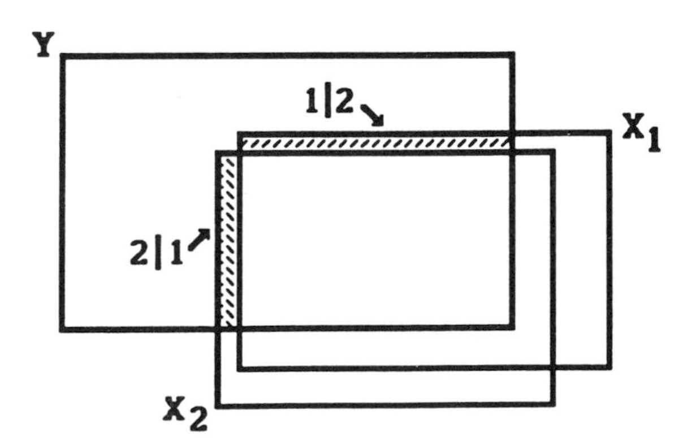

FIGURE 9.3. Partial sums of squares.

but not both. If either TRANS or GEARS is dropped from the model, the partial t test for the other becomes highly significant.

In this section we have learned how to compare two models, as long as one is a sub-model of the other. We have also seen that conclusions from such tests are most useful if the two models differ by only one predictor. If we drop a variable from the model or add a variable to the model, partial t tests for the other variables may be substantially altered. Thus, changes to the model should be made one predictor at a time. Each time the model is changed, every predictor in the new model should be reexamined. In a sense, the practical importance of predictors is a relative concept. The contribution of any variable to the overall regression depends on which other variables are also included in the model.

9.7 MODEL SELECTION

A model with substantial multicollinearity is said to be **overfit**. In Section 9.4 we learned that the effects of overfitting include overestimation of population coefficients, difficulty in interpreting regression coefficients, and inflation of variances for regression coefficients and predictions. If the full model shows evidence of multicollinearity, we should seek a sub-model with less redundancy among predictors.

But we must also be careful not to **underfit**. If we omit predictors that contribute significantly to the regression sum of squares, the quality of predictions suffers. In this case, the regression sum of squares is too small, and therefore the residual sum of squares is inflated. As a result, mean square residual overestimates random variability. In addition, the regression coefficients are biased estimates of population coefficients.

Our objective is to strike a compromise between overfitting and underfitting. We seek a sub-model with minimum multicollinearity, but with predictive accuracy that is essentially as good as the full model. We should recognize that this search does not have a unique answer. In most practical applications, several adequate sub-models exist with comparable prediction properties. The goal of model selection is to find a good prediction equation, not the "correct" one.

Several statistics may be useful in evaluating sub-models. For comparing sub-models of the same size (i.e., with the same number of predictors), R^2 is adequate. But adding more predictors tends to increase R^2, and it is not appropriate to compare models of different sizes by choosing the one with the largest R^2. The full model always has larger R^2 than any sub-model.

Recall from Section 8.8 that R^2 measures the proportion of variability accounted for *in the sample*, while adjusted R^2 estimates the proportion of variability *in the population* that our model would be expected to explain. Since we intend to use the prediction equation for additional observations in the population, it is better to choose a sub-model by maximizing adjusted R^2 rather than R^2. When comparing two models of the same size, the one with larger R^2 also has larger adjusted R^2. But if the two models are of different sizes, it is possible that the one with smaller R^2 has *larger* adjusted R^2. One of the elements used in computing adjusted R^2 is the number of predictors. (In the formula for adjusted R^2, (8.8.2), $df_{REGRESSION}$ is the number of predictors.) In this sense, adjusted R^2 accounts for the number of predictors in the model.

As noted above, an underfit model has inflated mean square residual—that is, too much of the Y variability is considered random error. Minimizing the residual mean square is also a logical method for selecting among candidate models. It can be shown that minimizing the residual mean square is equivalent to maximizing adjusted R^2. These two criteria produce identical decisions in comparing sub-models, regardless of whether the sub-models are of the same or different sizes.

A model that is overfit has too much variance in the regression coefficients and predicted values, while one that is underfit has bias in these quantities. A statistic that includes both variance and bias is called C_p. If C_p is small, the model is neither substantially overfit nor underfit. In model selection, adjusted R^2, $MS_{RESIDUAL}$, and C_p should all be considered. A good prediction equation will have large adjusted R^2, small $MS_{RESIDUAL}$, and small C_p.

Although the search for a satisfactory subset of predictors could be carried out by trial and error, most computer programs contain algorithms for stepping through various sub-models in search of a good solution. These procedures are called **stepwise regression routines**. Several logical search heuristics are available. Confusingly, however, the term "stepwise" is also used to refer to a particular one of these routines. The three most commonly used procedures are forward selection, backward elimination, and stepwise regression.

Forward selection begins with no predictors in the model and adds variables one at a time until further additions would no longer significantly improve prediction. To determine which variable should be entered first, each predictor is used separately in simple linear regression to predict Y. The predictor with the smallest p-value in simple linear regression is the first to enter the model.

At this stage, the additional predictive value of each of the other variables is evaluated by fitting a series of two-predictor models, each including the first predictor and one other variable. In each model, the partial t (or partial F) test for the new predictor is computed. The predictor with the smallest p-value is added to the model as the second variable. Thus, the second predictor is the variable whose addition to the model provides the greatest improvement in prediction, beyond what can be predicted from the first variable alone. Now each subset of size three that includes these two variables is examined, and the partial t test for the third predictor is computed. Again, the new predictor with the smallest p-value is added to the model, now as the third variable. The process continues until addition of further predictors is unjustified because all partial t tests for variables not already in the model are nonsignificant. At this stage, no variable can be added that significantly improves prediction beyond what can be predicted from the existing set of variables.

Forward selection does not necessarily produce the "best" terminal model. At any step, the model may not have the highest R^2 among all models of that size. For example, when the second predictor is added, the model may not be the best among subsets of size two. The only two-predictor models examined by this routine are those that include the first predictor entered. But the best subset of size two may not include the best single predictor.

Special consideration must be given to the α level used in forward selection. Typically, an α value larger than 5 percent should be employed with this technique. This is because mean square residual is inflated in an underfit model. In the early steps of forward selection, the model does not yet include enough predictors and is underfit. Since mean square residual is used in the denominator of partial F tests, this inflated residual term leads to F values that are too small. Predictors that are worth adding to the model may not be significant, due to small F values. To remedy this problem, a larger α level is used in forward selection, generally 25 to 50 percent. Unfortunately, default α levels in computer programs vary considerably. Generally, the user is ill-advised to alter default values in computer programs. Forward selection in regression may be an exception to this rule.

When forward selection is used for the data of Example 9.1 with an α value of 0.25, a terminal model with four variables results. In order, the predictors entered are WEIGHT, TRANS, HP, and GEARS. The final prediction equation is

$$\widehat{\text{MPG}} = 43.33 + (-0.00538)\text{WEIGHT} + (-2.06)\text{TRANS} + (-0.0307)\text{HP} + (1.15)\text{GEARS}.$$

These coefficients are the same as would be obtained if least squares were applied without a stepwise approach to a model that contained these four predictors. This model has $R^2 = 0.838$, adjusted $R^2 = 0.824$, $MS_{RESIDUAL} = 3.58$, and $C_p = 4.55$.

Backward elimination begins with the full model and removes variables one at a time until further deletions would significantly reduce predictability. In each step, partial t (or partial F) tests are computed for each predictor currently in the model, and the variable with the largest p-value is removed. After each deletion, the partial t tests must be recomputed, since removal of any variable alters the relative usefulness of remaining predictors. The process continues until all variables left in the model are significant at a specified α level.

As with forward selection, backward elimination does not necessarily produce the "best" terminal model. Once a variable has been deleted from the equation, it cannot reenter. After deletion of two or more predictors, the model may not have the highest R^2 among all models with the same number of predictors.

Since this routine begins with the full model, mean square residual is not underestimated, and a conventional α level, such as 5 percent, is appropriate. As in forward selection, computer programs vary considerably in default α levels.

When backward elimination is used for the data of Example 9.1 with an α value of 0.05, the terminal model contains three predictors. In order, variables deleted from the model are GEARS, DISP, and CYLIN. The final prediction equation is

$$\widehat{MPG} = 49.38 + (-0.00487)\text{WEIGHT} + (-3.90)\text{TRANS} + (-0.0280)\text{HP}.$$

Again, these coefficients are the same as would be obtained if least squares were applied without a stepwise approach to a model containing these three predictors. For this model, $R^2 = 0.832$, adjusted $R^2 = 0.821$, $MS_{RESIDUAL} = 3.65$, and $C_p = 4.42$.

Stepwise regression combines the search patterns used in forward selection and backward elimination. Like forward selection, stepwise regression begins with no variables in the model and adds predictors one at a time. But, as in backward elimination, variables no longer needed are deleted from the model. In each step the algorithm first tries to delete a variable by performing partial t (or partial F) tests for each predictor currently in the model. If any of these are nonsignificant at a specified α level, the variable with the largest p-value is deleted from the model, and a new step begins. If all variables currently in the model are significant, the routine tries to add a variable. Each new model that could be formed by the addition of one predictor is examined, and the partial t test for the new predictor is computed. If any of these are significant at a specified α level, the variable with the smallest p-value is added to the model, and a new step begins.

In each step the routine changes the model by either deleting a variable or adding a variable. As soon as the model is changed, a new step begins. Each step first tries to delete a variable and then, if none are deleted, to add a variable. The routine stops when the model is unchanged through a complete cycle—that is, when all variables currently in the model are significant, and no other predictors are worth adding.

Clearly, stepwise regression performs a more thorough search than either forward selection or backward elimination. But it still does not examine all possible subsets of predictors. Therefore, we cannot be sure that the terminal model produced by stepwise regression is "best" in terms of any of the criteria used for model evaluation.

Most stepwise regression programs allow the user to specify different α levels for adding and deleting predictors. Since, like forward selection, the routine begins with no predictors in the model, mean square residual is inflated in the first few steps. Accordingly, we should use a large α level in tests to add variables to the model. But, after several predictors are added to the model, this is no longer the case, and a small α level should be used in tests to delete variables from the model. However, using a large α for adding and a small α for deleting can cause difficulties. A variable may be significant enough to add to the model, but not significant enough to keep. The same variable may be alternately added and deleted repeatedly, and the program may get caught in a loop. To avoid this problem, the same α level should be employed for both types of decisions. Compromising between these objectives, an α in the range of 15 percent is reasonable.

When stepwise regression is used for the data of Example 9.1 with an α level of 0.15 for both adding and deleting variables, a terminal model with three predictors results. In this example, once added, no variables are ever deleted from the model. In order, the predictors entered are WEIGHT, TRANS, and HP. Since these are the same variables remaining after backward elimination, the terminal model is also the same. Resulting values of R^2, adjusted R^2, $MS_{RESIDUAL}$, and C_p are also the same as in backward elimination.

Although stepwise regression routines, such as forward selection, backward elimination, and stepwise regression, are commonly used, the power of modern computers now allows even more thorough searches. Some programs examine all possible subsets of predictors, printing descriptive statistics for the models of each size that have the highest R^2 values. Unlike stepwise routines, however, these procedures do not settle on a terminal model. For Example 9.1, the model of each size with highest R^2 is shown in Table 9.4, along with its adjusted R^2, $MS_{RESIDUAL}$, and C_p.

From inspection of this table, we see that forward selection picked the best subset of size four as its terminal model. However, backward elimination and stepwise regression settled on a terminal model with three predictors that is not the best model of this size. On the other hand, the sub-model with WEIGHT, TRANS, and HP selected by these procedures has statistical characteristics (R^2, adjusted R^2, $MS_{RESIDUAL}$ and C_p) nearly identical to those of the best subset of size three.

In general, computer routines can be useful for suggesting good sub-models for consideration. But nonstatistical factors may also play a role in model choice.

TABLE 9.4 Sub-model of Each Size with Highest R^2 in Prediction of Gas Mileage

Size	R^2	Adj. R^2	MS_{RES}	C_p	Predictors Included
1	0.703	0.697	6.17	35.81	WEIGHT
2	0.816	0.808	3.90	6.72	WEIGHT, TRANS
3	0.832	0.821	3.65	4.41	WEIGHT, GEARS, HP
4	0.838	0.824	3.58	4.55	WEIGHT, TRANS, HP, GEARS
5	0.841	0.823	3.60	5.84	WEIGHT, TRANS, HP, CYLIN, DISP
6	0.844	0.822	3.62	7.00	WEIGHT, TRANS, HP, GEARS, CYLIN, DISP

Certain variables may be more difficult or costly to obtain than others, and models without these predictors may be preferred. Other variables may be so inherently logical as predictors of the dependent variable that any regression model excluding them is unpalatable. One model may be preferred over another because the signs of the regression coefficients seem to make more sense. In the final analysis, selection of a good sub-model is the researcher's prerogative.

9.8 MODEL SELECTION IN COMPUTER PROGRAMS

Each of the programs, SAS, SPSS, and MINITAB, performs forward selection, backward elimination, and stepwise regression. SAS and MINITAB also have routines for examining all possible subsets. The SAS and SPSS commands for model selection procedures are included in inputs shown in Section 9.5, but are repeated here for easy reference. For MINITAB, the transcript shown here was actually appended to the transcript presented in Section 9.5.

SAS Input

```
•
•  (data input as in Section 9.6;
•  PROC REG must precede MODEL statements shown below)
•
MODEL MPG = CYLIN WEIGHT DISP HP TRANS GEARS /
      SELECTION = FORWARD SLENTRY = 0.25;
MODEL MPG = CYLIN WEIGHT DISP HP TRANS GEARS /
      SELECTION = BACKWARD SLSTAY = 0.05;
MODEL MPG = CYLIN WEIGHT DISP HP TRANS GEARS /
      SELECTION = STEPWISE SLENTRY = 0.15 SLSTAY = 0.15;
MODEL MPG = CYLIN WEIGHT DISP HP TRANS GEARS /
     SELECTION = RSQUARE BEST = 3 ADJRSQ CP MSE;
```

Discussion of SAS Input

The SAS procedure REG performs various routines for stepwise regression by using the SELECTION option on the MODEL statement. Each MODEL statement begins with the dependent variable on the left, and all predictors available for selection listed on the right. Only one SELECTION option can be used on each MODEL statement. The first MODEL statement shown here requests FORWARD selection, with a significance level for entry into the model (SLENTRY) of 0.25. BACKWARD elimination is requested in the next model statement, with a significance level for deletion (SLSTAY) of 0.05. In the next model statement, STEP-

WISE regression is requested, using α = 0.15 for both adding and deleting variables. SAS default α levels are 0.50 for entry in forward selection, 0.10 for deletion in backward elimination, and 0.15 for both entry and deletion in stepwise regression. The final MODEL statement uses the RSQUARE method of selection to examine all possible subsets. For each model size, the three sub-models with highest R^2 are printed (BEST = 3). Adjusted R^2, C_p, and $MS_{RESIDUAL}$ are requested for each of these models.

SAS Output

```
                    EXAMPLE 9.1 - GAS MILEAGE                          5
     Forward Selection Procedure for Dependent Variable MPG

Step 1    Variable WEIGHT Entered    R-square = 0.70344545
                                     C(p) = 35.80848005

                DF    Sum of Squares     Mean Square         F    Prob>F

Regression       1      702.09483874    702.09483874    113.86    0.0001
Error           48      295.98516126      6.16635753
Total           49      998.08000000

                Parameter     Standard            Type II
Variable         Estimate        Error     Sum of Squares         F   Prob>F

INTERCEP      49.11360433   2.26104271      2909.47602818    471.83   0.0001
WEIGHT        -0.00838090   0.00078543       702.09483874    113.86   0.0001

Bounds on condition number:      1,      1
-------------------------------------------------------------------------

Step 2    Variable TRANS Entered    R-square = 0.81612294
                                    C(p) = 6.72490993

                DF    Sum of Squares     Mean Square         F    Prob>F

Regression       2      814.55598590    407.27799295    104.30    0.0001
Error           47      183.52401410      3.90476626
Total           49      998.08000000

                Parameter     Standard            Type II
Variable         Estimate        Error     Sum of Squares         F   Prob>F

INTERCEP      50.51059497    1.81798456     3014.25250028    771.94   0.0001
WEIGHT        -0.00646198    0.00072007      314.47182790     80.54   0.0001
TRANS         -3.93908840    0.73399313      112.46114717     28.80   0.0001

Bounds on condition number:      1.327284,      5.309138
-------------------------------------------------------------------------

Step 3    Variable HP Entered    R-square = 0.83172070
                                 C(p) = 4.42206343
```

```
                DF    Sum of Squares       Mean Square          F   Prob>F

Regression       3      830.12379398      276.70793133      75.79   0.0001
Error           46      167.95620602        3.65122187
Total           49      998.08000000

              Parameter        Standard             Type II
Variable       Estimate           Error      Sum of Squares          F   Prob>F

INTERCEP    49.38171549      1.84101856       2626.96470192     719.48   0.0001
WEIGHT      -0.00486761      0.00103972         80.02693759      21.92   0.0001
HP          -0.02797354      0.01354730         15.56780807       4.26   0.0446
TRANS       -3.89807015      0.71004131        110.04499543      30.14   0.0001

Bounds on condition number:     2.959447,     20.88565
------------------------------------------------------------------------------
```

```
                         EXAMPLE 9.1 - GAS MILEAGE                            6

Step 4    Variable GEARS Entered       R-square = 0.83849671
                                       C(p) = 4.55281042

                DF    Sum of Squares       Mean Square          F   Prob>F

Regression       4      836.88679889      209.22169972      58.41   0.0001
Error           45      161.19320111        3.58207114
Total           49      998.08000000

              Parameter        Standard             Type II
Variable       Estimate           Error      Sum of Squares          F   Prob>F

INTERCEP    43.33038295      4.76659837        296.00710833      82.64   0.0001
WEIGHT      -0.00537765      0.00109468         86.44550693      24.13   0.0001
HP          -0.03068542      0.01356277         18.33586438       5.12   0.0285
TRANS       -2.06483843      1.50819368          6.71416670       1.87   0.1778
GEARS        1.15073895      0.83747926          6.76300492       1.89   0.1762

Bounds on condition number:     6.108789,     67.47293
------------------------------------------------------------------------------

No other variable met the 0.2500 significance level for entry into
the model.

  Summary of Forward Selection Procedure for Dependent Variable MPG

       Variable  Number  Partial    Model
Step   Entered       In     R**2     R**2        C(p)           F    Prob>F

   1   WEIGHT         1   0.7034   0.7034     35.8085    113.8589    0.0001
   2   TRANS          2   0.1127   0.8161      6.7249     28.8010    0.0001
   3   HP             3   0.0156   0.8317      4.4221      4.2637    0.0446
   4   GEARS          4   0.0068   0.8385      4.5528      1.8880    0.1762
```

EXAMPLE 9.1 - GAS MILEAGE 7

Backward Elimination Procedure for Dependent Variable MPG

Step 0 All Variables Entered R-square = 0.84412563
C(p) = 7.00000000

	DF	Sum of Squares	Mean Square	F	Prob>F
Regression	6	842.50490648	140.41748441	38.81	0.0001
Error	43	155.57509352	3.61802543		
Total	49	998.08000000			

Variable	Parameter Estimate	Standard Error	Type II Sum of Squares	F	Prob>F
INTERCEP	43.36121717	4.90649066	282.57425528	78.10	0.0001
CYLIN	0.76110234	0.61378047	5.56328863	1.54	0.2217
WEIGHT	-0.00488471	0.00142435	42.55128935	11.76	0.0013
DISP	-0.01762116	0.01647747	4.13770184	1.14	0.2909
HP	-0.03134912	0.01452760	16.84746670	4.66	0.0366
TRANS	-2.61442190	1.58119797	9.89123375	2.73	0.1055
GEARS	0.81175754	0.88522342	3.04241525	0.84	0.3643

Bounds on condition number: 12.27041, 246.824

Step 1 Variable GEARS Removed R-square = 0.84107736
C(p) = 5.84090488

	DF	Sum of Squares	Mean Square	F	Prob>F
Regression	5	839.46249124	167.89249825	46.57	0.0001
Error	44	158.61750876	3.60494338		
Total	49	998.08000000			

Variable	Parameter Estimate	Standard Error	Type II Sum of Squares	F	Prob>F
INTERCEP	47.18904548	2.57394648	1211.66289183	336.11	0.0001
CYLIN	0.93103458	0.58407887	9.15982318	2.54	0.1181
WEIGHT	-0.00441867	0.00132819	39.89865039	11.07	0.0018
DISP	-0.02191065	0.01577097	6.95812261	1.93	0.1717
HP	-0.02958981	0.01437430	15.27598580	4.24	0.0455
TRANS	-3.91105167	0.70642986	110.49622958	30.65	0.0001

Bounds on condition number: 11.28153, 140.6231

```
                         EXAMPLE 9.1 - GAS MILEAGE                          8

Step 2   Variable DISP Removed     R-square = 0.83410585
                                   C(p) = 5.76408722

                DF    Sum of Squares    Mean Square          F     Prob>F

Regression       4      832.50436862   208.12609216      56.56     0.0001
Error           45      165.57563138     3.67945848
Total           49      998.08000000

             Parameter       Standard          Type II
Variable      Estimate          Error   Sum of Squares         F    Prob>F

INTERCEP   49.65949185     1.88011174    2566.97195906    697.65    0.0001
CYLIN       0.29403651     0.36555465       2.38057465      0.65    0.4254
WEIGHT     -0.00530504     0.00117691      74.76082443     20.32    0.0001
HP         -0.03185108     0.01442871      17.92988258      4.87    0.0324
TRANS      -3.87698404     0.71326346     108.71062085     29.55    0.0001

Bounds on condition number:    3.762852,    44.03408
------------------------------------------------------------------------------

Step 3   Variable CYLIN Removed    R-square = 0.83172070
                                   C(p) = 4.42206343

                DF   Sum of Squares     Mean Square          F     Prob>F

Regression       3     830.12379398    276.70793133      75.79     0.0001
Error           46     167.95620602      3.65122187
Total           49     998.08000000

             Parameter       Standard          Type II
Variable      Estimate          Error   Sum of Squares         F    Prob>F

INTERCEP   49.38171549     1.84101856    2626.96470192    719.48    0.0001
WEIGHT     -0.00486761     0.00103972      80.02693759     21.92    0.0001
HP         -0.02797354     0.01354730      15.56780807      4.26    0.0446
TRANS      -3.89807015     0.71004131     110.04499543     30.14    0.0001

Bounds on condition number:    2.959447,    20.88565
------------------------------------------------------------------------------

All variables left in the model are significant at the 0.0500 level.

          Summary of Backward Elimination Procedure for
                    Dependent Variable MPG

       Variable   Number    Partial     Model
Step   Removed        In       R**2      R**2      C(p)         F    Prob>F

 1     GEARS           5     0.0030    0.8411    5.8409    0.8409    0.3643
 2     DISP            4     0.0070    0.8341    5.7641    1.9302    0.1717
 3     CYLIN           3     0.0024    0.8317    4.4221    0.6470    0.4254
```

EXAMPLE 9.1 - GAS MILEAGE 9

Stepwise Procedure for Dependent Variable MPG

Step 1 Variable WEIGHT Entered R-square = 0.70344545
C(p) = 35.80848005

	DF	Sum of Squares	Mean Square	F	Prob>F
Regression	1	702.09483874	702.09483874	113.86	0.0001
Error	48	295.98516126	6.16635753		
Total	49	998.08000000			

Variable	Parameter Estimate	Standard Error	Type II Sum of Squares	F	Prob>F
INTERCEP	49.11360433	2.26104271	2909.47602818	471.83	0.0001
WEIGHT	-0.00838090	0.00078543	702.09483874	113.86	0.0001

Bounds on condition number: 1, 1

Step 2 Variable TRANS Entered R-square = 0.81612294
C(p) = 6.72490993

	DF	Sum of Squares	Mean Square	F	Prob>F
Regression	2	814.55598590	407.27799295	104.30	0.0001
Error	47	183.52401410	3.90476626		
Total	49	998.08000000			

Variable	Parameter Estimate	Standard Error	Type II Sum of Squares	F	Prob>F
INTERCEP	50.51059497	1.81798456	3014.25250028	771.94	0.0001
WEIGHT	-0.00646198	0.00072007	314.47182790	80.54	0.0001
TRANS	-3.93908840	0.73399313	112.46114717	28.80	0.0001

Bounds on condition number: 1.327284, 5.309138

Step 3 Variable HP Entered R-square = 0.83172070
C(p) = 4.42206343

	DF	Sum of Squares	Mean Square	F	Prob>F
Regression	3	830.12379398	276.70793133	75.79	0.0001
Error	46	167.95620602	3.65122187		
Total	49	998.08000000			

Variable	Parameter Estimate	Standard Error	Type II Sum of Squares	F	Prob>F
INTERCEP	49.38171549	1.84101856	2626.96470192	719.48	0.0001
WEIGHT	-0.00486761	0.00103972	80.02693759	21.92	0.0001
HP	-0.02797354	0.01354730	15.56780807	4.26	0.0446
TRANS	-3.89807015	0.71004131	110.04499543	30.14	0.0001

Bounds on condition number: 2.959447, 20.88565

All variables left in the model are significant at the 0.1500 level.

No other variable met the 0.1500 significance level for entry into the model.

EXAMPLE 9.1 - GAS MILEAGE 10

Summary of Stepwise Procedure for Dependent Variable MPG

Step	Variable Entered	Number In	Partial R**2	Model R**2	C(p)	F	Prob>F
1	WEIGHT	1	0.7034	0.7034	35.8085	113.8589	0.0001
2	TRANS	2	0.1127	0.8161	6.7249	28.8010	0.0001
3	HP	3	0.0156	0.8317	4.4221	4.2637	0.0446

EXAMPLE 9.1 - GAS MILEAGE 11

N = 50 Regression Models for Dependent Variable: MPG

In	R-square	Adjusted R-square	C(p)	MSE	Variables in Model
1	0.7034455	0.6972672	35.8085	6.16636	WEIGHT
1	0.5558902	0.5466379	76.5136	9.23452	HP
1	0.5513155	0.5419679	77.7755	9.32965	DISP
2	0.8161229	0.8082984	6.7249	3.90477	WEIGHT TRANS
2	0.8096711	0.8015720	8.5047	4.04178	WEIGHT GEARS
2	0.7515398	0.7409670	24.5410	5.27624	HP TRANS
3	0.8317696	0.8207981	4.4086	3.65016	WEIGHT HP GEARS
3	0.8317207	0.8207460	4.4221	3.65122	WEIGHT HP TRANS
3	0.8201256	0.8083946	7.6207	3.90281	WEIGHT TRANS GEARS
4	0.8384967	0.8241409	4.5528	3.58207	WEIGHT HP TRANS GEARS
4	0.8341059	0.8193597	5.7641	3.67946	CYLIN WEIGHT HP TRANS
4	0.8326890	0.8178169	6.1550	3.71088	CYLIN WEIGHT HP GEARS

```
5  0.8410774  0.8230180   5.8409  3.60494  CYLIN WEIGHT DISP
                                           HP TRANS
5  0.8399800  0.8217959   6.1436  3.62984  CYLIN WEIGHT HP
                                           TRANS GEARS
5  0.8385516  0.8202052   6.5377  3.66224  WEIGHT DISP HP
                                           TRANS GEARS
---------------------------------------------------------------
6  0.8441256  0.8223757   7.0000  3.61803  CYLIN WEIGHT DISP HP
                                           TRANS GEARS
```

Discussion of SAS Output

Output pages 1 through 4 were discussed in Section 9.5. The results of forward selection are reported on pages 5 and 6. After each step, SAS shows the analysis of variance table for predictors included so far, along with estimated regression coefficients and partial F tests. A Type II Sum of Squares is a partial sum of squares, and reflects the additional predictive value of each variable above and beyond all other predictors currently included in the model. After GEARS is entered in step 4, we are told that no other variables meet the 0.25 significance level criterion for entry into the model. In the summary table, Partial R**2 is the increase in R^2 for each step, and the F test printed for each variable is its partial F when it was entered into the model.

Pages 7 and 8 chronicle the backward elimination procedure. SAS labels the full model as step 0, since no variables have been deleted from the model yet. As in forward selection, SAS shows the analysis of variance table after each step, along with estimated regression coefficients and partial F tests. After CYLIN is removed in the third step, all variables remaining in the model are significant at the 0.05 level. The F test printed for each variable in the summary table is its partial F when it was deleted from the model.

Results of stepwise regression, reported on pages 9 and 10, parallel those of forward selection, since no variables are deleted in this example. The terminal model differs from that of forward selection, due to use of a different α level for entry. After HP is entered in the third step, we are told that all variables remaining in the model are significant at the 0.15 level, and no other variables meet the 0.15 significance level criterion for entry into the model.

Page 11 of the output reports the results of the RSQUARE method of selection. For each model size, the best three models in terms of R^2 are identified, along with their adjusted R^2, C_p, and $MS_{RESIDUAL}$ values. (Only one model of size 6 exists.)

SPSS Input

```
•
• (data input as in Section 9.6)
•
```

```
REGRESSION VARIABLES = MPG TO GEARS
     / CRITERIA = PIN (0.25) POUT (0.99)
     / DEPENDENT = MPG
     / METHOD = FORWARD CYLIN TO GEARS
     / CRITERIA = PIN (0.01) POUT (0.05)
     / DEPENDENT = MPG
     / METHOD = BACKWARD CYLIN TO GEARS
     / CRITERIA = PIN (0.15) POUT (0.15001)
     / DEPENDENT = MPG
     / METHOD = STEPWISE CYLIN TO GEARS
FINISH
```

Discussion of SPSS Input

Stepwise regression routines are performed by the SPSS procedure REGRESSION. CRITERIA subcommands are used to set significance levels for entry and deletion. To avoid infinite loops, the significance level for entry (PIN) must be less than that for deletion (POUT). Default α levels are 0.05 for entry and 0.10 for deletion.

Each METHOD subcommand specifies the desired stepwise routine and lists the variables available for inclusion in the model. Before performing forward selection, the significance level for entry is set to 0.25, and for deletion to 0.99. Since no variables can be deleted using forward selection, the POUT value will not be used. Before requesting backward elimination, the significance level for entry is set to 0.01, and for deletion to 0.05. PIN will not be used during backward elimination. Although we would like to use $\alpha = 0.15$ for both entry and deletion in stepwise regression, a slightly higher value is specified for deletion to satisfy the requirement that PIN be less than POUT.

SPSS Output

```
                 EXAMPLE 9.1 - GAS MILEAGE                Page 6

    * * * * M U L T I P L E   R E G R E S S I O N * * * *
Listwise Deletion of Missing Data
Equation Number 1   Dependent Variable..   MPG

Block Number 1.   Method:   Forward    Criterion  PIN  .2500
   CYLIN    WEIGHT    DISP    HP    TRANS    GEARS

Variable(s) Entered on Step Number
   1..   WEIGHT
```

```
Multiple R              .83872
R Square                .70345
Adjusted R Square       .69727
Standard Error         2.48322

Analysis of Variance
                   DF        Sum of Squares        Mean Square
Regression          1             702.09484          702.09484
Residual           48             295.98516            6.16636

F =    113.85892         Signif F =    .0000

- - - - - - - - - -  Variables in the Equation - - - - - - - -
Variable                B           SE B       Beta          T   Sig T

WEIGHT          -.008381    7.8543E-04  -.838717    -10.670   .0000
(Constant)     49.113604      2.261043                21.722   .0000

- - - - - - - - -  Variables not in the Equation - - - - - - -
Variable    Beta In     Partial    Min Toler          T      Sig T

CYLIN        .020563     .023510     .387678       .161      .8726
DISP        -.081073    -.075935     .260158      -.522      .6041
HP          -.219422    -.246495     .374249    -1.744      .0878
TRANS       -.386724    -.616405     .753418    -5.367      .0000
GEARS        .328594     .598498     .983808     5.122      .0000
```

```
                  EXAMPLE 9.1 - GAS MILEAGE                 Page 7
    * * * * M U L T I P L E   R E G R E S S I O N * * * *
Equation Number 1       Dependent Variable..    MPG

Variable(s) Entered on Step Number
   2..    TRANS

Multiple R              .90340
R Square                .81612
Adjusted R Square       .80830
Standard Error         1.97605

Analysis of Variance
                   DF        Sum of Squares        Mean Square
Regression          2             814.55599          407.27799
Residual           47             183.52401            3.90477
```

```
F =    104.30278    Signif F =    .0000

- - - - - - - - -Variables in the Equation - - - - - - - - -
Variable               B          SE B        Beta        T   Sig T

WEIGHT          -.006462  7.2007E-04   -.646681   -8.974   .0000
TRANS          -3.939088     .733993   -.386724   -5.367   .0000
(Constant)     50.510595    1.817985               27.784   .0000

- - - - - - - -Variables not in the Equation - - - - - - - -
Variable     Beta In     Partial     Min Toler        T   Sig T
CYLIN        .006917     .010040       .339628     .068   .9460
DISP        -.106365    -.126426       .235319    -.864   .3919
HP          -.204231    -.291251       .337901   -2.065   .0446
GEARS        .135461     .147540       .167050    1.012   .3170
```

```
                 EXAMPLE 9.1 - GAS MILEAGE            Page 8

    * * * * M U L T I P L E   R E G R E S S I O N * * * *
Equation Number 1   Dependent Variable..   MPG

Variable(s) Entered on Step Number
   3..    HP

Multiple R             .91199
R Square               .83172
Adjusted R Square      .82075
Standard Error        1.91082

Analysis of Variance
                         DF      Sum of Squares        Mean Square
Regression                3           830.12379          276.70793
Residual                 46           167.95621            3.65122

F =    75.78502    Signif F =    .0000

- - - - - - - - - - Variables in the Equation- - - - - - - - -
Variable               B         SE B        Beta        T   Sig T

WEIGHT          -.004868     .001040   -.487125   -4.682   .0000
TRANS          -3.898070     .710041   -.382697   -5.490   .0000
HP              -.027974     .013547   -.204231   -2.065   .0446
(Constant)     49.381715    1.841019               26.823   .0000
```

```
- - - - - - - - - Variables not in the Equation - - - - - - -
Variable     Beta In     Partial     Min Toler        T     Sig T

CYLIN        .083246     .119054      .265756      .804     .4254
DISP        -.027856    -.032634      .207736     -.219     .8276
GEARS        .178145     .200665      .163699     1.374     .1762
```

```
                 EXAMPLE 9.1 - GAS MILEAGE                   Page 9

    * * * * M U L T I P L E   R E G R E S S I O N * * * *
Equation Number 1      Dependent Variable..   MPG

Variable(s) Entered on Step Number
   4..    GEARS

Multiple R               .91569
R Square                 .83850
Adjusted R Square        .82414
Standard Error          1.89264

Analysis of Variance
                     DF        Sum of Squares          Mean Square
Regression            4             836.88680            209.22170
Residual             45             161.19320              3.58207

F =    58.40802         Signif F =    .0000

- - - - - - - - - - Variables in the Equation - - - - - - - -
Variable                  B         SE B          Beta         T    Sig T

WEIGHT            -.005378      .001095     -.538167    -4.913    .0000
TRANS            -2.064838     1.508194     -.202717    -1.369    .1778
HP                -.030685      .013563     -.224030    -2.262    .0285
GEARS             1.150739      .837479      .178145     1.374    .1762
(Constant)       43.330383     4.766598                  9.090    .0000

- - - - - - - - -Variables not in the Equation - - - - - - -
Variable     Beta In     Partial     Min Toler        T     Sig T

CYLIN        .066203     .095833      .162138      .639     .5264
DISP        -.015462    -.018441      .162436     -.122     .9032

End Block Number   1   PIN =   .250 Limits reached.
```

```
                    EXAMPLE 9.1 - GAS MILEAGE          Page 10

   * * * * M U L T I P L E   R E G R E S S I O N * * * *
Equation Number 2     Dependent Variable..   MPG

Block Number  1.   Method:   Enter

Variable(s) Entered on Step Number
   1..    GEARS
   2..    CYLIN
   3..    HP
   4..    WEIGHT
   5..    TRANS
   6..    DISP

Multiple R              .91876
R Square                .84413
Adjusted R Square       .82238
Standard Error         1.90211

Analysis of Variance
                    DF       Sum of Squares         Mean Square
Regression           6            842.50491           140.41748
Residual            43            155.57509             3.61803

F =    38.81053        Signif F =   .0000

- - - - - - - - - Variables in the Equation - - - - - - - -
Variable               B         SE B        Beta        T   Sig T


GEARS            .811758      .885223     .125668     .917   .3643
CYLIN            .761102      .613780     .215479    1.240   .2217
HP              -.031349      .014528    -.228875   -2.158   .0366
WEIGHT          -.004885      .001424    -.488836   -3.429   .0013
TRANS          -2.614422     1.581198    -.256673   -1.653   .1055
DISP            -.017621      .016477    -.225541   -1.069   .2909
(Constant)     43.361217     4.906491                8.838   .0000

End Block Number   1   All requested variables entered.
```

```
                    EXAMPLE 9.1 - GAS MILEAGE             Page 11

      * * * * M U L T I P L E   R E G R E S S I O N * * * *
Equation Number 2   Dependent Variable..   MPG

Block Number  2.  Method:  Backward   Criterion  POUT  .0500
   CYLIN     WEIGHT      DISP     HP     TRANS     GEARS

Variable(s) Removed on Step Number
   7..   GEARS

Multiple R              .91710
R Square                .84108
Adjusted R Square       .82302
Standard Error         1.89867

Analysis of Variance
                 DF        Sum of Squares          Mean Square
Regression        5             839.46249            167.89250
Residual         44             158.61751              3.60494

F =    46.57285       Signif F =    .0000

- - - - - - - - - Variables in the Equation - - - - - - - -
Variable               B         SE B         Beta          T  Sig T

CYLIN            .931035     .584079      .263590      1.594  .1181
HP              -.029590     .014374     -.216031     -2.059  .0455
WEIGHT          -.004419     .001328     -.442197     -3.327  .0018
TRANS          -3.911052     .706430     -.383971     -5.536  .0000
DISP            -.021911     .015771     -.280445     -1.389  .1717
(Constant)     47.189045    2.573946                  18.333  .0000

- - - - - - - - Variables not in the Equation - - - - - - -
Variable      Beta In     Partial    Min Toler        T     Sig T

GEARS         .125668    .138495      .081497      .917     .3643
```

```
                     EXAMPLE 9.1 - GAS MILEAGE            Page 12

      * * * * M U L T I P L E   R E G R E S S I O N * * * *
Equation Number 2     Dependent Variable..   MPG

Variable(s) Removed on Step Number
   8..   DISP
```

```
Multiple R              .91329
R Square                .83411
Adjusted R Square       .81936
Standard Error         1.91819

Analysis of Variance
                  DF         Sum of Squares          Mean Square
Regression         4             832.50437             208.12609
Residual          45             165.57563               3.67946

F =    56.56433      Signif F =    .0000

- - - - - - - - - Variables in the Equation - - - - - - - -
Variable                B          SE B          Beta          T   Sig T

CYLIN            .294037      .365555      .083246        .804   .4254
HP              -.031851      .014429     -.232540      -2.207   .0324
WEIGHT          -.005305      .001177     -.530900      -4.508   .0000
TRANS          -3.876984      .713263     -.380626      -5.436   .0000
(Constant)     49.659492     1.880112                   26.413   .0000

- - - - - - - - Variables not in the Equation - - - - - - -
Variable       Beta In     Partial    Min Toler          T     Sig T

DISP          -.280445    -.204997      .088640     -1.389     .1717
GEARS          .167272     .188172      .162138      1.271     .2104
```

```
                   EXAMPLE 9.1 - GAS MILEAGE               Page 13

   * * * * M U L T I P L E   R E G R E S S I O N * * * *
Equation Number 2     Dependent Variable..   MPG

Variable(s) Removed on Step Number
   9..    CYLIN

Multiple R                .91199
R Square                  .83172
Adjusted R Square         .82075
Standard Error           1.91082
```

```
Analysis of Variance
                    DF        Sum of Squares        Mean Square
Regression           3             830.12379          276.70793
Residual            46             167.95621            3.65122

F =    75.78502       Signif F =   .0000

- - - - - - - - - - Variables in the Equation - - - - - - - -
Variable                B          SE B         Beta         T   Sig T

HP              -.027974      .013547    -.204231   -2.065   .0446
WEIGHT          -.004868      .001040    -.487125   -4.682   .0000
TRANS          -3.898070      .710041    -.382697   -5.490   .0000
(Constant)     49.381715     1.841019                26.823   .0000

- - - - - - - - - Variables not in the Equation - - - - - - -
Variable     Beta In     Partial     Min Toler        T    Sig T

CYLIN        .083246     .119054      .265756      .804    .4254
DISP        -.027856    -.032634      .207736     -.219    .8276
GEARS        .178145     .200665      .163699     1.374    .1762

End Block Number 2   POUT =     .050 Limits reached.
```

```
                 EXAMPLE 9.1 - GAS MILEAGE               Page 14

   * * * * M U L T I P L E   R E G R E S S I O N * * * *
Equation Number 3   Dependent Variable..   MPG

Block Number   1.  Method:   Stepwise   Criteria   PIN.1500
   POUT.1500   CYLIN   WEIGHT   DISP   HP   TRANS   GEARS

Variable(s) Entered on Step Number
   1..   WEIGHT

Multiple R                .83872
R Square                  .70345
Adjusted R Square         .69727
Standard Error           2.48322
```

```
Analysis of Variance
                    DF        Sum of Squares        Mean Square
Regression           1             702.09484          702.09484
Residual            48             295.98516            6.16636

F =   113.85892       Signif F =    .0000

- - - - - - - - - Variables in the Equation - - - - - - - -
Variable               B          SE B       Beta          T  Sig T

WEIGHT          -.008381  7.8543E-04  -.838717    -10.670  .0000
(Constant)     49.113604    2.261043                21.722  .0000

- - - - - - - - Variables not in the Equation - - - - - - -
Variable      Beta In     Partial     Min Toler         T   Sig T

CYLIN         .020563     .023510       .387678      .161   .8726
DISP         -.081073    -.075935       .260158     -.522   .6041
HP           -.219422    -.246495       .374249    -1.744   .0878
TRANS        -.386724    -.616405       .753418    -5.367   .0000
GEARS         .328594     .598498       .983808     5.122   .0000
```

```
                 EXAMPLE 9.1 - GAS MILEAGE              Page 15

   * * * * M U L T I P L E   R E G R E S S I O N * * * *
Equation Number 3     Dependent Variable..    MPG

Variable(s) Entered on Step Number
   2..    TRANS

Multiple R              .90340
R Square                .81612
Adjusted R Square       .80830
Standard Error         1.97605

Analysis of Variance
                    DF        Sum of Squares        Mean Square
Regression           2             814.55599          407.27799
Residual            47             183.52401            3.90477

F =    104.30278   Signif F =    .0000
```

```
- - - - - - - - - - Variables in the Equation - - - - - - - -
Variable              B          SE B        Beta          T   Sig T

WEIGHT        -.006462  7.2007E-04   -.646681     -8.974   .0000
TRANS        -3.939088     .733993   -.386724     -5.367   .0000
(Constant)   50.510595    1.817985                27.784   .0000

- - - - - - - - Variables not in the Equation - - - - - - -
Variable     Beta In     Partial    Min Toler          T   Sig T

CYLIN       .006917     .010040     .339628        .068   .9460
DISP       -.106365    -.126426     .235319       -.864   .3919
HP         -.204231    -.291251     .337901     -2.065   .0446
GEARS       .135461     .147540     .167050      1.012   .3170
```

```
                EXAMPLE 9.1 - GAS MILEAGE              Page 16

    * * * * M U L T I P L E   R E G R E S S I O N * * * *
Equation Number 3      Dependent Variable..   MPG
Variable(s) Entered on Step Number
   3..   HP

Multiple R                  .91199
R Square                    .83172
Adjusted R Square           .82075
Standard Error             1.91082

Analysis of Variance
                       DF      Sum of Squares       Mean Square
Regression              3           830.12379         276.70793
Residual               46           167.95621           3.65122

F =     75.78502       Signif F =    .0000

- - - - - - - - - - Variables in the Equation - - - - - - - -
Variable              B          SE B        Beta          T   Sig T

WEIGHT        -.004868     .001040   -.487125     -4.682   .0000
TRANS        -3.898070     .710041   -.382697     -5.490   .0000
HP            -.027974     .013547   -.204231     -2.065   .0446
(Constant)   49.381715    1.841019                26.823   .0000

- - - - - - - - Variables not in the Equation - - - - - - -
Variable     Beta In     Partial    Min Toler          T   Sig T

CYLIN       .083246     .119054     .265756       .804   .4254
DISP       -.027856    -.032634     .207736      -.219   .8276
```

```
GEARS          .178145    .200665      .163699    1.374     .1762
End Block Number   1   PIN =    .150 Limits reached.
```

Discussion of SPSS Output

Output pages 1 through 5 were discussed in section 9.5. The results of forward selection are reported in pages 6 through 9. After each step, SPSS shows the analysis of variance table for predictors included so far, along with information relevant to each variable in the equation and each variable not in the equation. For each predictor in the model, the estimated regression coefficient (B), its standard error (SE B), the standardized coefficient (Beta), and partial t test for deletion are printed. For each variable not in the model, SPSS shows the standardized regression coefficient if the variable were to enter in the next step (Beta In), its partial correlation with the dependent variable controlling for the predictors currently in the model (Partial), its minimum tolerance (Min Toler), and the partial t test for entry. Minimum tolerance for a variable is the smallest tolerance that any predictor already in the equation would have if the variable were entered in the next step. The variable with the smallest p-value is added to the model. After GEARS is entered in step 4, the procedure ends, since no other variables meet the 0.25 significance level criterion for entry into the model.

Pages 10 through 13 chronicle the backward elimination procedure, beginning with the full model on page 10. As in forward selection, SPSS shows the analysis of variance table after each step, along with information for each variable in the equation and each variable not in the equation. The step numbers appear strange, since the first six steps are allocated to entering all predictors to form the full model. After CYLIN is removed in step 9, the algorithm stops because all variables remaining in the model are significant at the 0.05 level.

Results of stepwise regression, reported on pages 14 through 16, parallel those of forward selection, since no variables are deleted in this example. The terminal model differs from that of forward selection, due to use of a different α level for entry. After HP is entered in the third step, we are told that no other variables meet the 0.15 significance level criterion for entry into the model. It is also true that all variables remaining in the model are significant at the 0.15001 level.

MINITAB Transcript

```
•
• (data input as in Section 9.6)
•
MTB > STEPWISE 'MPG' C3-C8;
SUBC > FREMOVE = 0.0;
SUBC > FENTER = 1.36.
```

```
STEPWISE REGRESSION OF MPG ON 6 PREDICTORS, WITH N = 50

    STEP           1          2          3          4
CONSTANT       49.11      50.51      49.38      43.33

WEIGHT      -0.00838   -0.00646   -0.00487   -0.00538
T-RATIO       -10.67      -8.97      -4.68      -4.91

TRANS                     -3.94      -3.90      -2.06
T-RATIO                   -5.37      -5.49      -1.37

HP                                  -0.028     -0.031
T-RATIO                              -2.06      -2.26

GEARS                                            1.15
T-RATIO                                          1.37

S               2.48       1.98       1.91       1.89
R-SQ           70.34      81.61      83.17      83.85
  MORE? (YES, NO, SUBCOMMAND, OR HELP)
SUBC > NO

MTB > STEPWISE 'MPG' C3-C8;
SUBC > ENTER C3-C8;
SUBC > FENTER = 100000.0;
SUBC > FREMOVE = 4.07.

STEPWISE REGRESSION OF MPG ON 6 PREDICTORS, WITH N = 50

    STEP           1          2          3          4
CONSTANT       43.36      47.19      49.66      49.38

CYLIN           0.76       0.93       0.29
T-RATIO         1.24       1.59       0.80

WEIGHT       -0.0049    -0.0044    -0.0053    -0.0049
T-RATIO        -3.43      -3.33      -4.51      -4.68

DISP          -0.018     -0.022
T-RATIO        -1.07      -1.39

HP            -0.031     -0.030     -0.032     -0.028
T-RATIO        -2.16      -2.06      -2.21      -2.06
```

```
TRANS          -2.61       -3.91       -3.88       -3.90
T-RATIO        -1.65       -5.54       -5.44       -5.49

GEARS           0.81
T-RATIO         0.92

S               1.90        1.90        1.92        1.91
R-SQ           84.41       84.11       83.41       83.17
  MORE? (YES, NO, SUBCOMMAND, OR HELP)
SUBC > NO

MTB > STEPWISE 'MPG' C3-C8;
SUBC > FENTER = 2.14;
SUBC > FREMOVE = 2.14.

STEPWISE REGRESSION OF MPG ON 6 PREDICTORS, WITH N = 50

STEP               1           2           3
CONSTANT       49.11       50.51       49.38

WEIGHT      -0.00838    -0.00646    -0.00487
T-RATIO       -10.67       -8.97       -4.68

TRANS                      -3.94       -3.90
T-RATIO                    -5.37       -5.49

HP                                    -0.028
T-RATIO                                -2.06

S               2.48        1.98        1.91
R-SQ           70.34       81.61       83.17
  MORE? (YES, NO, SUBCOMMAND, OR HELP)
SUBC > NO

MTB > BREG 'MPG' C3-C8;
SUBC > BEST 3.

Best Subsets Regression of MPG
```

Vars	R-sq	Adj. R-sq	C-p	s	CYLIN	WEIGHT	DISP	HP	TRANS	GEARS
1	70.3	69.7	35.8	2.4832		X				
1	55.6	54.7	76.5	3.0388				X		
1	55.1	54.2	77.8	3.0544			X			
2	81.6	80.8	6.7	1.9760		X			X	
2	81.0	80.2	8.5	2.0104		X				X
2	75.2	74.1	24.5	2.2970				X	X	
3	83.2	82.1	4.4	1.9105		X		X		X
3	83.2	82.1	4.4	1.9108		X		X	X	
3	82.0	80.8	7.6	1.9756		X			X	X
4	83.8	82.4	4.6	1.8926		X		X	X	X
4	83.4	81.9	5.8	1.9182	X	X		X	X	
4	83.3	81.8	6.2	1.9264	X	X		X		X
5	84.1	82.3	5.8	1.8987	X	X	X	X	X	
5	84.0	82.2	6.1	1.9052	X	X		X	X	X
5	83.9	82.0	6.5	1.9137		X	X	X	X	X
6	84.4	82.2	7.0	1.9021	X	X	X	X	X	X

```
MTB > STOP
```

Discussion of MINITAB Transcript

In MINITAB, the STEPWISE command is used to perform all stepwise regression routines. Instead of α levels, cutoff values for partial F statistics are used as decision criteria. Default critical values for partial F ratios are 4.0 for both adding and deleting variables. By carefully choosing critical F values for entry (FENTER) and deletion (FREMOVE), we can force the program to perform forward selection, backward elimination, or stepwise regression.

By setting a critical F value of 0.0 for removal, we guarantee that the program performs forward selection. Once a variable is in the equation, it will not be deleted unless its partial F statistic falls below this level. Since partial F values are always positive, no variables will ever be dropped out, and the routine performs forward selection. To roughly parallel using $\alpha = 0.25$ for forward selection in Example 9.1, the critical F value of 1.36 is chosen for entry. In the first step of forward selection for this example, partial F ratios for entry have 1 degree of freedom in the numerator and 48 in the denominator. The value 1.36 cuts off the upper 25% area for this F distribution. However, in subsequent steps the degrees of freedom change, and the upper 25% critical value is slightly different from 1.36. Thus, the specification of 1.36 for FENTER corresponds exactly to using $\alpha = 0.25$ in the first step, but only approximately for subsequent steps.

Each step in forward selection is summarized in a column of the resulting display. Strangely, despite requiring input of critical values for *partial F tests*, MINITAB reports results in terms of *partial t tests*. After each variable is entered, MINITAB prints the regression coefficient and partial t test for each variable currently in the model, and the square root of mean square residual (S) and R^2 for the model at this stage.

When no additional variables qualify for entry, we are asked if we wish to submit more subcommands. At this point, we could force inclusion of other predictors by using a FORCE subcommand, remove variables currently in the model with a REMOVE subcommand, ask for alternatives to the last step with a BEST subcommand, or change the FENTER and FREMOVE criteria. Responding NO terminates the algorithm.

Backward elimination is performed by starting with all predictors in the model and ensuring that, once deleted, no variable can reenter the equation. The ENTER subcommand accomplishes the first of these objectives, and setting FENTER to a very high value accomplishes the second. To roughly parallel using $\alpha = 0.05$ for backward elimination in Example 9.1, the critical F value of 4.07 is chosen for removal. In the first step of backward elimination for this example, partial F ratios for deletion have 1 degree of freedom in the numerator and 43 in the denominator. The value 4.07 cuts off the upper 5% area for this F distribution. However, as with forward selection, subsequent steps have slightly different degrees of freedom, and the use of 4.07 for FREMOVE corresponds only approximately to using $\alpha = 0.05$ in subsequent steps.

Again, each step is summarized in a column of the output display, including regression coefficients and partial t tests for each predictor currently in the model, and the square root of mean square residual (S) and R^2 for the model at this stage. After deletion of CYLIN in the fourth step, we are asked if we wish to continue by submitting additional subcommands.

Stepwise regression using α of approximately 0.15 for both entry and deletion is performed by setting both FENTER and FREMOVE to 2.14. In the first step of stepwise regression for this example, partial F ratios have 1 degree of freedom in the numerator and 48 in the denominator. The value 2.14 cuts off the upper 15% area for this F distribution. As with forward selection and backward elimination, this criterion for partial F tests corresponds only approximately to using $\alpha = 0.15$ in subsequent steps. Results parallel those of forward selection, since no variables are deleted in this example. The terminal model differs from that of forward selection, due to use of a different α level for entry.

The MINITAB command BREG examines all possible subsets. For each model size, the three sub-models with highest R^2 are requested by the subcommand BEST 3. In the output, predictors included in each model are indicated with an X in the columns on the right. For each model, the values of R^2 and adjusted R^2 (each expressed as a percentage), and C_p and S are printed. (Only one model of size 6 exists.)

9.9 CROSS VALIDATION

Model selection techniques pose substantial hazards for prediction of additional observations. Methods such as stepwise regression routines are designed to find a subset of predictors that yields excellent results in the original sample. But a prediction equation derived from such an approach may produce totally inadequate predictions when applied to new data.

To see the reason for this difficulty, recall from Section 8.8 the logic behind adjusted R^2. Least squares estimates of regression coefficients are selected to minimize residuals in the original sample. For the sample, R^2 is the proportion of variability accounted for by the fitted equation. But coefficients from the sample are probably not optimal for the whole population. Thus, the proportion of variation likely to be explained by our regression equation in the population is overestimated by R^2. The more coefficients we have estimated from the sample, the greater this problem is. Adjusted R^2 estimates the proportion of variability that our fitted model can predict in the population.

This difficulty is compounded by using model selection techniques. In addition to estimating the coefficients from sample data, we are also using the sample to determine *which variables to include* in the model. It is entirely possible that we will tailor the model too precisely to the sample, and the resulting equation will yield inaccurate predictions for additional observations.

To guard against this possibility, we strongly recommend that **cross validation** be used anytime a model selection strategy is employed. In essence, we fit the model using one sample, and check it on another. To obtain two separate samples, we split the available data in half before beginning analysis. Typically, a random division into halves is appropriate. In Example 9.1, for instance, we would not want to fit a model using all the cars with automatic transmissions, and then attempt to check it on cars with manual transmissions. In some situations, however, a nonrandom division may be in order. When the variables are time dependent and predictions constitute extrapolation into the future, it may be reasonable to fit a model using early observations and check it on more recent data. This strategy might simulate predicting future observations.

After splitting the data in half, one half is used for model selection, possibly by exploration in a variety of stepwise methods. Model choice may also be influenced by judgements based on cost or effort of obtaining variables, logical value of predictors, or signs of regression coefficients. Through some combination of these techniques, the researcher determines a sub-model believed to be most appropriate for its intended use. For this final model, the estimated regression equation is recorded.

Now the other half of the data is employed to check the accuracy of the predictions. For each observation, values of the independent variables are inserted in the equation and a predicted score, $\hat{Y}$, is obtained. These values are compared to the actual Y values for the corresponding observations. Typically, a statistic such as $\Sigma(Y-\hat{Y})^2$ is computed. If the proposed model is adequate, this sum should not be appreciably larger than the sum of squared residuals for the original fitting sample. Unsuccessful cross validation is usually a symptom of tailoring the model too precisely to the fitting sample. A prediction equation containing fewer predictors should be investigated.

If the cross validation yields satisfactory results, regression coefficients for the chosen subset of predictors should be reestimated, now using all the available data. To obtain these final regression coefficients, the fitting and checking samples are combined into a single set of data. Regression is performed on this combined data file, with a model specifying inclusion of the chosen subset of predictors, not as a stepwise regression routine, but as if the subset were a full model to be evaluated. This last step should produce a final prediction equation with more stable estimates of the population regression coefficients, since they are now based on the entire original sample.

9.10 SUMMARY

Multiple linear regression is used to predict a single dependent variable from a set of independent, or predictor, variables. As in simple linear regression, values of the independent variables are presumed to be known exactly, while the dependent variable is assumed to be normally distributed with constant variance. The multiple linear regression model is enormously general in that predictors can be powers or other functions of independent variables. Categorical predictors can be included by means of dummy variables. The model to be estimated, however, must be linear in the regression coefficients.

The least squares criterion produces estimated regression coefficients by minimizing the residual sum of squares. Relative importance of predictors cannot be determined from the size of their coefficients because the coefficients are scale dependent, are heavily influenced by correlations among predictors, and may be affected by unusual observations.

Analysis of variance is used to test the null hypothesis that no prediction is possible from the full model containing all the predictors. As in simple linear regression, the total sum of squares is partitioned into components for regression and residual. A significant result implies that some prediction is possible, but it does not tell us which independent variables are contributing to this prediction.

R^2 measures the proportion of variability in the sample explained by the model, and adjusted R^2 estimates the proportion of population variability that can be predicted. Partial regression plots examine the roles of individual predictors in the overall model. These plots reveal the amount of unique information provided by each predictor, and whether outliers heavily influence individual regression coefficients.

Multicollinearity refers to correlations among predictors. Effects of high multicollinearity include overestimation of population coefficients, difficulty in interpreting regression coefficients, and inflation of variances for regression coefficients and predictions. Experiments can sometimes be designed with uncorrelated predictors by including all possible combinations of values for predictors. A large condition number signifies the presence of unacceptably high multicollinearity. Tolerances and variance inflation factors for individual predictors help determine which variables are instrumental in causing high multicollinearity.

A full model may be compared to a sub-model by use of the partial F test. This technique assesses the additional predictive value of the variables included in the full model, but not the sub-model. The test may be interpreted as a test of the significance of increase in R^2 when the additional predictors are included in the model. The test is valid for comparison of any two models, as long as one contains a subset of the predictors in the other. To maximize the practical utility of such tests, the full and sub-models should differ by only one predictor. In this case, the square root of the partial F test can be interpreted as a partial t test. When partial t tests are examined for several variables, correlations among the variables must be considered in drawing conclusions about the importance of the various predictors.

A model with high multicollinearity is overfit because it contains predictors that provide redundant information. But a model that omits important independent variables is underfit and may yield poor predictions. The objective of model selection is to compromise between these extremes and fit a model with low multicollinearity, but with prediction characteristics nearly as good as the full model. For models of the same size, maximizing R^2 is an adequate decision criteria, but for models of different sizes we should attempt to maximize adjusted R^2. A good subset has high adjusted R^2, low mean square residual, and low C_p.

Stepwise regression routines search through a sequence of models and terminate with a reasonable subset of predictors. Forward selection begins with no predictors in the model and adds variables one at a time until further additions would no longer improve prediction. Backward elimination starts with the full model and removes variables one at a time until further deletions would significantly reduce predictability. Stepwise regression is similar to forward selection, but also allows predictors to drop out of the equation if they become redundant in later steps. Each of these methods produces a single terminal model, but none is guaranteed to find the best subset of predictors.

Some programs examine all possible subsets of predictors and locate the best few models of each size, but do not attempt to find a single terminal model. Decisions regarding inclusion of variables may also be made by the researcher, based on logical or practical grounds.

Whenever model selection routines are employed, cross validation of the proposed model may reveal if the fitted model is too precisely tailored to the fitting sample. The available data are split in half, with one half used to fit the model and the other used to check it. Multiple regression and model selection algorithms are demonstrated on SAS, SPSS, and MINITAB programs.

APPENDIX

Statistical Tables

Table 1: Upper tail areas for standard normal distribution
Table 2: Upper tail areas for t distribution
Table 3a: Lower tail areas for chi-square distribution
Table 3b: Upper tail areas for chi-square distribution
Table 4a: Upper tail area = 0.05 for F distribution
Table 4b: Upper tail area = 0.025 for F distribution
Table 4c: Upper tail area = 0.01 for F distribution

(All tables generated on MINITAB by the author.)

TABLE 1 Upper Tail Areas for Standard Normal Distribution

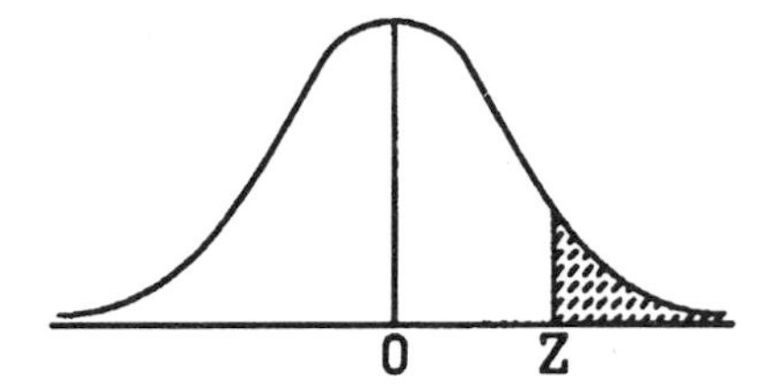

left margin - first two digits of Z
top margin - second decimal place of Z
body of table - area to right of Z

Z	.00	.01	.02	.03	.04	.05	.06	.07	.08	.09
0.00	.5000	.4960	.4920	.4880	.4840	.4801	.4761	.4721	.4681	.4641
0.10	.4602	.4562	.4522	.4483	.4443	.4404	.4364	.4325	.4286	.4247
0.20	.4207	.4168	.4129	.4090	.4052	.4013	.3974	.3936	.3897	.3859
0.30	.3821	.3783	.3745	.3707	.3669	.3632	.3594	.3557	.3520	.3483
0.40	.3446	.3409	.3372	.3336	.3300	.3264	.3228	.3192	.3156	.3121
0.50	.3085	.3050	.3015	.2981	.2946	.2912	.2877	.2843	.2810	.2776
0.60	.2743	.2709	.2676	.2643	.2611	.2578	.2546	.2514	.2483	.2451
0.70	.2420	.2389	.2358	.2327	.2296	.2266	.2236	.2206	.2177	.2148
0.80	.2119	.2090	.2061	.2033	.2005	.1977	.1949	.1922	.1894	.1867
0.90	.1841	.1814	.1788	.1762	.1736	.1711	.1685	.1660	.1635	.1611
1.00	.1587	.1562	.1539	.1515	.1492	.1469	.1446	.1423	.1401	.1379
1.10	.1357	.1335	.1314	.1292	.1271	.1251	.1230	.1210	.1190	.1170
1.20	.1151	.1131	.1112	.1093	.1075	.1056	.1038	.1020	.1003	.0985
1.30	.0968	.0951	.0934	.0918	.0901	.0885	.0869	.0853	.0838	.0823
1.40	.0808	.0793	.0778	.0764	.0749	.0735	.0721	.0708	.0694	.0681
1.50	.0668	.0655	.0643	.0630	.0618	.0606	.0594	.0582	.0571	.0559
1.60	.0548	.0537	.0526	.0516	.0505	.0495	.0485	.0475	.0465	.0455
1.70	.0446	.0436	.0427	.0418	.0409	.0401	.0392	.0384	.0375	.0367
1.80	.0359	.0351	.0344	.0336	.0329	.0322	.0314	.0307	.0301	.0294
1.90	.0287	.0281	.0274	.0268	.0262	.0256	.0250	.0244	.0239	.0233
2.00	.0228	.0222	.0217	.0212	.0207	.0202	.0197	.0192	.0188	.0183
2.10	.0179	.0174	.0170	.0166	.0162	.0158	.0154	.0150	.0146	.0143
2.20	.0139	.0136	.0132	.0129	.0125	.0122	.0119	.0116	.0113	.0110
2.30	.0107	.0104	.0102	.0099	.0096	.0094	.0091	.0089	.0087	.0084
2.40	.0082	.0080	.0078	.0075	.0073	.0071	.0069	.0068	.0066	.0064
2.50	.0062	.0060	.0059	.0057	.0055	.0054	.0052	.0051	.0049	.0048
2.60	.0047	.0045	.0044	.0043	.0041	.0040	.0039	.0038	.0037	.0036
2.70	.0035	.0034	.0033	.0032	.0031	.0030	.0029	.0028	.0027	.0026
2.80	.0026	.0025	.0024	.0023	.0023	.0022	.0021	.0021	.0020	.0019
2.90	.0019	.0018	.0018	.0017	.0016	.0016	.0015	.0015	.0014	.0014
3.00	.0013	.0013	.0013	.0012	.0012	.0011	.0011	.0011	.0010	.0010

TABLE 2 Upper Tail Areas for t Distribution

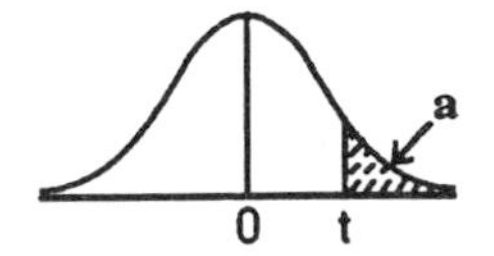

left margin = degrees of freedom
top margin = upper tail area
body of table = t cutoff point on axis

df	a=.1	a=.05	a=.025	a=.01	a=.005	a=.001
1	3.078	6.314	12.71	31.82	63.66	318.3
2	1.886	2.920	4.303	6.965	9.925	22.33
3	1.638	2.353	3.182	4.541	5.841	10.21
4	1.533	2.132	2.776	3.747	4.604	7.173
5	1.476	2.015	2.571	3.365	4.032	5.893
6	1.440	1.943	2.447	3.143	3.707	5.208
7	1.415	1.895	2.365	2.998	3.499	4.785
8	1.397	1.860	2.306	2.896	3.355	4.501
9	1.383	1.833	2.262	2.821	3.250	4.297
10	1.372	1.812	2.228	2.764	3.169	4.144
11	1.363	1.796	2.201	2.718	3.106	4.025
12	1.356	1.782	2.179	2.681	3.055	3.930
13	1.350	1.771	2.160	2.650	3.012	3.852
14	1.345	1.761	2.145	2.624	2.977	3.787
15	1.341	1.753	2.131	2.602	2.947	3.733
16	1.337	1.746	2.120	2.583	2.921	3.686
17	1.333	1.740	2.110	2.567	2.898	3.646
18	1.330	1.734	2.101	2.552	2.878	3.610
19	1.328	1.729	2.093	2.539	2.861	3.579
20	1.325	1.725	2.086	2.528	2.845	3.552
21	1.323	1.721	2.080	2.518	2.831	3.527
22	1.321	1.717	2.074	2.508	2.819	3.505
23	1.319	1.714	2.069	2.500	2.807	3.485
24	1.318	1.711	2.064	2.492	2.797	3.467
25	1.316	1.708	2.060	2.485	2.787	3.450
26	1.315	1.706	2.056	2.479	2.779	3.435
27	1.314	1.703	2.052	2.473	2.771	3.421
28	1.313	1.701	2.048	2.467	2.763	3.408
29	1.311	1.699	2.045	2.462	2.756	3.396
30	1.310	1.697	2.042	2.457	2.750	3.385
40	1.303	1.684	2.021	2.423	2.704	3.307
60	1.296	1.671	2.000	2.390	2.660	3.232
120	1.289	1.658	1.980	2.358	2.617	3.160
infinite	1.282	1.645	1.960	2.326	2.576	3.090

TABLE 3a Lower Tail Areas for Chi-Square Distribution

left margin = degrees of freedom
top margin = lower tail area
body of table = χ^2 cutoff point on axis

df	a=.001	a=.005	a=.01	a=.025	a=.05	a=.1
1	.0000	.0000	.0002	.0010	.0039	.0158
2	.0020	.0100	.0201	.0506	.1026	.2107
3	.0243	.0717	.1148	.2158	.3518	.5844
4	.0908	.2070	.2971	.4844	.7107	1.064
5	.2102	.4117	.5543	.8312	1.146	1.610
6	.3811	.6757	.8721	1.237	1.635	2.204
7	.5985	.9893	1.239	1.690	2.167	2.833
8	.8571	1.344	1.647	2.180	2.733	3.490
9	1.152	1.735	2.088	2.700	3.325	4.168
10	1.479	2.156	2.558	3.247	3.940	4.865
11	1.834	2.603	3.054	3.816	4.575	5.578
12	2.214	3.074	3.571	4.404	5.226	6.304
13	2.617	3.565	4.107	5.009	5.892	7.042
14	3.041	4.075	4.660	5.629	6.571	7.790
15	3.483	4.601	5.229	6.262	7.261	8.547
16	3.942	5.142	5.812	6.908	7.962	9.312
17	4.416	5.697	6.408	7.564	8.672	10.09
18	4.905	6.265	7.015	8.231	9.391	10.86
19	5.407	6.844	7.633	8.907	10.12	11.65
20	5.921	7.434	8.260	9.591	10.85	12.44
21	6.447	8.034	8.897	10.28	11.59	13.24
22	6.983	8.643	9.543	10.98	12.34	14.04
23	7.529	9.260	10.20	11.69	13.09	14.85
24	8.085	9.886	10.86	12.40	13.85	15.66
25	8.649	10.52	11.52	13.12	14.61	16.47
26	9.222	11.16	12.20	13.84	15.38	17.29
27	9.803	11.81	12.88	14.57	16.15	18.11
28	10.39	12.46	13.56	15.31	16.93	18.94
29	10.99	13.12	14.26	16.05	17.71	19.77
30	11.59	13.79	14.95	16.79	18.49	20.60
40	17.92	20.71	22.16	24.43	26.51	29.05
50	24.67	27.99	29.71	32.36	34.76	37.69
60	31.74	35.53	37.48	40.48	43.19	46.46
70	39.04	43.28	45.44	48.76	51.74	55.33
80	46.52	51.17	53.54	57.15	60.39	64.28
90	54.16	59.20	61.75	65.65	69.13	73.29
100	61.92	67.33	70.06	74.22	77.93	82.36

TABLE 3b Upper Tail Areas for Chi-Square Distribution

left margin = degrees of freedom
top margin = upper tail area
body of table = χ^2 cutoff point on axis

df	a=.1	a=.05	a=.025	a=.01	a=.005	a=.001
1	2.706	3.841	5.024	6.635	7.880	10.83
2	4.605	5.991	7.378	9.210	10.60	13.82
3	6.251	7.815	9.348	11.35	12.84	16.27
4	7.779	9.488	11.14	13.28	14.86	18.47
5	9.236	11.07	12.83	15.09	16.75	20.52
6	10.65	12.59	14.45	16.81	18.55	22.46
7	12.02	14.07	16.01	18.48	20.28	24.32
8	13.36	15.51	17.54	20.09	21.96	26.12
9	14.68	16.92	19.02	21.67	23.59	27.88
10	15.99	18.31	20.48	23.21	25.19	29.59
11	17.28	19.68	21.92	24.73	26.76	31.26
12	18.55	21.03	23.34	26.22	28.30	32.91
13	19.81	22.36	24.74	27.69	29.82	34.53
14	21.06	23.69	26.12	29.14	31.32	36.12
15	22.31	25.00	27.49	30.58	32.80	37.70
16	23.54	26.30	28.85	32.00	34.27	39.25
17	24.77	27.59	30.19	33.41	35.72	40.79
18	25.99	28.87	31.53	34.81	37.16	42.31
19	27.20	30.14	32.85	36.19	38.58	43.82
20	28.41	31.41	34.17	37.57	40.00	45.31
21	29.62	32.67	35.48	38.93	41.40	46.80
22	30.81	33.92	36.78	40.29	42.80	48.27
23	32.01	35.17	38.07	41.64	44.18	49.70
24	33.20	36.42	39.36	42.98	45.56	51.17
25	34.38	37.65	40.65	44.31	46.93	52.61
26	35.56	38.89	41.92	45.64	48.29	54.05
27	36.74	40.11	43.19	46.96	49.64	55.46
28	37.92	41.34	44.46	48.28	50.99	56.86
29	39.09	42.56	45.72	49.59	52.33	58.29
30	40.26	43.77	46.98	50.89	53.66	59.66
40	51.81	55.76	59.34	63.69	66.76	73.36
50	63.17	67.51	71.42	76.15	79.49	86.66
60	74.40	79.08	83.30	88.38	91.95	99.59
70	85.53	90.53	95.02	100.4	104.2	112.3
80	96.58	101.9	106.6	112.3	116.3	124.8
90	107.6	113.1	118.1	124.1	128.3	137.2
100	118.5	124.3	129.6	135.8	140.2	149.4

TABLE 4a Upper Tail Area = 0.05 for F Distribution

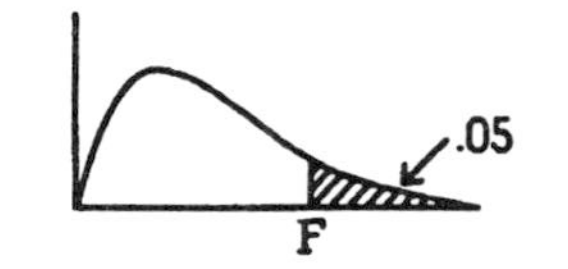

top margin = df_1 = numerator degrees of freedom
left margin = df_2 = denominator degrees of freedom
body of table = F cutoff point on axis with .05 upper area

df_2 \ df_1	1	2	3	4	5	6	7	8	9	10	12	15	20	24	30	40	60	120
1	161.4	199.5	215.7	224.6	230.2	234.0	236.8	238.9	240.5	241.9	243.9	246.0	248.0	249.0	250.1	251.1	252.2	253.3
2	18.51	19.00	19.16	19.25	19.30	19.33	19.35	19.37	19.39	19.40	19.41	19.43	19.45	19.45	19.46	19.47	19.48	19.49
3	10.13	9.55	9.28	9.12	9.01	8.94	8.89	8.85	8.81	8.79	8.75	8.70	8.66	8.64	8.62	8.59	8.57	8.55
4	7.71	6.94	6.59	6.39	6.26	6.16	6.09	6.04	6.00	5.96	5.91	5.86	5.80	5.77	5.75	5.72	5.69	5.66
5	6.61	5.79	5.41	5.19	5.05	4.95	4.88	4.82	4.77	4.74	4.68	4.62	4.56	4.53	4.50	4.46	4.43	4.40
6	5.99	5.14	4.76	4.53	4.39	4.28	4.21	4.15	4.10	4.06	4.00	3.94	3.87	3.84	3.81	3.77	3.74	3.70
7	5.59	4.74	4.35	4.12	3.97	3.87	3.79	3.73	3.68	3.64	3.58	3.51	3.44	3.41	3.38	3.34	3.30	3.27
8	5.32	4.46	4.07	3.84	3.69	3.58	3.50	3.44	3.39	3.35	3.28	3.22	3.15	3.12	3.08	3.04	3.01	2.97
9	5.12	4.26	3.86	3.63	3.48	3.37	3.29	3.23	3.18	3.14	3.07	3.01	2.94	2.90	2.86	2.83	2.79	2.75
10	4.97	4.10	3.71	3.48	3.33	3.22	3.14	3.07	3.02	2.98	2.91	2.85	2.77	2.74	2.70	2.66	2.62	2.58

df_2	df_1 1	2	3	4	5	6	7	8	9	10	12	15	20	24	30	40	60	120
11	4.84	3.98	3.59	3.36	3.20	3.10	3.01	2.95	2.90	2.85	2.79	2.72	2.65	2.61	2.57	2.53	2.49	2.45
12	4.75	3.89	3.49	3.26	3.11	3.00	2.91	2.85	2.80	2.75	2.69	2.62	2.54	2.51	2.47	2.43	2.38	2.34
13	4.67	3.81	3.41	3.18	3.03	2.92	2.83	2.77	2.71	2.67	2.60	2.53	2.46	2.42	2.38	2.34	2.30	2.25
14	4.60	3.74	3.34	3.11	2.96	2.85	2.76	2.70	2.65	2.60	2.53	2.46	2.39	2.35	2.31	2.27	2.22	2.18
15	4.54	3.68	3.29	3.06	2.90	2.79	2.71	2.64	2.59	2.54	2.48	2.40	2.33	2.29	2.25	2.20	2.16	2.11
16	4.49	3.63	3.24	3.01	2.85	2.74	2.66	2.59	2.54	2.49	2.43	2.35	2.28	2.24	2.19	2.15	2.11	2.06
17	4.45	3.59	3.20	2.97	2.81	2.70	2.61	2.55	2.49	2.45	2.38	2.31	2.23	2.19	2.15	2.10	2.06	2.01
18	4.41	3.56	3.16	2.93	2.77	2.66	2.58	2.51	2.46	2.41	2.34	2.27	2.19	2.15	2.11	2.06	2.02	1.97
19	4.38	3.52	3.13	2.90	2.74	2.63	2.54	2.48	2.42	2.38	2.31	2.23	2.16	2.11	2.07	2.03	1.98	1.93
20	4.35	3.49	3.10	2.87	2.71	2.60	2.51	2.45	2.39	2.35	2.28	2.20	2.12	2.08	2.04	1.99	1.95	1.90
21	4.33	3.47	3.07	2.84	2.69	2.57	2.49	2.42	2.37	2.32	2.25	2.18	2.10	2.05	2.01	1.97	1.92	1.87
22	4.30	3.44	3.05	2.82	2.66	2.55	2.46	2.40	2.34	2.30	2.23	2.15	2.07	2.03	1.98	1.94	1.89	1.84
23	4.28	3.42	3.03	2.80	2.64	2.53	2.44	2.38	2.32	2.28	2.20	2.13	2.05	2.01	1.96	1.91	1.87	1.81
24	4.26	3.40	3.01	2.78	2.62	2.51	2.42	2.36	2.30	2.26	2.18	2.11	2.03	1.98	1.94	1.89	1.84	1.79
25	4.24	3.39	2.99	2.76	2.60	2.49	2.41	2.34	2.28	2.24	2.17	2.09	2.01	1.96	1.92	1.87	1.82	1.77
26	4.23	3.37	2.98	2.74	2.59	2.47	2.39	2.32	2.27	2.22	2.15	2.07	1.99	1.95	1.90	1.85	1.80	1.75
27	4.21	3.35	2.96	2.73	2.57	2.46	2.37	2.31	2.25	2.20	2.13	2.06	1.97	1.93	1.88	1.84	1.79	1.73
28	4.20	3.34	2.95	2.71	2.56	2.45	2.36	2.29	2.24	2.19	2.12	2.04	1.96	1.92	1.87	1.82	1.77	1.71
29	4.18	3.33	2.93	2.70	2.55	2.43	2.35	2.28	2.22	2.18	2.10	2.03	1.95	1.90	1.85	1.81	1.75	1.70
30	4.17	3.32	2.92	2.69	2.53	2.42	2.33	2.27	2.21	2.17	2.09	2.02	1.93	1.89	1.84	1.79	1.74	1.68
40	4.09	3.23	2.84	2.61	2.45	2.34	2.25	2.18	2.12	2.08	2.00	1.92	1.84	1.79	1.74	1.69	1.64	1.58
60	4.00	3.15	2.76	2.53	2.37	2.25	2.17	2.10	2.04	1.99	1.92	1.84	1.75	1.70	1.65	1.59	1.53	1.47
120	3.92	3.07	2.68	2.45	2.29	2.18	2.09	2.02	1.96	1.91	1.83	1.75	1.66	1.61	1.55	1.50	1.43	1.35
inf.	3.84	3.00	2.60	2.37	2.21	2.10	2.01	1.94	1.88	1.83	1.75	1.67	1.57	1.52	1.46	1.39	1.32	1.22

Table 4b Upper Tail Area = 0.025 for F Distribution

top margin = df_1 = numerator degrees of freedom
left margin = df_2 = denominator degrees of freedom
body of table = F cutoff point on axis with .025 upper area

df_2 \ df_1	1	2	3	4	5	6	7	8	9	10	12	15	20	24	30	40	60	120
1	647.8	799.5	864.2	899.6	921.9	937.1	948.1	956.6	963.3	968.6	976.8	984.9	993.1	997.2	1002	1006	1010	1014
2	38.50	39.00	39.17	39.25	39.30	39.33	39.35	39.37	39.39	39.40	39.41	39.43	39.45	39.46	39.46	39.47	39.48	39.49
3	17.44	16.04	15.44	15.10	14.89	14.74	14.62	14.54	14.47	14.42	14.34	14.25	14.17	14.12	14.08	14.04	13.99	13.95
4	12.22	10.65	9.98	9.61	9.37	9.20	9.07	8.98	8.91	8.84	8.75	8.66	8.56	8.51	8.46	8.41	8.36	8.31
5	10.01	8.43	7.76	7.39	7.15	6.98	6.85	6.76	6.68	6.62	6.53	6.43	6.33	6.28	6.23	6.18	6.12	6.07
6	8.81	7.26	6.60	6.23	5.99	5.82	5.70	5.60	5.52	5.46	5.37	5.27	5.17	5.12	5.07	5.01	4.96	4.90
7	8.07	6.54	5.89	5.52	5.29	5.12	5.00	4.90	4.82	4.76	4.67	4.57	4.47	4.42	4.36	4.31	4.25	4.20
8	7.57	6.06	5.42	5.05	4.82	4.65	4.53	4.43	4.36	4.30	4.20	4.10	4.00	3.95	3.89	3.84	3.78	3.73
9	7.21	5.72	5.08	4.72	4.48	4.32	4.20	4.10	4.03	3.96	3.87	3.77	3.67	3.61	3.56	3.51	3.45	3.39
10	6.94	5.46	4.83	4.47	4.24	4.07	3.95	3.86	3.78	3.72	3.62	3.52	3.42	3.37	3.31	3.26	3.20	3.14

df_2	df_1 1	2	3	4	5	6	7	8	9	10	12	15	20	24	30	40	60	120
11	6.72	5.26	4.63	4.28	4.04	3.88	3.76	3.66	3.59	3.53	3.43	3.33	3.23	3.17	3.12	3.06	3.00	2.94
12	6.55	5.10	4.47	4.12	3.89	3.73	3.61	3.51	3.44	3.37	3.28	3.18	3.07	3.02	2.96	2.91	2.85	2.79
13	6.41	4.97	4.35	4.00	3.77	3.60	3.48	3.39	3.31	3.25	3.15	3.05	2.95	2.89	2.84	2.78	2.72	2.66
14	6.30	4.86	4.24	3.89	3.66	3.50	3.38	3.29	3.21	3.15	3.05	2.95	2.84	2.79	2.73	2.67	2.61	2.55
15	6.20	4.77	4.15	3.80	3.58	3.42	3.29	3.20	3.12	3.06	2.96	2.86	2.76	2.70	2.64	2.59	2.52	2.46
16	6.12	4.69	4.08	3.73	3.50	3.34	3.22	3.13	3.05	2.99	2.89	2.79	2.68	2.63	2.57	2.51	2.45	2.38
17	6.04	4.62	4.01	3.67	3.44	3.28	3.16	3.06	2.99	2.92	2.83	2.72	2.62	2.56	2.50	2.44	2.38	2.32
18	5.98	4.56	3.95	3.61	3.38	3.22	3.10	3.01	2.93	2.87	2.77	2.67	2.56	2.50	2.44	2.38	2.32	2.26
19	5.92	4.51	3.90	3.56	3.33	3.17	3.05	2.96	2.88	2.82	2.72	2.62	2.51	2.45	2.39	2.33	2.27	2.20
20	5.87	4.46	3.86	3.52	3.29	3.13	3.01	2.91	2.84	2.77	2.68	2.57	2.46	2.41	2.35	2.29	2.22	2.16
21	5.83	4.42	3.82	3.48	3.25	3.09	2.97	2.87	2.80	2.74	2.64	2.53	2.43	2.37	2.31	2.25	2.18	2.11
22	5.79	4.38	3.78	3.44	3.22	3.06	2.93	2.84	2.76	2.70	2.60	2.50	2.39	2.33	2.27	2.21	2.14	2.08
23	5.75	4.35	3.75	3.41	3.18	3.02	2.90	2.81	2.73	2.67	2.57	2.47	2.36	2.30	2.24	2.18	2.11	2.04
24	5.72	4.32	3.72	3.38	3.16	3.00	2.87	2.78	2.70	2.64	2.54	2.44	2.33	2.27	2.21	2.15	2.08	2.01
25	5.69	4.29	3.69	3.35	3.13	2.97	2.85	2.75	2.68	2.61	2.52	2.41	2.30	2.24	2.18	2.12	2.05	1.98
26	5.66	4.27	3.67	3.33	3.11	2.95	2.82	2.73	2.65	2.59	2.49	2.39	2.28	2.22	2.16	2.09	2.03	1.95
27	5.63	4.24	3.65	3.31	3.08	2.92	2.80	2.71	2.63	2.57	2.47	2.36	2.25	2.20	2.13	2.07	2.00	1.93
28	5.61	4.22	3.63	3.29	3.06	2.90	2.78	2.69	2.61	2.55	2.45	2.34	2.23	2.17	2.11	2.05	1.98	1.91
29	5.59	4.20	3.61	3.27	3.04	2.88	2.76	2.67	2.59	2.53	2.43	2.33	2.21	2.15	2.09	2.03	1.96	1.89
30	5.57	4.18	3.59	3.25	3.03	2.87	2.75	2.65	2.58	2.51	2.41	2.31	2.20	2.14	2.07	2.01	1.94	1.87
40	5.42	4.05	3.46	3.13	2.90	2.74	2.62	2.53	2.45	2.39	2.29	2.18	2.07	2.01	1.94	1.88	1.80	1.72
60	5.29	3.93	3.34	3.01	2.79	2.63	2.51	2.41	2.33	2.27	2.17	2.06	1.94	1.88	1.82	1.74	1.67	1.58
120	5.15	3.80	3.23	2.89	2.67	2.52	2.39	2.30	2.22	2.16	2.05	1.94	1.82	1.76	1.69	1.61	1.53	1.43
inf.	5.02	3.69	3.12	2.79	2.57	2.41	2.29	2.19	2.11	2.05	1.94	1.83	1.71	1.64	1.57	1.48	1.39	1.27

TABLE 4c Upper Tail Area = 0.01 for F Distribution

top margin = df_1 = numerator degrees of freedom
left margin = df_2 = denominator degrees of freedom
body of table = F cutoff point on axis with .01 upper area

df_2 \ df_1	1	2	3	4	5	6	7	8	9	10	12	15	20	24	30	40	60	120
1	4052	4999	5403	5624	5764	5859	5929	5981	6023	6055	6107	6157	6209	6235	6260	6287	6312	6339
2	98.51	99.00	99.17	99.25	99.30	99.33	99.35	99.38	99.39	99.40	99.41	99.43	99.44	99.45	99.47	99.47	99.48	99.49
3	34.12	30.82	29.46	28.71	28.24	27.91	27.67	27.49	27.35	27.23	27.05	26.87	26.69	26.60	26.51	26.41	26.32	26.22
4	21.20	18.00	16.69	15.98	15.52	15.21	14.98	14.80	14.66	14.55	14.37	14.20	14.02	13.93	13.84	13.75	13.65	13.56
5	16.26	13.27	12.06	11.39	10.97	10.67	10.46	10.29	10.16	10.05	9.89	9.72	9.55	9.47	9.38	9.29	9.20	9.11
6	13.74	10.92	9.78	9.15	8.75	8.47	8.26	8.10	7.98	7.87	7.72	7.56	7.40	7.31	7.23	7.14	7.06	6.97
7	12.25	9.55	8.45	7.85	7.46	7.19	6.99	6.84	6.72	6.62	6.47	6.31	6.16	6.07	5.99	5.91	5.82	5.74
8	11.26	8.65	7.59	7.01	6.63	6.37	6.18	6.03	5.91	5.81	5.67	5.52	5.36	5.28	5.20	5.12	5.03	4.95
9	10.56	8.02	6.99	6.42	6.06	5.80	5.61	5.47	5.35	5.26	5.11	4.96	4.81	4.73	4.65	4.57	4.48	4.40
10	10.04	7.56	6.55	5.99	5.64	5.39	5.20	5.06	4.94	4.85	4.71	4.56	4.41	4.33	4.25	4.17	4.08	4.00

df_2	df_1 1	2	3	4	5	6	7	8	9	10	12	15	20	24	30	40	60	120
11	9.65	7.21	6.22	5.67	5.32	5.07	4.89	4.74	4.63	4.54	4.40	4.25	4.10	4.02	3.94	3.86	3.78	3.69
12	9.33	6.93	5.95	5.41	5.06	4.82	4.64	4.50	4.39	4.30	4.16	4.01	3.86	3.78	3.70	3.62	3.54	3.45
13	9.07	6.70	5.74	5.21	4.86	4.62	4.44	4.30	4.19	4.10	3.96	3.82	3.66	3.59	3.51	3.43	3.34	3.25
14	8.86	6.51	5.56	5.04	4.69	4.46	4.28	4.14	4.03	3.94	3.80	3.66	3.51	3.43	3.35	3.27	3.18	3.09
15	8.68	6.36	5.42	4.89	4.56	4.32	4.14	4.00	3.89	3.80	3.67	3.52	3.37	3.29	3.21	3.13	3.05	2.96
16	8.53	6.23	5.29	4.77	4.44	4.20	4.03	3.89	3.78	3.69	3.55	3.41	3.26	3.18	3.10	3.02	2.93	2.84
17	8.40	6.11	5.18	4.67	4.34	4.10	3.93	3.79	3.68	3.59	3.46	3.31	3.16	3.08	3.00	2.92	2.83	2.75
18	8.29	6.01	5.09	4.58	4.25	4.01	3.84	3.71	3.60	3.51	3.37	3.23	3.08	3.00	2.92	2.84	2.75	2.66
19	8.19	5.93	5.01	4.50	4.17	3.94	3.77	3.63	3.52	3.43	3.30	3.15	3.00	2.92	2.84	2.76	2.67	2.58
20	8.10	5.85	4.94	4.43	4.10	3.87	3.70	3.56	3.46	3.37	3.23	3.09	2.94	2.86	2.78	2.69	2.61	2.52
21	8.02	5.78	4.87	4.37	4.04	3.81	3.64	3.51	3.40	3.31	3.17	3.03	2.88	2.80	2.72	2.64	2.55	2.46
22	7.95	5.72	4.82	4.31	3.99	3.76	3.59	3.45	3.35	3.26	3.12	2.98	2.83	2.75	2.67	2.58	2.50	2.40
23	7.88	5.66	4.76	4.26	3.94	3.71	3.54	3.41	3.30	3.21	3.07	2.93	2.78	2.70	2.62	2.54	2.45	2.35
24	7.82	5.61	4.72	4.22	3.90	3.67	3.50	3.36	3.26	3.17	3.03	2.89	2.74	2.66	2.58	2.49	2.40	2.31
25	7.77	5.57	4.68	4.18	3.85	3.63	3.46	3.32	3.22	3.13	2.99	2.85	2.70	2.62	2.54	2.45	2.36	2.27
26	7.72	5.53	4.64	4.14	3.82	3.59	3.42	3.29	3.18	3.09	2.96	2.81	2.66	2.58	2.50	2.42	2.33	2.23
27	7.68	5.49	4.60	4.11	3.78	3.56	3.39	3.26	3.15	3.06	2.93	2.78	2.63	2.55	2.47	2.38	2.29	2.20
28	7.64	5.45	4.57	4.07	3.75	3.53	3.36	3.23	3.12	3.03	2.90	2.75	2.60	2.52	2.44	2.35	2.26	2.17
29	7.60	5.42	4.54	4.04	3.73	3.50	3.33	3.20	3.09	3.00	2.87	2.73	2.57	2.49	2.41	2.33	2.23	2.14
30	7.56	5.39	4.51	4.02	3.70	3.47	3.30	3.17	3.07	2.98	2.84	2.70	2.55	2.47	2.39	2.30	2.21	2.11
40	7.31	5.18	4.31	3.83	3.51	3.29	3.12	2.99	2.89	2.80	2.66	2.52	2.37	2.29	2.20	2.11	2.02	1.92
60	7.08	4.98	4.13	3.65	3.34	3.12	2.95	2.82	2.72	2.63	2.50	2.35	2.20	2.12	2.03	1.94	1.84	1.73
120	6.85	4.79	3.95	3.48	3.17	2.96	2.79	2.66	2.56	2.47	2.34	2.19	2.03	1.95	1.86	1.76	1.66	1.53
inf.	6.63	4.61	3.78	3.32	3.02	2.80	2.64	2.51	2.41	2.32	2.18	2.04	1.88	1.79	1.70	1.59	1.47	1.32

Glossary

Numbers in parentheses are sections in which glossary terms are introduced.

Alpha level Probability of rejecting the null hypothesis when it is true; type one error rate (3.2).
Alternative hypothesis Statement that change has occurred or treatment has had an effect (3.1).
Analysis of variance Partition of variability into components corresponding to influences on the data; abbreviated anova (5.0).
Anova Abbreviation for analysis of variance (5.0).
Autocorrelation Time dependence in observations (8.8).

Backward elimination Stepwise regression routine that starts with the full model and removes variables one at a time until further deletions would significantly reduce predictability (9.7).
Balanced Analysis of variance design with equal sample sizes in every cell; orthogonal (6.1).
Batch mode Submission of a set of instructions to a computer program for which output is received after all commands have been performed (2.3).
Between groups Component of variability that measures differences among group averages (5.2).
Biased estimate Estimate whose probability distribution is not centered at the value of the population parameter that it is estimating (9.4).
Binomial distribution Discrete distribution for categorical data with only two categories; used in test of a proportion (3.18).
Blocking Technique for reducing error variance in anova designs (5.9).
Blocks Matched sets of observations containing one measurement for each treatment (5.9).
Bonferroni procedure Method of controlling experimentwise type one error rate for a set of tests (6.12).
Box-Cox transformation Class of transformations to achieve linearity (8.14).

Box plot Graphical representation of sample that shows each outlier separately (2.12).

Categorical variable Variable for which observed values are category names (1.2).
Centered scale Scale with a mean of 0 (1.5).
Chi-square distribution Asymmetric distribution with degrees of freedom as its only parameter; standardized distribution of sample variance (2.9).
Coefficient of determination R^2; measure of practical importance in analysis of variance (5.3) and regression (8.8).
Coefficient of variation Spread relative to the size of the numbers in a sample (2.2).
Comparisonwise Method of controlling type one error rate for a set of tests in which each test (comparison) has a 5% chance of producing a type one error (5.6).
Concordant A pair of observations ordered in the same direction on two ordered categorical variables (7.4).
Condition number Value summarizing the degree of multicollinearity among predictors in multiple regression (9.4).
Confidence bands Graph of confidence intervals for predicting the mean value of the dependent variable in regression (8.9).
Confidence interval Range of values that includes a population parameter with specified probability; interval estimate (2.5).
Confounded Two or more factors whose influence on results cannot be separated (4.6).
Contingency coefficient Measure of strength of association between two categorical variables (7.2).
Continuous random variable Variable that takes values along a continuum; has an infinite number of possible values (1.2).
Cook's D Regression diagnostic measuring how unusual the values of the independent and dependent variables are (8.12).
Correlation Relationship between two variables when order is relevant for each variable (7.4).
Cramer's V Measure of strength of association between two categorical variables; form of phi coefficient scaled to range from 0 to 1 (7.2).
Critical value Cutoff point on the distribution of test statistic marking location of rejection region (3.2).
Cross validation Method for examining adequacy of model selection in multiple regression that uses half of the data to select a sub-model and the other half to check it (9.9).

Decision rule Rule for determining when to reject the null hypothesis in terms of the value of the test statistic (3.2).
Degrees of freedom Number of values free to vary; number of independent pieces of information about the population; parameter of t, chi-square, and F distributions (2.7).
Dependent variable Variable that depends on the outcome of an experiment; random variable (1.2); target variable to be predicted in regression (8.0).
Descriptive statistics Use of statistics to describe the available data, typically a random sample from some population (1.1).
Design variable Variable built into the design of the experiment; independent variable (1.2).

Discordant A pair of observations ordered in opposite directions on two ordered categorical variables (7.4).

Discrete random variable Variable that is limited in the values it can assume, typically to a finite number of possible values (1.2).

Distribution-free Class of tests that do not assume normal data, usually based on ranks of observations; nonparametric (3.13).

Dummy variable Regression predictor with artificial values, used to represent a categorical variable (9.1).

Duncan's multiple range test Multiple range procedure that controls type one error rate between comparisonwise and experimentwise (5.6).

Durbin-Watson test Test for autocorrelation (8.8).

Error Unexplained variation; random fluctuation in data (5.9).

Error term Any estimate of random or unexplained variability; in analysis of variance, pooled estimate of random variability using within groups and nonsignificant interaction terms (6.6).

Expected count Number of observations expected in each combination of categories for two categorical variables when there is no relationship between the variables (7.1).

Experiment Any process of measurement or observation; method by which data are obtained (1.2).

Experimentwise Method of controlling type one error rate for a set of tests in which there is a 5% chance of *ever* producing a type one error (5.6).

External validity Ability to generalize results from sample to population (4.6).

Extreme outlier Value more than three interquartile ranges beyond box in box plot (2.12).

F distribution Asymmetric distribution with two separate degrees of freedom parameters; ratio of sample variances or mean squares (4.10).

Factor Classification variable in analysis of variance design (6.0).

Factorial design Analysis of variance experiment that includes every possible combination of levels, where one level is chosen from each factor (6.1).

Fisher's Least Significant Difference In analysis of variance, critical difference needed between two group averages to reject hypothesis of equality for corresponding population means, when controlling type one error rate on comparisonwise basis (5.5).

Forward selection Stepwise regression routine that starts with no predictors and adds variables one at a time until further additions would no longer significantly improve prediction (9.7).

Frequency distribution Condensation of sample into intervals with frequency shown for each interval (2.12).

Friedman's test Nonparametric test for equality of treatment means in randomized blocks design (5.9).

Full model Multiple regression model that includes all of the available predictors (9.0).

Gamma Measure of correlation between two ordered categorical variables (7.4).

Histogram Graphical representation of frequency distribution; display used to describe shape of population (2.12).
Homogeneity of variance Assumption of equal variances (4.2).
Hypothesis testing Procedure for decision making that controls the probability of reaching an incorrect decision; test of a claim concerning a population parameter (3.0).

Independent variable Variable whose values can be determined by the researcher; design variable (1.2); predictor variable in regression (8.0).
Inferential statistics Use of statistics to infer what we think is true in the population, based on what we observe in the sample (1.1).
Interaction effect In analysis of variance, pattern of cell means in which difference between levels of one factor varies over levels of another factor (6.1).
Interactive mode Submission of commands to a computer program in sequential fashion, with output received after each command (2.3).
Intercept Point where the regression line crosses the axis of the dependent variable (8.1).
Internal validity Ability to attribute causality to experimental manipulations (4.6).
Interquartile range Difference between third and first quartiles; describes spread of the middle half of sample (2.2).
Interval estimate Range of values that includes a population parameter with specified probability; confidence interval (2.0).

Kruskal-Wallis one-way analysis by ranks Nonparametric test for equality of several means (5.4).

Least squares Procedure for determining regression coefficients by minimizing the sum of squared residuals (8.3).
Leverage Regression diagnostic measuring how unusual the value of the independent variable is (8.12).
Linear contrast Combination of means in which coefficients sum to 0 (5.7).
Linear transformation Change of scale obtained by multiplying every value by a constant and then adding a constant (7.8).

Main effect In analysis of variance, difference in means between levels of a single factor (6.1).
Mean square Ratio of sum of squares to degrees of freedom; variance or variance-like term (5.2).
Median test Nonparametric test for comparing centers (means or medians) of two populations that does not assume identical shapes for the two distributions (4.4).
Midranking procedure Technique for ranking in the presence of ties; assigns average ranks to tied observations (3.13).
Mild outlier Value more than 1.5, but less than 3 interquartile ranges beyond box in box plot (2.12).
Model Representation of the influences acting on individual observations (5.2).

Multicollinearity Correlations among predictors in multiple regression; measure of redundancy among predictors (9.4).
Multiple linear regression Prediction of a dependent variable from several independent variables by using a model linear in the regression coefficients (9.0).
Multiple range procedure In analysis of variance, method for presenting results of all pairwise comparisons among means (5.6).

Natural log transformation Transformation applied to the dependent variable in regression to stabilize variance when variance increases with the mean (8.8).
Nonparametric Class of tests that do not assume normal data, usually based on ranks of observations; distribution-free (3.13).
Normal distribution Unimodal, symmetric, continuous probability distribution; approximate distribution for the sample mean (1.5).
Null hypothesis Statement that no change has occurred or treatment has had no effect; specifies value of a population parameter (3.1).
Numeric variable Variable whose values are numeric quantities (1.2).

Observed count Number of observations falling in each combination of categories for two categorical variables (7.1).
One-sided test Test with direction predicted; test in which only one direction of departure from null hypothesis is of interest (3.1).
One-way analysis of variance Test for equality of several means (5.0).
Ordered category variable Categorical variable whose categories have a natural ordering (1.2).
Orthogonal Analysis of variance design with equal sample sizes in every cell; balanced (6.1).
Outlier Extreme value in a sample (2.1); in correlation, observation that does not follow the pattern of most of the data (7.9); in regression, data point that differs from the bulk of the sample in location, pattern, or both (8.12).
Overfit Multiple regression model with high multicollinearity (9.7).

p-value Tail area beyond test statistic; measure of statistical significance (3.10).
Paired t test Test for comparing two means, based on paired or matched samples; one sample test on difference scores (4.7).
Parameter Quantity that characterizes a population; unknown constant (1.1).
Partial F test In regression, test on part of the model; comparison of full model against sub-model (9.6).
Partial t test Test of contribution of each predictor above and beyond all other variables in the model; square root of partial F test when full and sub-models differ by a single predictor (9.6).
Partial regression plot For each predictor in multiple regression, graph of residuals when dependent variable is predicted from all other predictors against residuals when this predictor is predicted from all other predictors (9.3).
Partial sum of squares Difference between regression sums of squares for full model and sub-model (9.6).

Pearson's r Measure of correlation for two numeric variables; sample correlation (7.6).
Phi coefficient Measure of strength of association between two categorical variables, generally used only when the number of rows or columns is two (7.2).
Point estimate Single best guess at the value of a population parameter (2.0).
Pool Combine terms by adding sums of squares and dividing by sum of degrees of freedom; form weighted average of terms using degrees of freedom as weights (6.6).
Pooled t test Test for comparing two means, based on independent samples and employing a pooled variance estimate (4.2).
Pooled variance estimate Weighted average of sample variances, using sample sizes minus 1 (degrees of freedom) as weights (4.2).
Population mean Center of probability distribution describing the population; long-run average value of a random variable (1.4).
Population standard deviation Square root of variance; measure of spread in same units as original measurements (1.4).
Population variance Spread of probability distribution describing the population; long-run average squared distance from the mean (1.4).
Power Probability of rejecting the null hypothesis when it is false; chance of reaching a correct decision when the alternative hypothesis is true (3.5).
Practical importance Degree to which the null hypothesis is wrong; amount of error in hypothesized value of parameter (3.10).
Predicted value Value of the dependent variable estimated from the regression equation (8.3).
Prediction bands In regression, graph of prediction intervals for predicting individual observations (8.9).
Prediction interval Probability interval for predicting an individual observation using the regression equation (8.9).
PRESS residual Residual obtained by predicting an observation from the regression equation computed without the observation (8.12).
Principal components regression Regression on a smaller set of predictors, each of which is a linear function of the original predictors (9.4).
Probability distribution For discrete random variable, list of possible values and their associated probabilities; for continuous random variable, curve in which area corresponds to probability (1.3).
Pure error Estimate of variance in the dependent variable that is independent of the regression model, obtainable only when the data include replications (8.5).

Random assignment Division of experimental units into groups on random basis; method used to establish causality or internal validity (4.6).
Random variable Variable that takes on values according to the outcome of an experiment; dependent variable (1.2).
Randomized complete blocks Extension of paired t test to several groups; analysis of variance design allowing separation of variation associated with blocking variable (5.9).
Rank sum test Nonparametric test for comparing centers (means or medians) of two populations; equivalent to U test (4.4).

Regression Prediction (8.0).
Regression coefficients In simple regression, intercept and slope of the regression line (8.3); in multiple regression, intercept and set of multipliers for all independent variables (9.1).
Rejection region Range of values for test statistic leading to rejection of the null hypothesis; typically, 5 percent tail area (3.2).
Residual In regression, difference between the actual value of the dependent variable and the predicted value; error of prediction (8.3).
Ridge regression Biased regression technique designed to combat multicollinearity, but still include all the predictors (9.4).

Sample correlation Measure of correlation for two numeric variables; Pearson's r (7.6).
Sample covariance Measure of direction of relationship between two numeric variables; unbounded form of correlation (7.6).
Sample mean Average in the sample (1.6).
Sample median Middle value if sample size is odd, average of two middle values if sample size is even, after ordering data; estimate of population center that is insensitive to outliers (2.1).
Sample mode Most frequently occurring value in sample (2.1).
Sample range Largest value in sample minus smallest; estimate of population spread (2.2).
Sample standard deviation Square root of sample variance; estimate of population standard deviation (2.2).
Sample variance Average squared distance from sample mean (except using $n-1$ instead of n in denominator); estimate of population variance (2.2).
Scale-dependent Altered by linear transformation of the data (8.6).
Scale-free Not influenced by scale of measurements; scale-independent (3.10).
Scale-independent Unaffected by linear transformation of the data; scale-free (8.6).
Sign test Test on the median appropriate for asymmetric data (3.13).
Signed rank test Nonparametric test on the mean or median; Wilcoxon signed rank test (3.13).
Significance level Degree of certainty that the null hypothesis is wrong; p-value (3.10).
Significant Result leading to rejection of the null hypothesis (3.2).
Simple effects tests In two-way design, these are one-way anovas performed on rows (or columns); in three-way design, these are two-way anovas performed on levels of the third factor; similar extension to higher order designs (6.5).
Simple linear regression Prediction from a single variable using a linear model (8.0).
Slope Direction and degree of slant in the regression line; rise over run (8.1).
Spearman's r_s Correlation coefficient computed on ranks of values for two numeric variables (7.6).
Standard error Estimated standard deviation (2.2).
Standard error of prediction Estimated standard deviation of a predicted value (8.9).
Standard normal distribution Normal distribution with mean of 0 and variance of 1; Z curve (1.5).
Standardized residual Measure of how unusual the value of the dependent variable is for its associated value of the independent variable (8.12).

Standardized scale Scale with mean of 0 and standard deviation of 1 (1.5).
Statistic Quantity that characterizes a sample; all random variables are statistics (1.1).
Statistical significance Level of certainty that the null hypothesis is wrong; measured by p-value (3.10).
Stem-and-leaf diagram Condensation of sample obtained by dividing each observation into stem (first digit) and leaf (second digit) (2.12).
Stepwise regression Stepwise regression routine that starts with no predictors and adds variables one at a time until further additions would no longer significantly improve prediction, but also deletes predictors if they later become redundant (9.7).
Stepwise regression routines Collection of algorithms for stepping through sub-models in search of a subset that is neither overfit nor underfit (9.7).
Studentized residual Standardized residual obtained by predicting an observation from the regression equation computed without the observation; standardized PRESS residual (8.12).
Sum of squares Numerator of a variance or variance-like term (5.2).

t distribution Unimodal, symmetric distribution with degrees of freedom as its only parameter; standardized distribution of sample mean used when population standard deviation is unknown (2.7).
t test Any test whose test statistic follows a t distribution; in one sample case, test on the mean when variance is unknown (3.8).
Test statistic Standardized statistic used in hypothesis testing (3.2).
Three-way interaction In analysis of variance, effect arising when pattern of a two-way interaction changes over levels of the third factor (6.10).
Tolerance For each predictor in multiple regression, one minus R^2 for predicting this variable from all other predictors; measure of the degree of uniqueness of each predictor (9.4).
Total variability Overall sample variance of all observations; ratio of total sum of squares to total degrees of freedom (5.2).
Treatments Groups compared in analysis of variance designs (5.9).
Trimmed mean Average in a sample after discarding extremes; estimate of population center (2.1).
Tukey's Honestly Significant Difference In analysis of variance, critical difference needed between two group averages to reject hypothesis of equality for corresponding population means, when controlling type one error rate on experimentwise basis (5.6).
Two-sided test Test with direction not predicted; test in which departures from hypothesized value of parameter in either direction are of interest (3.7).
Type one error Rejection of the the null hypothesis when it is true (3.4).
Type one error rate Probability of rejecting the null hypothesis when it is true; alpha level (3.2).
Type two error Acceptance of the null hypothesis when it is false (3.4).

U test Nonparametric test for comparing centers (means or medians) of two populations; equivalent to rank sum test (4.4).

Unbalanced design Analysis of variance experiment with unequal sample sizes (6.8).
Underfit Multiple regression model that omits important predictors (9.7).
Unequal variance t test Test for comparing two means based on independent samples and employing separate estimates of the two variances (4.3).

Variance inflation factor For each predictor in multiple regression, reciprocal of tolerance; measure of the degree of uniqueness of each predictor (9.4).

Weighted average An average of two or more quantities obtained by multiplying each quantity by a weight, and then dividing by the sum of the weights (4.2).
Weighted regression Application of least squares procedure to weighted residuals (8.8).
Whiskers Extensions to box in box plot that span all non-outlying values in sample (2.12).
Within groups Component of variability in analysis of variance that measures random fluctuations within each group (5.2).

Z test Any test whose test statistic follows a standard normal distribution; in one sample case, test on the mean when variance is known (3.2).

Symbols

Numbers in parentheses are sections in which symbols are introduced.

a	Number of groups in one-way analysis of variance (5.1); number of rows in two-way analysis of variance (6.2).
adjusted R^2	Proportion of Y variability that can be predicted from X in the population by using the estimated regression equation (8.8, 9.3).
α	Alpha level; probability of type one error; probability of rejecting null hypothesis when it is true (3.2).
b	Number of blocks in randomized complete blocks design (5.9); number of columns in two-way analysis of variance (6.2).
b_0	Estimated intercept in regression (8.3, 9.2).
b_1	Estimated slope in simple linear regression (8.3).
$b_0, b_1, b_2, \ldots, b_k$	Estimated regression coefficients in multiple regression (9.2).
β	Probability of type two error; probability of accepting null hypothesis when it is false (3.4).
β_0	Population intercept in regression (8.2, 9.1).
β_1	Population slope in simple linear regression (8.2).
$\beta_0, \beta_1, \beta_2, \ldots, \beta_k$	Population regression coefficients in multiple regression (9.1).
c	Number of columns in categorical table (7.1).
C	Number of concordant pairs of observations for two ordered categorical variables (7.4).
C_p	Regression diagnostic that includes both variance and bias of estimated regression coefficients (9.7).
CC	Contingency coefficient; measure of strength of relationship for categorical data (7.2).
χ^2	Variable that has chi-square distribution; standardized form of sample variance (2.9); statistic used in testing for association between two categorical variables (7.1).

D	Difference score for paired data (4.7); number of discordant pairs of observations for two ordered categorical variables (7.4).
$\overline{D}$	Sample mean (average) of difference scores (4.7).
df	Degrees of freedom; parameter of t distribution (2.7); parameter of chi-square distribution (2.9).
$df_{BETWEEN}$	Degrees of freedom for variability among group averages in analysis of variance (5.2).
df_{den}	Denominator degrees of freedom of F distribution (4.10).
df_{num}	Numerator degrees of freedom of F distribution (4.10).
$df_{REGRESSION}$	Degrees of freedom for regression sum of squares; number of parameters tested in null hypothesis (8.7, 9.3).
$df_{RESIDUAL}$	Degrees of freedom for residual sum of squares in regression; number of observations minus number of parameters estimated (8.7, 9.3).
df_{TOTAL}	Total degrees of freedom among all observations in analysis of variance (5.2); degrees of freedom for overall variability among Y values in regression (8.7).
df_{WITHIN}	Degrees of freedom for variability within groups in analysis of variance; denominator of pooled variance estimate (5.2).
F	Variable that has F distribution; ratio of sample variances (4.11); test statistic in analysis of variance (5.3) and regression (8.7, 9.3).
γ	Gamma; sample correlation between two ordered categorical variables (7.4).
H_0	Null hypothesis (3.1).
H_1	Alternative hypothesis (3.1).
HSD	Tukey's Honestly Significant Difference (5.6)
IQR	Interquartile range; difference between third and first quartiles (2.2).
k	Number of predictors in multiple regression (9.0).
LSD	Fisher's Least Significant Difference (5.5).
$MS_{BETWEEN}$	Measure of variability among group averages in analysis of variance (5.2).
MS_{TOTAL}	Overall sample variance among all observations in analysis of variance (5.2).
MS_{WITHIN}	Measure of variability within groups in analysis of variance; pooled variance estimate (5.2).
μ	Population mean; center of probability distribution; long-run average value of a random variable (1.4).
$\mu_{\overline{X}}$	Mean of probability distribution for $\overline{X}$ (1.6).
$\mu_{y\|x}$	Population mean value of Y for a specific X value in regression (8.2).
n	Number of observations in the sample for one-sample procedures (1.6); number of observations per group in balanced analysis of variance designs (5.1, 6.2).
N	Total number of observations in analysis of variance designs (5.1, 6.2).
p	p-value; tail area beyond observed value of test statistic; significance level (3.10, 3.14).
P	Sample proportion of successes; estimate of π (3.18).

φ — Phi coefficient; measure of strength of relationship for categorical data used for tables with two rows or two columns (7.2).

π — Probability of success in binomial distribution; hypothesized value of proportion (3.18).

r — Number of rows in cross-classification table for two categorical variables (7.1); Pearson's r; sample correlation for two numeric variables (7.6).

R^2 — Coefficient of determination; in analysis of variance, proportion of total variance related to differences among group averages (5.3); in regression, proportion of Y variability that can be predicted from X in the sample (8.8, 9.3).

S — Sample standard deviation (2.2).

S^2 — Sample variance (2.2).

S_D — Sample standard deviation of difference scores for paired data (4.7).

S_P^2 — Pooled variance estimate in two-sample t test; weighted average of two sample variances by using sample sizes minus 1 as weights (4.2).

S_{XY} — Sample covariance; measure of direction of relationship between two numeric variables (7.6).

$SE(\overline{X})$ — Standard error of $\overline{X}$; estimated standard deviation of sample mean (2.2).

$SS_{BETWEEN}$ — In analysis of variance, portion of SS_{TOTAL} related to differences among group averages (5.2).

SS_{COLS} — In two-way analysis of variance, portion of SS_{TOTAL} related to differences among column averages (6.2).

SS_{INT} — In two-way analysis of variance, portion of SS_{TOTAL} related to interaction between row and column factors (6.2).

$SS_{REGRESSION}$ — In regression, portion of SS_{TOTAL} we can predict from X (8.7).

$SS_{RESIDUAL}$ — In regression, portion of SS_{TOTAL} we cannot predict from X; term minimized to find least squares estimators (8.7, 9.2).

SS_{ROWS} — In two-way analysis of variance, portion of SS_{TOTAL} related to differences among row averages (6.2).

SS_{TOTAL} — Numerator of overall sample variance among all observations in analysis of variance (5.2); numerator of overall variability among Y values in regression (8.7, 9.3).

SS_{WITHIN} — In analysis of variance, portion of SS_{TOTAL} related to random variability within groups; numerator of pooled variance estimate (5.2).

σ — Population standard deviation; measure of spread in original units (1.4).

σ^2 — Population variance; spread of probability distribution; long-run average squared distance from the mean (1.4).

$\sigma_{\overline{X}}$ — Standard deviation of distribution for $\overline{X}$ (1.6).

$\sigma_{\overline{X}}^2$ — Variance of probability distribution for $\overline{X}$ (1.6).

t — Variable that has t distribution; statistic used in one-sample estimation and tests on the mean (2.7, 3.8) or for comparing two means (4.2, 4.3, 4.7); partial t test in multiple regression (9.6).

V	Cramer's V; measure of strength of relationship for categorical data used for tables with more than two rows and more than two columns (7.2).
X	Independent variable in simple linear regression; predictor (8.0).
$X_1, X_2, \ldots, X_k$	Independent variables in multiple regression; predictors (9.0).
$\overline{X}$	Sample mean; average (1.6).
Y	Dependent variable in regression; target of predictions (8.0, 9.0).
$\hat{Y}$	Estimated, or predicted, value of Y in regression (8.3, 9.2).
Y^*	Transformed Y variable (8.14).
$\widehat{Y^*}$	Predicted value for transformed Y variable in regression (8.14).
Z	Standard normal variable; normal variable with mean of 0 and variance of 1 (1.5).

Index